T0291909

Laboratory Animals

Laboratory Animals

Regulations and Recommendations for the Care and Use of Animals in Research

Second Edition

Edited by

Javier Guillén

Academic Press is an imprint of Elsevier
125 London Wall, London EC2Y 5AS, United Kingdom
525 B Street, Suite 1800, San Diego, CA 92101-4495, United States
50 Hampshire Street, 5th Floor, Cambridge, MA 02139, United States
The Boulevard, Langford Lane, Kidlington, Oxford OX5 1GB, United Kingdom

Copyright © 2018 Elsevier Inc. All rights reserved.

No part of this publication may be reproduced or transmitted in any form or by any means, electronic or mechanical, including photocopying, recording, or any information storage and retrieval system, without permission in writing from the publisher. Details on how to seek permission, further information about the Publisher's permissions policies and our arrangements with organizations such as the Copyright Clearance Center and the Copyright Licensing Agency, can be found at our website: www.elsevier.com/permissions.

This book and the individual contributions contained in it are protected under copyright by the Publisher (other than as may be noted herein).

Notices
Knowledge and best practice in this field are constantly changing. As new research and experience broaden our understanding, changes in research methods, professional practices, or medical treatment may become necessary.

Practitioners and researchers may always rely on their own experience and knowledge in evaluating and using any information, methods, compounds, or experiments described herein. In using such information or methods they should be mindful of their own safety and the safety of others, including parties for whom they have a professional responsibility.

To the fullest extent of the law, neither the Publisher nor the authors, contributors, or editors, assume any liability for any injury and/or damage to persons or property as a matter of products liability, negligence or otherwise, or from any use or operation of any methods, products, instructions, or ideas contained in the material herein.

Library of Congress Cataloging-in-Publication Data
A catalog record for this book is available from the Library of Congress

British Library Cataloguing-in-Publication Data
A catalogue record for this book is available from the British Library

ISBN: 978-0-12-849880-4

For information on all Academic Press publications visit our website at https://www.elsevier.com/books-and-journals

Working together to grow libraries in developing countries

www.elsevier.com • www.bookaid.org

Publisher: Shirle Decker-Lucke
Acquisition Editor: Mica Haley
Editorial Project Manager: Timothy Bennett
Production Project Manager: Kiruthika Govindaraju
Designer: Miles Hitchen

Typeset by TNQ Books and Journals

Contents

List of Contributors

Juan M. Baamonde Centro de Estudios Científicos, Valdivia, Chile

Michael Baar Canadian Council on Animal Care, Ottawa, ON, Canada

Kathryn Bayne AAALAC International, Frederick, MD, United States

Basil Taylor Bennett National Association for Biomedical Research, Washington, DC, United States

Ekaterina A. Botovchenco Rivera Federal University of Goiás, Goiania, Brazil

Tamara A. Botovchenco Rivera State of Goiás Public Ministry, Goiania, Brazil

John F. Bradfield University of Nebraska Medical Center, Omaha, NE, United States

Cecilia Carbone Universidad de La Plata, La Plata, Argentina

André S. Carissimi Federal University of Rio Grande do Sul, Porto Alegre, Brazil

Pradon Chatikavanij National Research Council of Thailand (NRCT), Bangkok, Thailand

Margaret Cooper The University of Kent, Canterbury, United Kingdom

Julie Dale Canadian Council on Animal Care, Ottawa, ON, Canada

Cheryl Inguito Daludado Agri-Food & Veterinary Authority of Singapore, Singapore, Singapore

Holl Davun Royal University of Agriculture, Phnom Penh, Kingdom of Cambodia

Anne-Dominique Degryse ESLAV&ECLAM (Past President), Puylaurens, France

Jade Donavanik Dhurakit Pundit University, Bangkok, Thailand

Maria Amelita C. Estacio University of the Philippines Los Banos, Laguna, Philippines

Idle O. Farah Institute of Primate Research, Karen, Kenya

Yacov Fischer Israeli Council for Animal Experimentation, Ministry of Health, Jerusalem, Israel

Montip Gettayacamin AAALAC International, Samutprakarn, Thailand

Cynthia S. Gillett University of Minnesota, Minneapolis, MN, United States

Richard Grant University of Washington, Seattle, WA, United States

Gilly Griffin Canadian Council on Animal Care, Ottawa, ON, Canada

Faisal A. Guhad Jigjiga Export Slaughter House (JESH), Jigjiga, Ethiopia

Javier Guillén AAALAC International, Pamplona, Spain/Frederick, MD, United States

Marcel Gyger Private Consultant, Lausanne, Switzerland

Alon Harmelin The Weizmann Institute of Science, Rehovot, Israel

Amanda R. Hau Administrative Court, Malmo, Sweden

Jann Hau University of Copenhguen, Copenhaguen, Denmark

John E. Heidrich Central New Mexico College, Albuquerque, NM, United States

Rafael Hernández-González Universidad Nacional Autónoma de México, Mexico City, Mexico

Chou-Chu Hong Level Biotechnology, Inc., New Taipei City, Taiwan

Bryan Howard Private Consultant, Preston, United Kingdom

Rony Kalman Hebrew University, Jerusalem, Israel

Pattamarat Kunjara National Research Council of Thailand (NRCT), Bangkok, Thailand

Tsutomu M. Kurosawa Kagoshima University, Kagoshima, Japan

Goh Yong Meng University Putra Malaysia, Serdang, Malaysia

Bert J. Mohr University of Cape Town, Cape Town, South Africa

Abdul Rahim Mutalib University Putra Malaysia, Serdang, Malaysia

Denise Noonan The University of Adelaide, Adelaide, Australia

Wanyong Pang Sanofi R & D, Sanofi China, Beijing, P.R. China

Yasmina Arditi Paramastri National University of Singapore, Singapore, Singapore

Jae-Hak Park Seoul National University, Seoul, Korea

Chumpol Pholpramool Mahidol University, Bangkok, Thailand

Jan-Bas Prins Leiden University Medical Centre and Leiden University, Leiden, The Netherlands

Syed S.Y.H. Qadri National Institute of Nutrition (ICMR), Hyderabad, India

Nguyen Bao Quoc Nong Lam University, Ho Chi Minh City, Vietnam

Subbaraya G. Ramachandra Indian Institute of Science, Bangalore, India

Ranillo Rodrigo G. Resuello Simian Conservation Breeding and Research Center, Inc. (SICONBREC), Makati City, Philippines

Dondin Sajuthi Bogor Agricultural University (IPB), Bogor, Indonesia

Sorn San Ministry of Agricultural, Phnom Penh, Kingdom of Cambodia

Ouajdi Souilem National School of Veterinary Medicine, Sidi Thabet, Tunisia

Nguyen Ba Tiep Vietnam National University of Agriculture, Hanoi, Vietnam

Joel Villa Tuplano Simian Conservation Breeding and Research Center, Inc. (SICONBREC), Makati City, Philippines

Patri Vergara Universitat Autonoma de Barcelona, Bellaterra, Spain

Jianfei Wang GlaxoSmithKline R&D China, Pudong, P.R. China

Virginia Williams Virginia Williams Consulting, North Canterbury, New Zealand

Imelda Liunanita Winoto Bogor Agricultural University (IPB), Bogor, Indonesia

Hon Mun Wong Agri-Food & Veterinary Authority of Singapore, Singapore, Singapore

Ehud Ziv Hadassah University Hospital, Jerusalem, Israel

Preface

This book is a revised second edition of the book first published in 2014, entitled “Laboratory Animals: Regulations and Recommendations for Global Collaborative Research.” That first edition was the first ever worldwide compilation of regulations and some major professional recommendations for the protection, oversight, care, and use of animals in research.

Regulatory frameworks and professional recommendations evolve constantly, although not all at the same time. While some of the chapters have only minor modifications, others have had to be more extensively revised to take account of recent developments in the field, which evidence, first, the increasing number of countries that are implementing legislation on the protection of animals used in research; and second, the common international approach applied to the care and use of these animals. This common approach is described in detail in the first chapter, which has been significantly extended, to show how the same principles are embedded in all pieces of legislation and international recommendations, and how this makes the harmonization of animal use and welfare a practical reality.

The aim of this book is to provide an easy-to-search summary of how laws and international recommendations refer to all the different areas of an animal care and use program. For this purpose, chapters are distributed in similar sections addressing all relevant areas. I hope this publication will be useful for all professionals in laboratory animal science, including personnel responsible for the direct care of the animals, such as: care takers, technicians, and veterinarians; investigators using animals in all kind of research disciplines; research institution administrators; members of Institutional Animal Care and Use Committees, Ethics Committees, or other equivalent oversight bodies; and course organizers and students.

The only geographical area, which is not represented in this book, is part of the Middle East, due to the difficulty of obtaining accurate information. Hopefully, future editions of this book will provide information on a larger number of countries both in respect of the development and implementation of legislation and the harmonization of international recommendations and practices.

Preface

[illegible]

[illegible]

[illegible]

[illegible]

Chapter 1

Global Guiding Principles: A Tool for Harmonization

Javier Guillén[1], Patri Vergara[2]
[1]*AAALAC International, Pamplona, Spain/Frederick, MD, United States;* [2]*Universitat Autonoma de Barcelona, Bellaterra, Spain*

GLOBAL PRINCIPLES

Animals play an essential role in the development of many areas of science. Science, as in many other activities of human life, is being globalized, and the care and use of animals in science is being subjected to this transformation. Globalization of science implies an increase in the level of collaborative research, which may be affected by different cultural traditions, religious beliefs, financial opportunities, and legal requirements. Societal ethical concerns on animal use for scientific purposes have led to the development and implementation of regulatory frameworks in many geographical areas. Also, even in those areas where specific regulations do not exist, practices in the field are significantly influenced by the existing regulations in other areas. Communication and the spread of science, as well as increasing interinstitutional collaboration and outsourcing, are also important factors contributing to globalization.

When reviewing the existing regulations described in this book, one easily notes that although there are differences in the way they are implemented, they share the same ethical core principles. The same principles also underlie most nonstatutory guidelines, or recommendations, on the care and use of animals used in research.

These principles are based on the principle of the Three Rs, first developed by Russell and Burch in 1959, that proposed the implementation of replacement (to avoid or replace the use of animals), reduction (to minimize the number of animals used), and refinement (to minimize animal suffering and increase animal well-being) to the use of animals in experimental research.[1]

- *Replacement* refers to the avoidance or replacement of the use of animals in experiments where otherwise they would have been used. Currently, considerable efforts are being made to develop alternative methods, and although a number of specialized centers have made significant progress, the process of validation in certain areas (i.e., regulated toxicology) has

Laboratory Animals. http://dx.doi.org/10.1016/B978-0-12-849880-4.00001-5
Copyright © 2018 Elsevier Inc. All rights reserved.

been slow. Complete replacement of animal use is already an objective in modern legislation: for example, Recital 10 in European Directive 2010/63/EU states that "this Directive represents an important step towards achieving the final goal of full replacement of procedures on live animals for scientific and educational purposes as soon as it is scientifically possible to do so."[2] However, as absolute replacement of animals is still a utopia in many research areas, the concept of relative replacement derived from approaches leading to reduction of animals in the same type of experiments and/or refinement of the experimental procedures is also valid, as it contributes to minimize the overall pain/distress/discomfort of animals used in research.

- *Reduction* refers to minimizing the number of animals needed to obtain the desired research objectives. Reduction can be achieved by improving the experimental design (including selection of the right animal model) and biostatistical approaches, effective sharing of scientific results (including negative results), or implementation of new techniques (e.g., imaging). Reduction is not only a matter of using a lower number of animals in a certain experiment but also of using the right number, as using too few could invalidate the results obtained from them.
- *Refinement* refers to minimizing the pain/distress of animals subjected to research experiments. Refinement applies not only to the implementation of pain-alleviation measures, such as anesthesia or analgesia during experimental procedures (which Russell and Burch called "direct inhumanity"), but also to the animal's environment during the research period (which Russell and Burch called "contingent inhumanity"), including housing conditions, husbandry, and care. Much important progress has been made in refinement thanks to the evolution of regulatory frameworks that usually define the minimum requirements for animal housing, care, and environment, as well as to scientific progress in different disciplines (anesthesia, analgesia, health monitoring, disease control, etc.).

Another concept that has had some influence in some regulations and recommendations is the "Five Freedoms." This concept was created in the United Kingdom in response to a 1965 UK government report on livestock husbandry[3] and later was proposed to be applicable to animals used in research.[4] The Five Freedoms include the following:

- *Freedom from hunger or thirst* by ready access to freshwater and a diet to maintain full health and vigor
- *Freedom from discomfort* by providing an appropriate environment including shelter and a comfortable resting area
- *Freedom from pain, injury, or disease* by prevention or rapid diagnosis and treatment
- *Freedom to express (most) normal behavior* by providing sufficient space, proper facilities, and company of the animal's own kind
- *Freedom from fear and distress* by ensuring conditions and treatment, which avoid mental suffering

Although not as significant as the Three Rs, the concept of the Five Freedoms has been used in some of the documents described in other chapters of this book, especially in Asia.

The importance of these principles is that they have been, and still are, used as the basis for the definition of legal requirements across the world for institutions performing research and breeding or supplying animals intended to be used in research. These requirements refer to designation of responsible persons at the institution (at administrative, animal care, veterinary, and research levels); definition of necessary training for personnel caring for or using research animals; implementation of oversight systems and ethical evaluation of research proposals; animal housing and husbandry conditions; animal care and veterinary needs; and animal facilities. Therefore, they could be claimed to be the global guiding principles that serve as the basis for harmonization.

However, there are still differences in the way these principles are defined within the regulations and implemented in practice. When principles are enshrined in legislation, legislators inevitably feel the need to provide definition and to establish clear boundaries between what is legal and what is not. This is the reason for the variation in standards that we so frequently find across countries or geopolitical areas. Why do they differ? We can identify a number of reasons including tradition, politics, financial implications, pragmatism, and culture. The outcome of this is the large variation in requirements and standards in respect of such things as regulatory bodies, personal responsibilities, or cage and enclosure dimensions. We face a situation where all members of the scientific community speak of the same aims and follow the same principles, but where animals may be treated differently in practice. Differences can have an adverse effect on collaborative research.

But there is also a very positive side to this development. The increased use of laboratory animal welfare legislation has created new regulations in areas where they did not exist before and has prompted the revision and improvement of existing laws and guidelines. While it is true that the same outcomes can be achieved by different means, the use of performance standards based on the same principles is spreading, thus providing a very powerful tool for harmonization. And this is mainly thanks to the interest, energy, and enthusiasm of the laboratory animal professionals who are continuously seeking to spread knowledge, ethical principles, and good practices around the world. International organizations are the key to achieving these, and fortunately there are very good examples: The International Council for Laboratory Animal Science (ICLAS), with its national, scientific, institutional, associate, and affiliate members, is playing an important role. ICLAS has collaborated with the Council for International Organizations for Medical Sciences (CIOMS) to revise the old International Guiding Principles for Biomedical Research Involving Animals that are now in a joint CIOMS-ICLAS document.[5] These principles, which are in accordance with the principles in the regulations described in the rest of the chapters of this book, can facilitate the advancement of international collaboration in biomedical sciences. The World Organization for Animal Health has focused on animal welfare and the use of animals in research and education in different chapters of the *Terrestrial Animal Health Code*.[6] The principles described in Chapter 7.1 of the Code are perfectly aligned with the principles in all laboratory animal regulations and are developed for animals used in research and education in Chapter 7.8.

At a more regional level, national associations have established umbrella organizations that contribute to the same aims. The recommendations of the Federation of European Laboratory Animal Science Associations (FELASA) are well known and recognized internationally,[7] and similar organizations such as the Asian Federation of Laboratory Animal Science Associations (AFLAS), the Federation of South American Laboratory Animal Science Associations exist in Asia and South America (FESSACAL), the Federation of Hispanic Laboratory Animal Societies and Associations from North America, Central America and Caribe (FESAHANCCCAL), and the Australian and New Zealand Council for the Care of Animals in Research and Teaching Ltd (ANZCCART) are working in their respective geographical areas. It is noteworthy to mention the initiative of FELASA and the American Association for Laboratory Animal Science (AALAS) to create the AALAS-FELASA liaison body, which is also producing common sets of recommendations.

From another practical perspective, AAALAC International has accredited almost 1000 animal care and use programs in 44 countries around the globe, serving as a unique tool for the improvement and harmonization of animal care and use standards and practices.

THE WORLD ORGANIZATION FOR ANIMAL HEALTH (OFFICE INTERNATIONAL DES EPIZOOTIES) TERRESTRIAL CODE CHAPTER 7.8

The Office International des Epizooties (OIE) was created in 1924 to fight animal diseases at a global level. The OIE became the World Organization for Animal Health in 2003 but kept the historical acronym OIE. The OIE has a total of 180 member countries and permanent relations with 71 other international and regional organizations. The OIE is mainly focused on animal health, but because of the close relationship between animal health and animal welfare, the OIE has also become a leading organization for animal welfare. The latest edition of the OIE *Terrestrial Animal Health Code* includes Section 7 dedicated to animal welfare.[6] Chapter 7.1 recognizes the Five Freedoms as valuable guidance in animal welfare and the Three Rs as valuable guidance for the use of animals in science. Although the Code mainly focuses on livestock issues, Chapter 7.8 is dedicated to the use of animals in research and education, with the aim "to provide advice and assistance for OIE Members to follow when formulating regulatory requirements, or other form of oversight, for the use of live *animals* in research and education". The Three Rs are clearly described in Article 7.8.3 and also mentioned as one of the key points to implement during the ethical review of the use of animals. The chapter defines how the oversight framework should be established, including the role of the Competent Authority and the process of ethical review, which may also be performed by a local committee. Oversight responsibilities should include the review of project proposals, facility inspections, and ethical evaluation (this refers to general policies and practices of the institution addressing all animal care and use program areas). Ensuring that personnel are appropriately trained and competent to work with the species used and the procedures to be performed is another important item in the OIE recommendation that training and competency should be considered for scientific staff, veterinarians, animal care staff, students, and members of the local oversight committee or others involved with oversight. Interestingly, occupational health and safety training is also listed as one of the training topics to be covered.

The provision of veterinary care and the position of the veterinarian are especially highlighted. The veterinarian should have authority and responsibility for making judgments concerning animal welfare, and veterinary advice should be available at all times. Veterinary functions should include clinical responsibilities; postmortem examinations; medical records; advice on zoonotic risks and notifiable diseases; advice on surgery and postoperative care; advice on analgesia, anesthesia, and euthanasia; and advice on humane end points.

OIE Chapter 7.8 also gives guidance on the source of animals (transport, documentation, health status), including genetically altered animals and animals captured in the wild and endangered species. Article 7.8.8 describes the recommendations for the physical facility and environmental conditions, and Article 7.8.9 focuses on husbandry aspects. Within this article, provision of environmental enrichment and social environment are also proposed.

The participation of the OIE in global harmonization is not restricted to Chapter 7.8 of the Terrestrial Code as in 2008; the OIE signed an agreement with ICLAS with the aim of collaborating in the following areas:

- The promotion of high standards of animal care and use in education, research, testing, and diagnostic work
- Cooperation in the development and revision of international animal welfare standards and guidelines relevant for both organizations
- The role and responsibilities of the veterinary profession in the welfare of animals used in experimental work
- Relationships between private veterinarians and the Veterinary Authority

The OIE and ICLAS are currently considering further collaboration to produce guiding principles to ensure integrity and the translatability of animal-based research.

In summary, the OIE is playing an important role in harmonization in the animal research field. In addition to collaborations with other global organizations, such as ICLAS, the recommendations of Chapter 7.8 on the establishment of regulatory and oversight systems are addressed to 180 countries, many of which may lack such systems, and include the basic guiding principles already implemented by most of the regulatory frameworks already established.

THE INTERNATIONAL GUIDING PRINCIPLES FOR BIOMEDICAL RESEARCH INVOLVING THE USE OF ANIMALS

The revised International Guiding Principles for Biomedical Research Involving Animals[5] were the result of a partnership between the CIOMS and the ICLAS. CIOMS is formed by international, national, and associate member organizations, representing many of the biomedical disciplines, national academies of sciences, and medical research councils. One of the important long-term programs in CIOMS is bioethics with contributions such as the International Ethical Guidelines for Biomedical Research Involving Human Subjects.[8]

The International Guiding Principles for Biomedical Research Involving Animals document is the result of the international collaboration of scientists, veterinarians, and other experts from around the world. It aims to reach global acceptance and to serve as a framework for all countries, particularly those that are still in the process of developing legislation on the use of experimental animals.

The document is structured with a preamble defining the rationale, scope, and methodology used, followed by the principles themselves:

i) Recognition of the value of scientific knowledge for the advance of human and animal welfare and the role of the animals played in the advance of the scientific knowledge.
ii) Recognition that the use of animals is a privilege and therefore there is the obligation of those involved to demonstrate respect for the animals and responsibility for their welfare.
iii) Adoption of the principle of the Three Rs.
iv) Recognition of the importance of the genetic, microbiological, and health status of animals.
v) Consideration to the health and welfare of the animals must be taken in any part of the processes where animals are involved to contribute to their well-being.
vi) The need to monitor the welfare of the animals and the need to provide veterinary care to animals.
vii) Establishes the analogy principle for pain and distress between humans and animals and the obligation to alleviate it applying analgesia and anesthesia.
viii) Unnecessary suffering must be avoided establishing humane end points.
ix) Recognizes the role of education and training to guarantee animal welfare.
x) Proposes that a system adapted to each country must be developed to guarantee that the principles are accomplished. This system should include an ethical review process, including a harm–benefit analysis for animal use.

CIOMS and ICLAS are actively promoting the use of these principles internationally.

THE INTERNATIONAL COUNCIL FOR LABORATORY ANIMAL SCIENCE

This organization had its origin, under the auspices of the CIOMS, the International Unions of Biological Sciences (IUBS), and the UNESCO in 1956 with the main objective of raising standards in the use of laboratory animals on a global basis. Currently, ICLAS maintains relationship with these organizations and in 2008 also signed an agreement of collaboration with OIE.

ICLAS aims can be summarized as the promotion and coordination of the development of Laboratory Animal Science throughout the world, particularly in developing countries and in the promotion of quality definition and monitoring of laboratory animals.

ICLAS seeks to achieve these objectives by working in collaboration with other laboratory animal science organizations to promote harmonization in the care and use of animals and education and training in the ethical use of animals.

Members of ICLAS belong to several categories: (1) national members, countries represented by a person appointed by an appropriate national body, concerned with the direction and encouragement of scientific research; (2) scientific/union members, laboratory animal science associations, scientific societies/scientific unions; (3) institutional members, universities, research institutes, or other noncommercial organizations, (4) associate members, organizations in sympathy with the aims of ICLAS, and (5) affiliate members, regional or global organization that shares common objectives with ICLAS.

ICLAS develops its programs by means of specific task and regional committees. Currently ICLAS is focused in the following areas: (1) harmonization of guidelines and principles, recognizing those guidelines that are suitable for implementation internationally and[9,10] (2) ethics and animal welfare, developing guidelines for researchers as well as for editors and reviewers to promote the ethical use of animals in research and the acceptance of the Three Rs principle.

A particular area of interest for ICLAS is the promotion of education and training in the ethical use of experimental animals and basic knowledge in laboratory animal science. To accomplish this, ICLAS has a specific committee for education and training, which searches for programs and scholarship opportunities for those living and working in countries where opportunities for education and training in laboratory animal science are limited.

Furthermore, to understand and address the need for resources in different geographical areas, ICLAS has regional committees. Currently there are committees for Africa, Americas, Asia, Europe, and Australia–New Zealand. Specific resources are approved every year for each of these regions.

As mentioned above, an important aim of ICLAS has always been the promotion of the quality of animals used in research. Only by using high-quality animals can we be sure of the reliability of experimental results. From its origin, ICLAS had specific programs for laboratory animal quality. In the beginning, monitoring and reference centers for both genetic and health monitoring were established. However, the rapid development of scientific institutions and laboratories around the world made it necessary to develop a functional network that could integrate more laboratories interested in the promotion of the quality of the animals used in research. As a result, in 2006, ICLAS set up the ICLAS Laboratory Animal Quality Network that has developed two programs: (1) the Performance Evaluation Program (PEP) for diagnostic laboratories and (2) Genetic Monitoring Reference Program.

PEP started in 2007 as a self-assessment program for any diagnostic laboratory that wanted to participate. Under the program, serum and microbiological specimens are prepared and confirmed by internationally recognized laboratories and then sent to subscribing participating laboratories worldwide for analysis. This program provides participating laboratories with a scientifically robust program to monitor their diagnostic performance. Currently, more than 30 laboratories around the world in all continents have joined this program.

The Genetic Monitoring Reference Program provides reference DNA from the most commonly used rodent breeds to enable research institutions worldwide to check whether the specific strains of research animals they have developed are genetically sound and truly representative of their assumed genetic background. This program became operational in 2015.

In summary, ICLAS is a global organization promoting the general acceptance of the Three Rs in the use of animals in research, promoting and facilitating education and training in laboratory animal science and providing tools to maintain and improve the quality of animals used in research.

THE ROLE OF INTERNATIONAL PROFESSIONAL ORGANIZATIONS

Professionals in the laboratory animal field normally group together in national and international associations where they establish contact with colleagues; share, disseminate, and promote scientific information and good practice; defend their interests and collaborate with competent authorities; and produce recommendations on specific topics of interest. Such professionals may include animal care takers, technicians, veterinarians, biologists, animal welfare officers, researchers, and any other group involved in animal care and use. Very often, associations at a national level come together to constitute wider international organizations that have a more significant impact in the implementation of shared principles.

The Federation of European Laboratory Animal Science Associations (FELASA) was established in 1978 and has 20 national members (national associations), representing more than 3000 professionals from 25 European countries. FELASA (www.felasa.eu) is the major representative of the interests of the laboratory animal community in Europe and, thanks to the many working groups formed through the years, has published several recommendations on the care and use of experimental animals, of which those dedicated to training, education, and health monitoring are the most popular internationally.[7] The FELASA Accreditation Board on Education and Training in Laboratory Animal Science evaluates and accredits training courses across Europe in accordance with the newly required functions under Directive 2010/63/EU. The importance of FELASA in education and training is recognized by the European Commission, which considers FELASA a very significant stakeholder in promoting education and training in Europe. More recently, FELASA has established with the AALAS the AALAS-FELASA liaison body, to promote joint working groups to produce guidance documents to help harmonize practices on both sides of the Atlantic. The first results of this collaboration have focused on health monitoring for rodent transportation and harm–benefit analysis of animal experiments.[11–13] These joint activities between FELASA and AALAS (which is also an ICLAS scientific member) are an excellent example of how professional organizations try to harmonize practices in animal care and use in different geographical areas through the use of shared principles. As regards education and training in laboratory animal science, AALAS offers extensive courses and a very complete e-learning library, accessible, which is also a very useful harmonization tool. The AALAS annual meeting is the largest in the world in laboratory animal science, where professionals from all around the world gather to share and learn about the latest scientific developments in the field.

FESSACAL and *FESAHANCCCAL* play a similar role in Latin American regions. FESSACAL is an ICLAS affiliate member and has representation from the national associations of Argentina, Brazil, Chile, Colombia, Uruguay, and Venezuela and has working groups on legislation, education, and institutional animal care and use committees. FESAHANCCCAL (www.fesahancccal.com) is a younger organization with representatives from the national associations of Costa Rica, Cuba, Dominican Republic, Guatemala, and Mexico. FESAHANCCCAL has already organized scientific congresses and is working on the promotion of high standards of animal care and use. Several of the national associations in Latin America are members of ICLAS.

In Asia, the *Asian Federation of Laboratory Animal Science Associations (AFLAS)* has 11 national associations as members. AFLAS (www.aflas-office.org) has the aim of promoting regional congresses for the purpose of reviewing scientific, technical, and educational problems in Laboratory Animal Science, to develop other relevant activities in the interest of Laboratory Animal Science and to contribute to animal welfare. AFLAS has already held seven international congresses since 2004 and is also an affiliate ICLAS member.

The Australian and New Zealand Council for the Care of Animals in Research and Teaching Ltd (ANZCCART) was established in 1987. The main role of ANZCCART (https://www.adelaide.edu.au/ANZCCART/) is to provide leadership in developing community consensus on ethical, social, and scientific issues relating to the use and well-being of animals in research and teaching. ANZCCART has published numerous documents pertaining to ethics committees and ethical approaches, euthanasia, housing, farm animals in research and the use of wildlife among others. The importance of ANZCCART in the area is evidenced by the presence of governmental departments from Australia and New Zealand.

In summary, international professional associations play one of the most important roles in terms of harmonization, as they represent the main force on the ground that can implement the general principles on laboratory animal care and use.

More specifically in the laboratory animal medicine field, laboratory animal medicine colleges in several geographical areas have been established. Examples include the United States (ACLAM; https://www.aclam.org/), Europe (ECLAM; http://eslav-eclam.org/), Japan (JCLAM; http://plaza.umin.ac.jp/~jclam/_AAAAAA_/index.html), and South Korea (KCLAM; http://www.kclam.org/). Laboratory animal veterinarians can become diplomates by these colleges that establish training programs in their respective areas. These listed colleges together conform the International Association of Colleges of Laboratory Animal Medicine (IACLAM; http://www.iaclam.org/), which provides a common platform at the global level for communication by, and representation of, these colleges and their diplomates. IACLAM assists achieving the shared objectives of all these laboratory animal medicine specialty organizations to promote the welfare and responsible use of laboratory animals through the certification of veterinary specialists, education, the dissemination of information relevant to the field, and serving as research partners. IACLAM has collaborated with ICLAS in the global organization, and support of Three Rs advances in laboratory animal science.[14]

AAALAC INTERNATIONAL: GLOBAL ACCREDITATION OF ANIMAL CARE AND USE PROGRAMS

AAALAC International (AAALAC) is a private, nonprofit organization that promotes the humane treatment of animals in science through voluntary accreditation and assessment programs. AAALAC (www.aaalac.org) was established in 1965 in the United States and, for many years, was focused exclusively in that country. However, when the international interest in accreditation was evident, AAALAC expanded its activities internationally, and nowadays there are approximately 1000 institutional programs accredited in more than 40 countries. This, along with the fact that AAALAC is governed by more than 60 scientific organizations from around the world, is what gives AAALAC its privileged position to promote global harmonization.

The scope of the assessment AAALAC performs includes institutional responsibilities (key institutional responsible persons; ethical review process; training, education, and competence of personnel; occupational health and safety program); the animal environment, housing, and management (physical environment, environmental enrichment, and social housing; husbandry; etc.); the veterinary care program (from procurement of animals to euthanasia); and the physical plant.

AAALAC uses three primary standards for the evaluation process: the *Guide for the Care and Use of Laboratory Animals* (*Guide*),[15] the European Convention ETS 123,[16] and the *Guide for the Care and Use of Agricultural Animals in Research and Teaching* (*Ag Guide*).[17] These documents are discussed in this book in the US chapter (*Guide* and *Ag. Guide*) and the European chapter (ETS 123), which explain how they basically follow the same principles. The recommendations of these documents are applied in addition to the legislation implemented in the accredited program's country or region. Where there is no implemented legislation, the minimum standards necessary to achieve accreditation are those specified in *Guide*.

In view of the diversity of research environments, the AAALAC Council on Accreditation (independent professionals in the laboratory animal field) applies these standards using a performance-based approach, which allows more flexibility. Performance standards focus on the outcome rather on the process used to achieve it. This allows institutions to implement different procedures, according to particular situations, to achieve the same outcome. For example, AAALAC may accept different approaches for ethical evaluation of research proposals (local committees, external or government committees, combination of any of them…), and the system is effective in producing a good ethical evaluation according to the Three Rs principle. While engineering standards (those measurable and nonflexible, such as minimum cage dimensions) must be met when they are legally required, AAALAC additionally promotes the use of performance standards.

The scheme described above is based on standards that follow very similar global principles and facilitates the dissemination of good practices and harmonization in institutions in all continents.

CONCLUSIONS

As can be seen in the rest of the chapters of this book, regulatory frameworks on the protection of animals used for scientific purposes around the world are based on different requirements for ethical review processes, the training and competence of personnel, animal housing and environment, veterinary care involvement, and animal facilities. However, when looking in detail into the core principles underlying these requirements, it is apparent that they are mostly based on the same tenets of replacement, reduction, and refinement, with the addition in some cases of other concepts applied to animal welfare such as the five freedoms.

Different regulatory frameworks can obstruct efficiency in achieving global collaborative research. Therefore, global organizations and international professional associations in the field use the same shared principles to establish a more common framework for new regulations on the care and use of experimental animals. Of particular importance is the participation of global organizations not directly related to laboratory animal professionals such as OIE and CIOMS.

These organizations can approach other professionals and regulators and promote the implementation of the same principles for laboratory animal care and use. Moreover, their engagement with other global organizations in the laboratory animal field, e.g., ICLAS, ensures the appropriateness of their recommendations. More specifically, in our field during the past decades, professionals have been grouping in regional/national associations, many of which have joined together to constitute larger international organizations. These organizations play a fundamental role in the expansion and harmonization of best practices. Examples shown include FELASA in Europe, AALAS in the United States, FESSACAL and FESAHANCCCAL in Latin America, AFLAS in Asia, and ANZCCART in Oceania. There is also increasing recognition of the need for these larger organizations to engage with each other, via ICLAS or through initiatives such as the AALAS-FELASA liaison body. In this way, laboratory animal professionals, regulators, animal users, and animal care personnel can become aware of the need to adopt a more common approach to laboratory animal care and use based on the same common principles already accepted in the community.

REFERENCES

1. Russell WMS, Burch RL. *The principles of humane experimental technique*. Special ed. Potters Bar, England: Universities Federation for Animal Welfare; 1959.
2. The European Parliament and the Council of the European Union. Directive 2010/63/EU of the European Parliament and of the Council of 22 September 2010 on the protection of animals used for scientific purposes. *Off J Eur Union* 2010;**L 276**:33–79.
3. Brambell R. *Report of the technical committee to enquire into the welfare of animals kept under intensive livestock husbandry systems, cmd*. Great Britain, Parliament: H.M. Stationery Office; 1965. p. 1–84.
4. Mellor D, Reid C. Concepts of animal well-being and predicting the impact of procedures on experimental animals. In: Baker RM, Jenkin G, Mellor DJ, editors. *Improving the well-being of animals in the research environment*. Australia: Australia and New Zealand Council for the Care of Animals in Research and Testing; 1994. p. 3–18.
5. Council for International Organizations for Medical Sciences; and International Council for Laboratory Animal Science. *International guiding principles for biomedical research involving animals*. 2012. Available from: www.iclas.org.
6. World Health Organization. *Terrestrial animal health code*. 2012. Available from: http://www.oie.int/international-standard-setting/terrestrial-code/access-online/.
7. Guillen J. FELASA guidelines and recommendations. *J Am Assoc Lab Anim Sci* 2012;**51**(3):311–21.
8. Council for International Organizations for Medical Sciences. *International ethical guidelines for biomedical research involving human subjects*. Geneva: CIOMS; 2002. ISBN: 92 9036 075 5. Available from: http://www.cioms.ch/publications/guidelines/guidelines_nov_2002_blurb.htm.
9. Demers G, Griffin G, De Vroey G, Haywood JR, Zurlo J, Bedard M. Harmonization of animal care and use guidance. *Science* May 2006;**5**(312):700–1.
10. Rose M, Everitt J, Hedrich H, Schofield J, Dennis M, Scott E, Griffin G. ICLAS Working Group on Harmonization: international guidance concerning the production care and use of genetically-altered animals. *Lab Anim* July 2013;**47**(3):142–52.

11. Pritchett-Corning KR, Prins JB, Feinstein R, Goodwin J, Nicklas W, Riley L. AALAS/FELASA working group on health monitoring of rodents for animal transfer. *J Am Assoc Lab Anim Sci* 2014;**53**(6):633–40.
12. Brønstad A, Newcomer CE, Decelle T, Everitt JI, Guillen J, Laber K. Current concepts of harm–benefit analysis of animal experiments – report from the AALAS–FELASA working group on harm–benefit analysis – part 1. *Lab Anim* 2016;**50**(1S):1–20.
13. Laber K, Newcomer CE, Decelle Y, Everitt JI, Guillen J, Brønstad A. Recommendations for addressing harm–benefit analysis and implementation in ethical evaluation – report from the AALAS–FELASA working group on harm–benefit analysis – part 2. *Lab Anim* 2016;**50**(1S):21–42.
14. Turner PV, Pekow C, Clark JM, Vergara P, Bayne K, White WJ, Kurosawa TM, Seok SH, Baneux P. Roles of the international Council for laboratory animal science (ICLAS) and international association of colleges of laboratory animal medicine (IACLAM) in the global organization and support of 3Rs advances in laboratory animal science. *J Am Assoc Lab Anim Sci* March 2015;**54**(2):174–80.
15. National Academy of Sciences (NAS). *Guide for the care and use of laboratory animals.* Washington, D.C.: National Academy Press; 2011.
16. Council of Europe. European convention for the protection of vertebrate animals used for experimental and other scientific purposes. European treaty series 123. Council of Europe. 2006. Appendix A of the European convention for the protection of vertebrate animals used for experimental and other scientific purposes (ETS No. 123). Guidelines for accommodation and care of animals (Article 5 of the convention). Approved by the multilateral consultation. *Cons* 1986;**123**(2006):3.
17. Federation of Animal Science Societies (FASS). *Guide for the care and use of agricultural animals in research and teaching*. 3rd ed. 2010.

Chapter 2

Oversight of Research Animal Welfare in the United States

John F. Bradfield[1], Basil Taylor Bennett[2], Cynthia S. Gillett[3]

[1]University of Nebraska Medical Center, Omaha, NE, United States; [2]National Association for Biomedical Research, Washington, DC, United States; [3]University of Minnesota, Minneapolis, MN, United States

INTRODUCTION

The United States (US) has many regulations and guidelines regarding the care and use of animals. A comprehensive review of all laws, regulations, and guidelines in the US regarding the welfare, care, and use of animals is beyond the scope of this chapter, but the key regulations and guidelines that impact the care and use of animals in research and teaching will be covered. In many ways the oversight framework for animal use is a reflection of societal values and priorities regarding the overall standing of animals in society. The regulatory landscape in the US, as in many other countries, is continuously evolving, and as one reviews the current standards it must be kept in mind that the implementation of animal use oversight is a rather dynamic process, which depends on our evolving understanding of animals, their needs, and our obligation to be a good steward when they are in our care.

The careful consideration of animal welfare guidelines and policies involves an international perspective. The US has animal welfare laws, guidelines, and policies that are by their very nature, US-based. However there is a global view of animal welfare that impacts the US landscape in tangible ways. While non–US-based laws and guidelines do not carry a statutory requirement in the US, they are often used as appropriate resources that impact animal care (AC) and use in this country.

The key regulations and guidelines that will be the focus of this chapter include the Animal Welfare Regulations (AWR) as mandated in the US law; the Animal Welfare Act (AWA) (PL-89-544)[1]; the Public Health Service Policy (PHS Policy) on Humane Care and Use of Laboratory Animals,[2] which is a statutory mandate of the Health Research Extension Act (HREA) of 1985; "Animals in Research" (PL-99-158)[3]; and the US Government Principles for

Laboratory Animals. http://dx.doi.org/10.1016/B978-0-12-849880-4.00002-7
Copyright © 2018 Elsevier Inc. All rights reserved.

the Utilization and Care of Vertebrate Animals Used in Testing, Research, and Training.[4] Apart from the AWR, the *Guide for the Care and Use of Laboratory Animals* (*Guide*, NRC, 2011)[5] further details the requirements of the PHS Policy and is used by institutions to comply with the PHS Policy. For agricultural animals used in research and teaching, a similar set of guidelines has been published by the Federation of Animal Science Societies, the "*Guide for the Care and Use of Agricultural Animals in Research and Teaching*" (*Ag Guide*, FASS, 2010).[6] Another set of guidelines with universal use is the Guidelines on Euthanasia (2013) published by the American Veterinary Medical Association (AVMA).[7]

The focus on these regulations and guidelines is not meant to imply that the many others that impact animal use and welfare are not important, but simply to direct attention to the core of animal welfare standards in the US, which focus on animals in research, teaching, and testing. The US has many relevant guidelines and regulations about animals and welfare, several of which are listed later. The breadth and scope of these laws and guidelines are simply too great to include in one chapter but these help provide the overall framework in the US regarding the consideration of animal welfare.

A partial list of additional regulations and guidelines in the US include: the Humane Methods of Slaughter Act,[8] Horse Protection Act,[9] Twenty-Eight Hour Law,[10] Guidelines for the Humane Transportation of Research Animals,[11] Guidelines of the American Society of Mammalogists for the Use of Wild Mammals in Research,[12] Guidelines to the Use of Wild Birds in Research,[13] Psychological Well-Being of Nonhuman Primates,[14] Guidelines for the welfare and use of animals in cancer research,[15] Guidelines for the Care and Use of Mammals in Neuroscience and Behavioral Research,[16] Guidelines for Use of Live Amphibians and Reptiles in Field and Laboratory Research,[17] and Environmental Enrichment for Nonhuman Primates Resource Guide[18].

These and many other guidelines are used throughout the US as they apply to a particular species, or discipline of research and are often developed by the various research societies engaged in animal-related research that consider the humane care and use of animals paramount in the conduct of high-quality science.

Brief overviews of the core documents that comprise the focus of this chapter are described.

The Animal Welfare Act (Public Law 89-544)[1]

The AWA originally enacted in 1966 and was entitled the "Laboratory Animal Welfare Act." Its title was changed to the current title when the Act was amended in 1970 (PL-91-579). The purposes of the original act were to protect against theft of pet dogs and cats, prevent the sale or use of dogs and cats that had been stolen, and ensure that research facilities provided humane animal care and use. The AWA covers all warm-blooded animals used, or intended for use in research, teaching, testing, experimentation, or exhibition purposes, or as a pet. This does not include birds, rats of the genus *Rattus*, and mice of the genus *Mus*, bred for use in research; horses not used for research purposes; and other farm animals used or intended for use for food or fiber, for improving animal nutrition, breeding, management, or production efficiency, or for improving the quality of food or fiber.

The AWA authorized the Secretary of Agriculture, the US Department of Agriculture (USDA), Animal and Plant Inspection Service (APHIS) to develop regulations to implement the Act. These AWRs are set forth in Title 9 of the Code of Federal Regulations (CFR) Chapter 1, Subchapter A—Animal Welfare, Part 1—Definitions, Part 2—Regulations, and Part 3—Standards.[19] Enforcement duties are the responsibility of the APHIS Deputy Administrator for Animal Care. To comply with the AWA, the USDA-APHIS requires all registered research facilities to adhere to the AWRs. As part of ensuring that institutions follow the regulations, USDA-APHIS conducts unannounced inspections of research facilities by Veterinary Medical Officers (VMO) who inspect at least annually and more often if deemed necessary by the agency. The USDA uses a risk-based inspection system to support and focus its inspection strategy, allowing more frequent and in-depth inspections at facilities experiencing problems and fewer at those that are consistently in compliance. The process uses several objective criteria including past compliance history to determine inspection frequency.

The AWA has been amended eight times since 1966 with the most significant changes made in 1985 in which an amendment entitled the "Improved Standards for Laboratory Animals Act" strengthened the standards for providing laboratory animal care, increased enforcement of the AWA, provided for collection and dissemination of information to reduce unintended duplication of experiments using animals and mandated training for those who handle animals. The 1985 amendment also included standards for exercise of dogs, psychological well-being of primates, limitation of multiple survival surgeries, and a requirement that investigators consult with a veterinarian in the design of experiments, which have the potential for causing pain, to ensure the proper use of anesthetics, analgesics, and tranquilizers. Investigators must also consider alternatives to procedures that may cause pain or distress. Each research facility must demonstrate upon inspection and report annually, that professional standards of animal care, treatment, and use are employed during research, teaching, or testing.

Perhaps most significantly, the 1985 amendment required the chief executive officer (CEO) of each research facility to appoint an Institutional Animal Care and Use Committee (IACUC) consisting of at least three members including a doctor of veterinary medicine and one member who is not affiliated with the institution. The IACUC is charged to act as an agent of the research facility in assuring compliance with the AWA. Central to this charge is the authority of the IACUC mandated in the AWRs. The role and authority of the IACUC in the care, use, and oversight of research animals cannot be overstated and it is this somewhat unique premise that is at the core of animal use oversight in the US. The AWRs stipulate that the IACUC is required to conduct a variety of activities to ensure compliance with the regulations. These responsibilities include inspection of all animal facilities and study areas at least once every 6 months, and review once every 6 months the research facility's program of animal care and use. The IACUC must file a report of its inspection with the Institutional Official (IO) of the research facility. If significant deficiencies or deviations are not corrected in accordance with the specific plan approved by the IACUC, the USDA and any Federal funding agencies must be notified in writing. The IACUC must also review and approve all proposed activities involving the care and use of animals in research, testing, or teaching procedures and all subsequent significant changes of ongoing activities. As part of this review, the IACUC must evaluate procedures that minimize discomfort, distress, and pain and that when an activity is likely to cause pain and that a veterinarian has been consulted in planning for the administration of anesthetics, analgesics, and tranquilizers and that paralytic agents are not employed except in the anesthetized animal. The IACUC must also determine that animals, which experience severe or chronic pain, are euthanatized consistent with the design of study, that the living conditions meet the species' needs, that necessary medical care will be provided, that all procedures will be performed by qualified individuals, that survival surgery will be performed aseptically, and that no animal will undergo more than one operative procedure unless justified and approved by the IACUC. Methods of euthanasia must be consistent with the definition contained in the regulations.

The IACUC must also assure on behalf of the research facility that the principal investigator (PI) considered alternatives to painful procedures and that the work being proposed does not unnecessarily duplicate previous experiments. To provide this assurance the IACUC must review the written narrative description provided by the investigator and the description must include the methods and sources used in determining that appropriate alternatives were not available. The IACUC can grant exceptions to the regulations and standards, if they have been justified in writing by the PI.

The research facility is required to provide training to scientists, animal technicians, and other personnel involved with the care and use of animals and training must include humane practices of experimentation; methods that minimize pain and distress; the proper use of anesthetics, analgesics, and tranquilizers; the use of appropriate information services such as the National Agricultural Library; and methods to report deficiencies in animal care and treatment.

Public Health Service Policy[2]

The National Institutes of Health (NIH) is a major component of the PHS of the US, Department of Health and Human Services (the largest funding source for biomedical research in the US). In 1971, the NIH issued its first Policy on the care and treatment of laboratory animals (NIH Policy, "Care and Treatment of

Laboratory Animals," [NIH 4206], NIH Guide for Grants and Contracts, No. 7, June 14, 1971). The Policy required that institutions receiving NIH grant awards provide written assurance that a committee of at least three members, one of which was a veterinarian, had been established or that the institution was accredited by a "professional accrediting body." The Policy was revised in 1973 and 1979 with a significant revision in 1985 as a result of the passage of the HREA (1985) (Public Law 99-158).[3] The HREA provided statutory authority of the PHS Policy for PHS-funded research involving live vertebrate animals used (or intended for use) in biomedical and behavioral research. An additional revision of the Policy was released in September 1986, which reflected the changes required by the HREA. In 2002, another revision occurred to reflect changes in the process of submitting IACUC approval and the information that must be included in the Animal Welfare Assurance. The most recent revision to the PHS Policy was in 2015 to incorporate changes to key references and internal administrative processes.

Under the PHS Policy, each institution using animals in PHS-sponsored projects must provide acceptable written assurance of its compliance with the Policy. In this Animal Welfare Assurance the institutions must describe:

1. The Institutional Program for the Care and Use of Animals
2. The Institutional Status
3. The IACUC

The Institutional Program must include a list of every branch and major component; the lines of authority for administering the program; the qualifications, authority, and responsibility of the veterinarian(s); the membership of the IACUC; and the procedures that they follow must be stated. The employee health program must be described for those who have frequent animal contact. A training or instruction program in the humane practices of animal care and use must be available to scientists, animal technicians, and other personnel involved in animal care, treatment, and use.

The Institutional Status must be stated as either Category one (1) AAALAC International accredited or Category two (2) nonaccredited. Institutions in Category two (2) must establish a reasonable plan with a specific timetable for correcting any departures from the recommendations in the *Guide for the Care and Use of Laboratory Animals* (NRC, 2011).

The IACUC must be appointed by the CEO, who can delegate that authority to the IO. The IACUC must consist of at least five members; one of whom is a veterinarian with program responsibility, a practicing scientist, an individual whose expertise is in a nonbiological science, and an individual who is not affiliated with the institution. The functions of this Committee are analogous to those required under the AWA with the exception that it must use the *Guide* to review the animal facilities and the institutional program for humane care and use of animals. The other major difference is that the PHS Policy covers all vertebrate animals.

The institution is responsible for maintaining all the necessary records to document compliance with the PHS Policy and for filing annual reports developed by the IACUC, which detail any changes in the program and indicate the dates of the semiannual inspections and programmatic reviews. Institutions must also report any serious or continuing noncompliance with the Policy, any serious deviations from the provisions of the *Guide* and any IACUC suspension of an activity.

The PHS Policy is intended to implement and supplement the "US Government Principles for the Utilization and Care of Vertebrate Animals in Testing, Research, and Training" (the US Government Principles). The nine principles are published in the PHS Policy and in the Appendix of the *Guide*.

The NIH, Office of Laboratory Animal Welfare (OLAW) is tasked with oversight of animal care and use for PHS-funded research. OLAW requires that institutions base their animal care and use programs on the *Guide* and as previously described, each institution receiving PHS funds for animal research must maintain a PHS Assurance Statement that has been approved by OLAW. OLAW endorses the performance based concepts of the *Guide* and provides guidance as to how institutions can best implement and comply with the PHS Policy and the *Guide*. Institutions with PHS Assurances must self-report instances of noncompliance with the *Guide* and PHS Policy and the means to correct them. In addition to this self-reporting to ensure compliance, OLAW sometimes conducts on-site evaluations of programs to ensure they comply with the *Guide*, PHS Policy, and the details of the institution's Assurance Statement.

The Guide for the Care and Use of Laboratory Animals (NRC 2011)[5]

The original "*Guide*" was first published in 1963 by the US Department of Health, Education, and Welfare, PHS, and was titled the *Guide for Laboratory Animal Facilities and Care*. It has been revised or updated seven times since in 1965, 1968, 1972, 1978, 1985, 1996, and 2011 with more recent editions published by the Institute for Laboratory Animal Resources (ILAR). Eventually it became the "*Guide for the Care and Use of Laboratory Animals*," commonly referred to as the *Guide*. The PHS Policy requires institutions that receive PHS funding adhere to the standards of the *Guide*. The *Guide* states on page xiii that it is intended to "*assist institutions in caring for and using animals in ways judged to be scientifically, technically, and humanely appropriate and is also intended to assist investigators in fulfilling their obligation to plan and conduct animal experiments in accord with the highest scientific, humane, and ethical principles.*" The 1996 version (seventh edition) stressed the importance of performance standards in developing acceptable standards for the care and use of laboratory animals. The performance standards approach requires clearly defined appropriate outcomes; the criteria for determining how outcomes are achieved; and the methods for evaluating the outcomes, without specifying exactly how the outcome is accomplished. The performance-based approach described in the 1996 *Guide* is in contrast to the engineering approach in which both the outcome and methods to achieve them are specified. The performance-based approach recognizes the wide array of scientific disciplines and animal models employed. It allows institutions to best determine how to achieve the standards described in the *Guide* based on the needs of the animals and the type of research conducted. The current, 2011 edition of the *Guide* (eighth edition) extends the performance-based approach and underscores the need for the IO, IACUC, attending veterinarian (AV), and PI to collaborate when implementing the standards of the *Guide*. The eighth edition contains five chapters: Key Concepts; Animal Care and Use Program; Environment, Housing, and Management; Veterinary Care; and Physical Plant. The purpose of the *Guide* is to assist institutions in laying the foundation for a comprehensive animal care and use program that relies on the use of performance standards and professional judgment to assure that such use is in accordance with the highest scientific, humane, and ethical principles. The *Guide* strongly affirms the principle that those who care for and use laboratory animal must assume responsibility for their well-being. Laboratory animals are defined as any vertebrate animal used in research, testing, and teaching. Key terms in the Guide that help the reader prioritize the importance of the recommendations are: "*must,*" which denotes an imperative or mandatory requirement; "*should*" is a string recommendation for achieving a goal; and "*may*," which is a suggestion for consideration.

Guide for the Care and Use of Agricultural Animals in Research and Teaching (FASS, 2010)[6]

The *Guide for the Care and Use of Agricultural Animals in Agricultural Research and Teaching* was published originally in 1988 and revised in 1999 and 2010 and is commonly referred to as the *Ag Guide*. The current third edition, published by the Federation of Animal Science Societies, is intended to provide standards for agricultural animals involved in research and teaching with the viewpoint stated in the preface that, "*Farm animals have certain needs and requirements and these needs and requirements do not necessarily change because of the objectives of the research or teaching activity. Therefore, regardless of the teaching or research objectives, the FASS Ag Guide should serve as a primary reference document for the needs and requirements of agricultural animals.*" Similar to the *Guide*, the *Ag Guide* emphasizes the use of performance standards and professional judgment when implementing institutional standards described in the *Ag Guide*, to be applicable among diverse institutions and their agricultural programs.

The third edition of the *Ag Guide* contains 11 chapters: Institutional Policies; Agricultural Animal Health Care; Husbandry, Housing, and Biosecurity; Environmental Enrichment; Animal Handling and Transport; Beef Cattle; Dairy Cattle; Horses; Poultry; Sheep and Goats; and Swine. There are many similarities between the standards of the *Ag Guide* and the *Guide* with the exception that standards unique to individual species of farm animals are described in separate chapters. Many aspects of animal care and use are described in this comprehensive document devoted to the care and use of agricultural animals.

US Government Principles for the Utilization and Care of Vertebrate Animals Used in Testing, Research, and Training (Federal Register, Vol. 50, No. 97, 1985)[21] (the US Government Principles)

The US Government Principles were drafted by the Interagency Research Animal Committee in response to public interest in the care and use of laboratory animals. This document stresses compliance with federal laws, policies, and guidelines and establishes overarching principles that should be applied when using animals, or sponsoring the use of animals, in research, teaching, and testing. There are nine principles:

I. The transportation, care, and use of animals should be in accordance with the Animal Welfare Act (7 U.S.C. 2131 et. seq.) and other applicable Federal Laws, guidelines, and policies.[1]

II. Procedures involving animals should be designed and performed with due consideration of their relevance to human or animal health, the advancement of knowledge, or the good of society.

III. The animals selected for a procedure should be of an appropriate species and quality and the minimum number required to obtain valid results. Methods such as mathematical models, computer stimulation, and in vitro biological systems should be considered.

IV. Proper use of animals, including the avoidance or minimization of discomfort, distress, and pain when consistent with sound scientific practices, is imperative. Unless the contrary is established, investigators should consider that procedures that cause pain or distress in human beings may cause pain or distress in other animals.

V. Procedures with animals that may cause more than momentary or slight pain or in distress should be performed with appropriate sedation, analgesia, or anesthesia. Surgical or other painful procedures should not be performed on unanesthetized animals paralyzed by chemical agents.

VI. Animals that would otherwise suffer severe or chronic pain or distress that cannot be relieved should be painlessly killed at the end of the procedure or, if appropriate, during the procedure.

VII. The living conditions of animals should be appropriate for their species and contribute to their health and comfort. Normally, the housing, feeding, and care of all animals used for biomedical purposes must be directed by a veterinarian or other scientist trained and experienced in the proper care, handling, and use of the species being maintained or studied. In any case, veterinary care shall be provided as indicated.

VIII. Investigators and other personnel shall be appropriately qualified and experienced for conducting procedures on living animals. Adequate arrangements shall be made for their in-service training, including the proper and humane care and use of laboratory animals.

IX. Where exceptions are required in relation to the provisions of these Principles, the decisions should not rest with the investigators directly concerned but should be made, with due regard to Principle II, by an appropriate review group such as an institutional animal research committee. Such exceptions should not be made solely for the purposes of teaching or demonstration.

The US Government Principles are featured prominently in the PHS Policy, the *Guide*, the *Ag Guide*, and many other animal care and use guidelines and have become the foundation for laboratory animal care and use in the US.

The American Veterinary Medical Association Guidelines on Euthanasia

The first report of the AVMA Panel on Euthanasia was published in the AVMA Journal in 1993. The Panel report was revised in 1972, 1978, 1986, 1993, 2000, and 2013. In 2007 the Guidelines were updated and renamed the *AVMA Guidelines on Euthanasia*. The Guidelines are intended for veterinarians who carry out or oversee euthanasia of animals and as a result have become widely used by research institutions in the US. Many IACUCs have developed institutional policies that require euthanasia procedures that conform to the recommendations of the AVMA Guidelines on Euthanasia. The 2013 Guidelines titled *AVMA Guidelines for the Euthanasia of Animals: 2013 Edition* provide a comprehensive discussion of many aspects of euthanasia for a variety of species.[7] The 2013 updated Guidelines include detailed, expanded explanation of methods, techniques, and agents of euthanasia to assist veterinarians in applying their professional judgment.

AAALAC International

The American Association for Accreditation of Laboratory Animal Care (AAALAC) was first established in 1965 as a private nonprofit organization that accredits animal care and use programs. The founders of AAALAC envisioned a voluntary, collaborative, peer-reviewed evaluation of animal care and use programs that engaged both the scientific and laboratory AC communities to promote high standards of animal care and use. AAALAC is governed by Member Organizations and a Board of Directors who represent scientific organizations and they rely upon members of the Council on Accreditation and Ad Hoc consultant specialists to conduct site visits to participating institutions. This triennial site visit process is the basis for rigorous, peer-reviewed assessments to ensure high standards of animal care and use are met by the participating institutions. In 1996, AAALAC was renamed the Association for the Assessment and Accreditation of Laboratory Animal Care, International to reflect its growth worldwide and even more recently it is known simply as AAALAC International. As primary standards of accreditation, AAALAC uses the *Guide, Ag Guide* for research programs that involve agricultural animals, and the European Convention for the Protection of Vertebrate Animals Used for Experimental and Other Scientific Purposes, Council of Europe, European Treaty Series 123 (ETS 123),[20] for accredited programs in Europe. In addition to these primary standards of accreditation, there are many other documents used by AAALAC International as reference resources to aid in the accreditation process. Accreditation by AAALAC International is voluntary but over the years, accreditation has come to represent a standard of excellence in proactive, continuous self-improvement in animal care. The peer review accrediting process fosters collaboration among stakeholders to ensure high-quality science through the promotion of high-quality animal welfare. The impact of the AAALAC International accreditation process in the US has been the establishment of a system of self-assessment and improvement that has become the benchmark of quality animal research programs and includes many institutions across the spectrum of animal care and use; industry, academia, hospitals, nonprofit, and governmental organizations.

The Cornerstones of Research Animal Oversight in the United States

The AWA and AWR, the PHS Policy, the *Guide*, and the *Ag Guide*, all describe the central role of the IACUC in providing the necessary institutional oversight of animals used in research. It is the IACUC that is responsible for oversight and monitoring to ensure the regulations and guidelines are fully implemented. The concept and importance of this committee is vital in animal use oversight in the US. The AWRs, PHS Policy, *Guide,* and *Ag Guide* also stipulate the key roles of the IO and the AV. The prominent role and authority given to the AV in the US may be considered somewhat unique by those in other countries. The AV and the program of veterinary care comprise key aspects of animal health, welfare, and oversight in the US. The *Guide* also describes the requirement for the IO, the IACUC, and the AV to collaborate in the implementation and oversight of the entire animal care and use program and it is this concept that provides the basis for effective animal care and use oversight. Although these three entities are crucial to the effective institutional oversight of animal care and use, one must not overlook the critical role of the investigator and research team. The *Guide* places a high degree of responsibility on those engaged in research to be good stewards of the animals with which they work. It is these four principal entities that form the fabric of research animal compliance in the US.

THE PRINCIPLES

Animal Welfare Act/Animal Welfare Regulations

Section 2131 of the AWA titled, "Congressional statement of policy" on page 1 states, Congress finds that, "(3) measures which eliminate or minimize the unnecessary duplication of experiments on animals can result in more productive use of Federal funds and (4) measures which help meet the public concern for laboratory animal care and treatment are important in assuring that research will continue to progress." The AWRs further state that the IACUC review of proposed animal activities will include consideration of the following principles: procedures will avoid or minimize discomfort, distress, and pain to the animals; ensure that the PI has considered alternatives to procedures that may cause more than momentary or slight pain or distress to the animals and has provided a written narrative description used to determine that alternatives were not available; and further that a written assurance be provided that activities do not unnecessarily duplicate previous experiments; procedures that may cause more than momentary or slight pain of distress will be performed with appropriate sedative, analgesics, or anesthetics unless withholding such agents is justified for scientific reasons and will continue for only the necessary period of time; that paralytics be used only with anesthetics; that a veterinarian be involved in the planning of the experiment; and that timely euthanasia be performed.

Public Health Service Policy and Office of Laboratory Animal Welfare

The HREA of 1985, Public Law 99–158, "Animals In Research," provides the statutory mandate for the PHS Policy. The introduction of the Policy states, "*It is the Policy of the Public Health Service (PHS) to require institutions to establish and maintain proper measures to ensure the appropriate care and use of all animals involved in research, research training, and biological testing activities (hereinafter referred to as activities) conducted or supported by the PHS. The PHS endorses the 'US Government Principles for the Utilization and Care of Vertebrate Animals Used in Testing, Research, and Training' developed by the Interagency Research Animal Committee. This Policy is intended to implement and supplement those Principles.*"

The introductory paragraph of the US Government Principles states, in part, "Whenever U.S. Government agencies develop requirements for testing, research, or training procedures involving the use of vertebrate animals, the following principles shall be considered; and whenever these agencies actually perform or sponsor such procedures, the responsible Institutional Official shall ensure that these principles are adhered to." The Principles were incorporated into the PHS Policy in 1986 and continue to provide a framework for conducting research in accordance with the Policy."

The Three Rs (3Rs) (Replacement, Refinement, and Reduction) are not specifically mentioned in the PHS Policy or the US Government Principles. However, several of the US Government Principles relate directly to the concept of the 3Rs. Principle III relates to using only the minimum number of animals necessary and Principle IV indicates that animal pain and distress should be minimized or avoided and procedures that cause pain or distress in humans may cause pain or distress in other animals. Principle V specifies that appropriate use of sedatives, anesthetics, and analgesics should be employed to minimize pain or distress. Principle VI states that humane euthanasia should be employed to relieve chronic or severe pain or distress. Also, the PHS Policy does require adherence to the USDA AWA, which determines that "*The principal investigator has considered alternatives to procedures that may cause more than momentary or slight pain or distress to the animals, and has provided a written narrative description of the methods and sources, e.g., the Animal Welfare Information Center, used to determine that alternatives were not available.*" OLAW has provided further clarification that the US Government Principles are a federal mandate and that Principles III and IV embody key aspects of the 3Rs. OLAW further states that "*consideration of the three* 'Rs' *should be incorporated into IACUC review, as well as other aspects of the institution's program.*"

Guide

The *Guide* is an internationally accepted primary reference on animal care and use, and its use is required in the US by the PHS Policy; the principles supporting The *Guide* are the same as for the PHS Policy. The *Guide*, on page 1, "*strongly affirms the principle that all who care for, use or produce animals for research, testing or teaching must assume responsibility for their well-being.*" and "*The Guide plays an important role in decision making regarding the use of vertebrate laboratory animals because it establishes the minimum ethical, practice, and care standards for researchers and their institutions.*" With reference to the US AWA and AWR and the PHS Policy, the *Guide* states, "*Taken together, the practical effect of these laws, regulations, and policies is to establish a system of self-regulation and regulatory oversight that binds researchers and institutions using animals.*" Ethics of animal use is of primary concern when making the decision to use animals in research and requires "*critical thought, judgment and analysis.*" The *Guide* specifies that using animals in research is, "*a privilege granted by society*" with the expectation that such use will lead to significant new knowledge or the improvement of human and/or animal well-being. The *Guide* endorses the US Government Principles and places the responsibility for their implementation on the research community. The concepts of 3Rs and their relevance to the ethical use of animals are discussed in detail in the first chapter of the *Guide* on "Key Concepts." In regard to the 3Rs, the *Guide* states on page 5 that "*the Three Rs have become an internationally accepted approach for researchers to apply when deciding to use animals in research and in designing humane animal research studies.*"

Ag Guide

While the *Ag Guide* does not reference the 3Rs directly, like the PHS Policy and the *Guide* it affirms the fundamental importance of the US Government Principles in the use of agricultural animals. The preface to the *Ag Guide* states, *"The US Government Principles for the Utilization and Care of Vertebrate Animals Used in Testing, Research, and Training of the IRAC (1985; Appendix 1) are endorsed in this guide as a basis for professional judgments about the appropriate treatment and use of agricultural animals in research and teaching activities."*

SCOPE/APPLICABILITY

Animal Welfare Act/Animal Welfare Regulations

The AWRs apply to animal activities including "research, teaching, testing, experimentation or exhibition purposes, or as a pet." The AWR further defines the species of animals that are covered and those not covered by the regulations. Those covered include, "any live or dead dog, cat, nonhuman primate, guinea pig, hamster, rabbit, or any other warm-blooded animal…being used for the purposes stated above. Animals excluded are birds, rats of the genus *Rattus*, mice of the genus *Mus* bred for use in research, and horses not used for research purposes and other farm animals, such as, but not limited to, livestock or poultry, used or intended for use as food or fiber, or livestock or poultry used or intended for use for improving management, or production efficiency, or for improving the quality of food or fiber."

Public Health Service Policy and Office of Laboratory Animal Welfare

In section II—Applicability, the PHS Policy states "*The PHS Policy is applicable to all PHS-conducted or supported activities involving animals, whether the activities are performed at a PHS agency, an awardee institution, or any other institution and conducted in the United States, the Commonwealth of Puerto Rico, or any territory or possession of the United States. Institutions in foreign countries receiving PHS support for activities involving animals shall comply with this Policy, or provide evidence to the PHS that acceptable standards for the humane care and use of the animals in PHS-conducted or supported activities will be met.*" It further states that, "*All institutions are required to comply, as applicable, with the Animal Welfare Act, and other Federal statutes and regulations relating to animals.*" OLAW provides further guidance that the PHS Policy is applicable to the generation of custom antibodies in vertebrate species and to larval forms of fish and amphibians, but is not applicable to avian and other egg-laying vertebrate species prior to hatching. The Policy defines "animal" as any live, vertebrate animal used or intended for use in research, research training, experimentation, or biological testing or for related purposes.

Guide

The PHS requires institutions to use the *Guide* as a basis for developing and implementing an institutional program for activities involving animals. In the section titled Applicability and Goals, the *Guide* states that it applies to all vertebrate animals but "*does not address in detail agricultural animals used in production, agricultural research or teaching, wildlife and aquatic species studied in natural settings, or invertebrate animals (e.g., cephalopods) used in research, but establishes general principles and ethical considerations that are also applicable to these species and situations.*" It further states that "*The goal of the Guide is to promote the humane care and use of laboratory animals by providing information that will enhance animal well-being, the quality of research and the advancement of scientific knowledge that is relevant to both humans and animals.*" The intended audience of the *Guide* includes: the scientific community; administrators; IACUCs; veterinarians; educators and trainers; producers of laboratory animals; accreditation bodies; regulators; and the public.

Ag Guide

The *Ag Guide* states that "*Farm animals have certain needs and requirements and these needs and requirements do not necessarily change because of the objectives of the research or teaching activity. Therefore, regardless of the teaching or research objective, the FASS Ag Guide should serve as a primary reference document for the needs and requirements of agricultural animals.*" There are detailed recommendations for specific species including beef and dairy cattle, horses, poultry, sheep, goats, and swine.

AUTHORIZATION OF USER-BREEDING INSTITUTIONS/ INSPECTIONS/PENALTIES BY COMPETENT AUTHORITIES

Animal Welfare Act/Animal Welfare Regulations

The AWRs require dealers/breeders to be licensed and research facilities and carriers to be registered. The AWR define three types of licensees. Type A licensees are breeders that only sell animals bred and raised on their premises. Type B licensees include business that purchase and resell animals. Type C licensees are involved in showing or displaying animals to the public. The type of license that an organization has is dependent on the type of activity constituting the majority of their business. Retail pet stores are not required to be licensed in most cases, as are individuals who have sales of less than the amount defined in the regulations or who do not maintain more breeding animals or sell more animals than defined in the regulations. Prior to obtaining a license each applicant must demonstrate their compliance with the standards and regulations and pay a small application fee. Licenses must be renewed annually and along with a licensing fee based upon their annual sales for dealer and number of animals exhibited for exhibitors.

For research facilities to be registered, they must have in place programs that meet the requirement contained in the regulations relative to composition and function of the IACUC, a program that assures that personnel who care for and use animals are properly qualified to do so, a program of adequate veterinary care, meeting the record-keeping requirements described in the regulations, filing an annual report, and making the premises and records available for unannounced inspections by the officials from the USDA's APHIS. By law inspection of research facilities must take place at least annually. Registration must be renewed every 3 years.

Responsibility for administering the AWA was delegated within the USDA to the Administrator of the APHIS. Enforcement duties are the responsibility of the APHIS Deputy Administrator for Animal Care. Inspections of research facilities are conducted by VMO working under one of the AC Regional Supervisors. They identify and report noncompliant items (NCIs). Depending upon the nature of the action, subsequent enforcement actions are handled by the Investigative and Enforcement Services (IES). The IES investigators follow up on the NCIs and prepare case reports. If a case warrants prosecution, APHIS takes legal action, usually through an administrative law process. Many cases are closed with an official warning, but sometimes IES issues stipulations, which may include a civil fine or other penalty. Fines may be as much as $10,000 per animal per day that the facility is found to be in noncompliance with the regulations and or standards. The more serious cases or those involving repeat violations are submitted to an administrative law judge (ALJ) who can suspend or revoke the violator's USDA license and impose a fine. If a violation is serious enough, IES will work with the Department of Justice to build a criminal case.

The first step in the enforcement process is a 90-day reinspection, which takes place if a facility is making clear progress toward compliance and the inspector found only a few minor NCIs with only a few repeat NCIs demonstrating no signs of animal health or welfare being in jeopardy. The next step is the issuance of an Official Warning Letter (7060). A 7060 is a notification to an individual or company regarding an alleged violation. It may be issued with or without an IES investigation. The reasons for issuing a 7060 include an inspector finding that a facility: is out of compliance after a 90-day reinspection, has multiple repeat NCIs, has a direct NCI (one that has a high potential to have a serious, adverse impact on the animals health and well-being), has incomplete documentation of a serious NCI, or is making slow progress toward compliance. The next step in the process is the issuance of a stipulation. A stipulation is an agreement in which the USDA gives notice of an alleged violation and agrees to accept a specified penalty to settle the matter. The settlement agreement form used by IES requires that the penalty be paid within a designated time frame and states that the payment constitutes a waiver of the alleged violator's right to a hearing and a finding that violations of the law have occurred. The fourth and final step in the process is an Office of General Counsel (OGC) Prosecution. An OGC Complaint gives notice to a facility of a formal allegation of possible violations of the AWA. The Complaint does not mean the facility is guilty of these violations, but serves as a notice that they must respond and either agree to the allegations in the Complaint, or seek a hearing date before a USDA ALJ. The ALJ will issue a Decision and Order based upon the evidence presented by APHIS and the facility. The facility has the right to appeal this decision. A copy of a Decision and Order is made available on the USDA website. This process is initiated following an IES investigation when there are serious NCIs and/or repeat direct or multiple direct NCIs with no progress toward compliance and where animal health and welfare have been impacted when the facility has usually had previous enforcement actions.

In addition to the enforcement actions described above the USDA has the authority to temporarily suspend the license of a dealer and to confiscate and destroy animals being held by a licensee or registered facility, if the animals are suffering due to a failure to comply with the regulations and/or standards.

Public Health Service Policy and Office of Laboratory Animal Welfare

The PHS Policy states on page 19 that "*Each awardee institution is subject to review at any time by PHS staff and advisors, which may include a site visit, in order to assess the adequacy or accuracy of the institution's compliance or expressed compliance with this Policy.*" The NIH OLAW is the office charged with general administration and coordination of the PHS Policy. OLAW will evaluate allegations of noncompliance with the Policy, which may include conducting site visits to selected institutions. OLAW does not have established penalties for noncompliance nor a schedule of regularly conducted inspections; however, OLAW may periodically conduct random site visits to institutions to assess compliance with PHS Policy and the *Guide*. Institutions with a PHS Assurance are required to report to OLAW, any activities that are judged by the IACUC to be noncompliant with approved protocols, protocol suspensions, and deviations from the *Guide* and PHS Policy. If warranted, OLAW can rescind an institution's PHS assurance, which may have direct consequences on the institution's eligibility to receive NIH grant awards.

Guide

The *Guide* itself does not authorize inspections or require reporting to competent authorities; OLAW is authorized to monitor and oversee all institutions that hold a PHS assurance for adherence to the PHS Policy and the *Guide* as outlined above. Given the prominent role of the IACUC and the level of institutional authority placed upon it, the *Guide's* importance in assisting the IACUC in protocol review, assessment, and oversight of the program is central to the animal oversight/regulatory process in the US.

Ag Guide

There are no statutory mandates or penalties associated with the *Ag Guide*. The standards of the *Ag Guide* are recommendations by the agricultural animal research community and are accepted as such.

AAALAC International

Although participation in AAALAC accreditation is voluntary, there are standards that must be met in order for an institution to attain or maintain accreditation. Apart from a requirement to adhere to the AWRs, AAALAC uses the *Guide* and the *Ag Guide* as primary standards for accreditation for the US institutions. AAALAC conducts site visits at least once every 3 years to assess animal care and use programs and determines whether the standards of the *Guide* and/or *Ag Guide* are met. Animal care and use programs must specify which standard they have implemented, the *Guide*, the *Ag Guide*, or both (Note: AAALAC International also uses ETS 123 as a primary standard of accreditation, but this standard is used mainly for institutions in Europe). In the US, accredited programs that do not meet one or more of the standards of the *Guide* or *Ag Guide* are placed on Deferred or Probationary accreditation, depending on the nature of the concern, until the concern is resolved. For programs that are unable to meet the standards of the *Guide* or *Ag Guide*, AAALAC International will withhold accreditation or revoke accreditation (if the program is currently accredited). Because accreditation by AAALAC International is widely recognized as a high standard, it has become valued by many institutions in the US and maintaining accreditation is respected.

NONHUMAN PRIMATES: SPECIAL CONSIDERATIONS AND RESTRICTIONS

Animal Welfare Act/Animal Welfare Regulations

Within the Standards promulgated under the AWA is a section on nonhuman primates entitled "Environment enhancement to promote psychological well-being." To comply with these requirements dealers, exhibitors, and research facilities must develop, implement, and document a plan that promotes the psychological well-being of these animals. This plan must be in accordance with currently accepted standards in the field and directed by the AV. The plan must include provisions to address social grouping of nonhuman primates known to exist in social groups, environmental enrichment to allow animals to express noninjurious species–typical behavior, special consideration such as occurs with infants and juveniles, animals showing signs of psychological distress, animals on research projects requiring IACUC-approved restricted activity, individually housed animals, and great apes weighing over 50 kg. In addition the plan must address the use of restraint devices. The component of a plan that addresses social grouping must include provisions for dealing with primates exhibiting vicious or overly aggressive behavior or are debilitated because of age or other conditions; animals suspected of having contagious diseases; and animals that are not compatible with other animals. The IACUC can exempt animals from the program for scientific reasons and these exemptions must be reviewed and approved annually. The AV can exempt animals from the plan for reasons of health or well-being. These exemptions must be reviewed every 30 days unless the basis for the exemption is a permanent condition.

Public Health Service Policy and Office of Laboratory Animal Welfare

The PHS Policy does not make specific mention of nonhuman primates. OLAW has guidance regarding nonhuman primates: one regarding social housing expectations and the other encouraging positive reinforcement training (OLAW, frequently asked questions, http://grants.nih.gov/grants/olaw/faqs.htm#useandmgmt_14).

Guide

Information regarding nonhuman primate environment, husbandry, minimum space requirements for cages and other primary enclosures, environmental enrichment, social housing, veterinary care, and occupational health and safety issues are discussed in the *Guide*. Use of chair restraint for nonhuman primate studies is specifically discouraged. Emphasis is placed on careful consideration for establishing pair or group housing of nonhuman primates and the need to consider important species-specific factors that can affect compatibility: age, behavioral repertoire, sex, natural social organization, breeding, and health status. When considering the cage space requirements for nonhuman primates, professional judgment is paramount and decisions should not be based on body weight alone. Further, the *Guide* states, "*Because of the many physical and behavioral characteristics of nonhuman primate species and the many factors to consider when using these animals in a biomedical research setting, species-specific plans for housing and management should be developed. Such plans should include strategies for environmental and psychological enrichment.*"

Ag Guide

The *Ag Guide* does not apply to research, teaching, or testing with nonhuman primates.

GENETICALLY ALTERED ANIMALS: SPECIAL CONSIDERATION

Animal Welfare Act/Animal Welfare Regulations

The AWRs and standards make no special provisions for genetically altered animals.

Public Health Service Policy and Office of Laboratory Animal Welfare

Neither the PHS Policy nor OLAW make specific mention of genetically altered animals.

Guide

Because genetically modified animals (GMAs), particularly mice and fish, are important animal models and the results of genetic manipulation can be unpredictable, there may be unexpected consequences that impact animal well-being. The *Guide* recommends that the first offspring of a newly generated GMA line be carefully observed from birth into early adulthood for signs of disease, pain, or distress and that when the initial characterization of a GMA reveals a condition that negatively affects animal well-being, this should be reported to the IACUC for review. Careful consideration of the phenotype is important during IACUC review of proposed studies to ensure that "*proactive measures can circumvent or alleviate the impact of the genetic modification on the animal's well-being and to establish humane endpoints specific to the GMA line.*"

Ag Guide

The *Ag Guide* provides careful consideration and recommendations for genetically altered farm animals and defines specific terms. A transgenic animal is one that carries a foreign gene that has been deliberately inserted into the genome. Genetic engineering is defined as the direct manipulation of an organism's genes including heritable and nonheritable recombinant DNA constructs. The genetic engineering of agricultural animals has been extensively reviewed (National Research Council, 2002; Council on Agricultural Science and Technology, 2003, 2007, 2009).[21] In many instances, research with genetically engineered farm animals is regulated by the USDA and NIH as most of these studies are considered biomedical activities. Also the US Food and Drug Administration has guidelines for the conduct of research with GMAs (http://www.fda.gov/AnimalVeterinary/DevelopmentApprovalProcess/GeneticEngineering/GeneticallyEngineeredAnimals/default.htm).[22] The *Ag Guide* describes cloning as one method of genetic engineering involving somatic cell nuclear transfer. Cloning is considered an assisted reproductive technology similar to artificial insemination, embryo transfer, and in vitro fertilization. The welfare of cloned animals is similar to that of other conventional agricultural animals with the recognition that there may be a rather low incidence animal health issues that may arise in cloned animals. Because of concerns related to the disposition of cloned agricultural animals in the food chain it is recommended that "*institutions and researchers participate in the Livestock Industry Clone Registry whereby and animal clones are registered in the database or registry* (www.livestockcloneregistry.com)." The *Ag Guide* indicates that health and welfare concerns, which may arise with transgenic and genetically engineered animals, are important to understand so that issues related to animal care and well-being are thoroughly considered. The *Ag Guide* states on page 14 that, "*The scientist is responsible for identifying physiologic and phenotypic changes and must have a plan to address changes that affect animal health to facilitate and ensure animal welfare.*" It is further acknowledged that the IACUC, AC staff, and veterinarians as well as researchers should be involved in the monitoring and care of genetically engineered animals because of unexpected phenotypes that can result. Protocols involving genetically engineered animals should receive careful review by the IACUC with consideration of such issues as proper animal identification, public safety, animal welfare, and final disposition of the animals. In several of the species-specific chapters there are recommendations regarding known welfare concerns associated with genetically altered animals, from dystocia and large offspring to retained placenta, hydrops and multiple births, and the emphasis to provide careful IACUC review, monitoring, and oversight, with recognition of potential welfare concerns that are at the forefront of these guidelines.

INSTITUTIONAL AND DESIGNATED PERSONNEL RESPONSIBILITIES

Animal Welfare Act/Animal Welfare Regulations

The AWR define the responsibilities of five key individuals or groups: the CEO, the IO, the AV, the PI, and the IACUC. The regulations require that members of the IACUC be appointed by the CEO. At a minimum, this committee must include a chair, a veterinarian with training and experience in laboratory science and medicine and with direct or delegated programmatic responsibilities for activities involving animals, and at least one person who is not affiliated with the research facility who is expected to provide representation for the general community interests in the proper care and treatment of animals. If the committee consists of more than the mandated three members, no more than three can come from the same administrative unit.

The IACUC is charged to act as an agent of the research facility in assuring compliance with the AWA. It is required to inspect all animal facilities and study areas at least once every 6 months, and to review the condition of the animals and the practices involving pain and distress to the animals to assure compliance with the regulations and standards promulgated under the

AWA. The IACUC is also required to review once every 6 months the research facility's program to assure the care and use of the animals conforms to the regulations and standards. The IACUC must file a report of its inspections with the IO of the research facility. The IO is an individual within the research facility who is legally authorized to commit on behalf of the facility that the regulations and standards will be followed. Failure to adhere to the plan and schedule that results in a deficiency, which may be or is a threat to animal health or safety, remaining uncorrected shall be reported in writing within 15 business days by the IACUC, through the IO, to APHIS and any Federal agency funding that activity.

The IACUC must also review and approve all proposed activities involving the care and use of animals in research, testing, or teaching procedures and all subsequent significant changes of ongoing activities. It is the responsibility of the PI to submit these proposals and for the design and implementation of the research involving animals.

Research facilities are required to file an annual report in which either the CEO or the IO must assure that professionally accepted standards governing the care, treatment, and use of animals were in place prior to, during and following the actual use of the animals; the PI considered alternatives to painful procedures; and the facility is adhering to the regulations and standards.

Public Health Service Policy and Office of Laboratory Animal Welfare

PHS Policy assigns each awardee institution the responsibility of providing verification of approval by the IACUC of those components of the PHS application or proposal related to the care and use of animals. Institutions must have on file with OLAW an approved written Animal Welfare Assurance document, which describes the institution's compliance with the PHS Policy. That Assurance must describe the lines of authority and responsibility for administering the program and ensuring compliance. There must be an IO named who has the authority to sign the institution's Assurance, making a commitment on behalf of the institution that the requirements of the PHS Policy will be met. The CEO acting as the IO, or a designated delegate must appoint an IACUC that consists of at least 5 members including: the veterinarian(s) with direct or delegated program authority and responsibility for animal activities; one practicing scientist experienced in research involving animals; one member whose primary concerns are in a nonscientific area; and one individual who is not affiliated with the institution in any way other than as a member of the IACUC.

OLAW further specifies that each institution must provide: personnel training necessary to comply with PHS Policy; an occupational health and safety program (OHSP) for personnel who have frequent contact with animals; and an animal facility disaster plan.

Guide

The *Guide* reflects the PHS Policy in regard to institutional responsibilities and required specific personnel. The *Guide*, on pages 13 and 14, states that the primary oversight responsibilities for the institutional animal care and use program rest with the IO, the AV, and the IACUC. Together they "*establish policies and procedures, ensure regulatory compliance, monitor Program performance, and support high-quality science and humane animal use.*" The IO bears ultimate responsibility for the Program and must have "the authority to allocate the resources needed to ensure the Program's overall effectiveness." The AV is responsible for the health and well-being of all laboratory animals used at the institution. The AV must have sufficient institutional authority and resources to manage the program of veterinary care. The IACUC (or institutional equivalent) is responsible for assessment and oversight of the institution's program components and facilities, and review and approval of proposed animal use and of proposed significant changes to animal use. The IACUC should have appropriate institutional authority and resources to fulfill its responsibilities. The "*IO is responsible for resource planning and ensuring alignment of Program goals of quality animal care and use with the institution's mission.*" Further, the IO ensures that the Program's overall effectiveness and the needs are met. The needs of the Program "*should be clearly and regularly communicated to the IO by the AV the IACUC and others associated with the program (e.g., facilities management staff, occupational health and safety personnel, scientists).*"

Ag Guide

The *Ag Guide* states on page 1 that there are critical components in an effective agricultural animal care and use program and include "*1) clearly established lines of authority and responsibility; 2) an active Institutional Animal Care and Use Committee (IACUC); 3) procedures for self monitoring of the IACUC through semi-annual review of programs and facility oversight by the institutional officer; 4) appropriately maintained facilities for proper management, housing and support of animals; 5) an adequate program of veterinary care; and 6) training and occupational health programs for individuals who work with animals.*" From these critical components, three relate to the important roles of the IO, the IACUC, and the AV. The *Ag Guide* furthers states that there should be clearly designated lines of authority within the animal care and use program. The CEO should appoint the IACUC, which is tasked with monitoring the program of animal care and use and specific responsibilities of the IACUC are outlined. The IACUC should be comprised of at least five members, each fulfilling a specific role. In addition to describing the responsibilities of the IACUC, their role of the AV is also provided on page 9. The AV "*must have authority to ensure that the provisions of the program are met. The Attending Veterinarian must be provided access to all research and teaching animals and to any related documents including health care records. The attending*

Veterinarian also must be involved in the development of and oversight of the veterinary care program, as well as other aspects of animal care and use such as protocol review, establishment of anesthetic and analgesic guidelines, study removal criteria, training of animal users, and responsible conduct of research activities."

OVERSIGHT AND ETHICAL REVIEW PROCESS

Animal Welfare Act/Animal Welfare Regulations

The IACUC is responsible for reviewing and approving and requiring modifications of proposed activities involving the use of animals in research, teaching, testing, and experimentation. As part of this review, the IACUC must evaluate procedures that minimize discomfort, distress, and pain. In addition, the IACUC must determine that a veterinarian has been consulted in planning for the administration of anesthetics, analgesics, and tranquilizers, and that paralytic agents are used only in anesthetized animals. The IACUC must also determine that animals, which experience severe or chronic pain, are euthanatized consistent with the design of the study, that the living conditions meet the species' needs, that necessary medical care will be provided, that all procedures will be performed by qualified individuals, that survival surgery will be performed aseptically, and that no animal will undergo more than one major operative procedure that is not justified and approved. The IACUC must also assure on behalf of the research facility that the PI considered alternatives to painful procedures and that the work being proposed does not unnecessarily duplicate previous experiments. To provide assurance of the former, the IACUC must review the written narrative description provided by the investigator. This description must include the methods and sources used in determining that alternatives were not available.

In reviewing proposed activities and modifications, the IACUC can grant exceptions to the regulations and standards, if they have been justified in writing by the PI.

The IACUC must conduct continuing review of activities at least annually and may suspend an activity that it has previously approved for failure to adhere to the description of the activity approved by the committee.

Public Health Service Policy and Office of Laboratory Animal Welfare

PHS Policy requirements for IACUC oversight and ethical review of animal activities are similar to the AWRs previously described. The PHS Policy does not explicitly require postapproval monitoring (PAM) procedures to compare the practices described in approved protocols against the manner in which they are actually conducted. IACUCs are charged, however, with program oversight and as such are responsible for conducting semiannual facility inspections and programmatic review, review of protocols and amendments, reporting noncompliance, ensuring that individuals who work with animals are appropriately trained and qualified, and addressing concerns involving the care and use of animals at the institution. Specific protocol elements to be reviewed include minimization of pain and distress; appropriate sedation, analgesia, or anesthesia; appropriate living conditions and veterinary care; adequately trained personnel; and appropriate euthanasia methods. The maximum duration for a protocol following IACUC approval is 3 years (i.e., a complete de novo review of protocols is required at least every 3 years).

Guide

The *Guide* supports and elaborates on PHS Policy expectations and procedures for IACUC review and oversight of the care and use of animals. Specifically, the *Guide* states that, "*The committee must meet as often as necessary to fulfill its responsibilities and records of committee meeting and results of deliberations should be maintained.*" The following topics should be included in protocols and reviewed by the IACUC:

- Rationale and purpose of the proposed use of animals
- A clear and concise sequential description of the procedures involving the use of animals that is easily understood by all members of the committee
- Availability or appropriateness of the use of less invasive procedures, other species, isolated organ preparation, cell or tissue culture, or computer simulation
- Justification of the species and number of animals proposed; whenever possible, the number of animals and experimental group sizes should be statistically justified
- Unnecessary duplication of experiments
- Nonstandard housing and husbandry requirements
- Impact of the proposed procedures on the animals' well-being
- Appropriate sedation, analgesia, and anesthesia
- Conduct of surgical procedures, including multiple operative procedures
- Postprocedural care and observation (e.g., inclusion of posttreatment or postsurgical animal assessment forms)
- Description and rationale for anticipated or selected endpoints
- Criteria and process for timely intervention, removal of animals from a study, or euthanasia if painful or stressful outcomes are anticipated
- Method of euthanasia or disposition of animals, including planning for care of long-lived species after study completion
- Adequacy of training and experience of personnel in the procedures used, and roles and responsibilities of the personnel involved
- Use of hazardous materials and provision of a safe working environment

The *Guide*, on pages 27–28, describes specific issues that require special consideration by the IACUC: experimental humane endpoints, unexpected outcomes, physical restraint of conscious animals, multiple survival surgical procedures, food and/or water regulation, the use of nonpharmaceutical grade chemicals and other substances, field investigations, and the use of agricultural animals in biomedical research. In addition to the specific topics to be included in animal use protocols, the IACUC is also tasked with evaluating scientific elements of the protocol as they relate to animal welfare (e.g., hypothesis testing, samples size, group numbers, and adequacy of controls). IACUC members named in protocols or who have other conflicts must recuse themselves from decisions concerning these protocols.

The *Guide* describes ongoing IACUC monitoring of approved activities as PAM and states on page 33 that, "*PAM helps ensure the well-being of the animals and may also provide opportunities to refine research procedures.*" Various methods of PAM are described: continuing protocol review; laboratory inspections; veterinary or IACUC observations of specific procedures; observations of animals by animal care, veterinary, and IACUC staff; and external regulatory inspections. In the US, institutions are required by the AWRs and PHS Policy, to conduct semiannual inspections and program reviews. Depending on the complexity and size of the animal care and use program, the conduct of these required activities may be sufficient to meet the expectations of the *Guide* for PAM. Effective monitoring includes as follows:

- Examination of surgical areas, including anesthetic equipment
- Use of appropriate aseptic technique, and handling and use of controlled substances
- Review of protocol-related health and safety issues
- Review of anesthetic and surgical records
- Regular review of adverse or unexpected experimental outcomes affecting the animals
- Observation of laboratory practices and procedures and comparison with approved protocols

Ag Guide

The *Ag Guide* specifies that an institutional IACUC shall be appointed by the CEO and should consist of at least five members; a scientist who has experience in agricultural animal research or teaching; an animal, dairy, or poultry scientist who has training and experience in the management of agricultural animals; a veterinarian who has training or experience in agricultural animal medicine and who is licensed or eligible to be licensed to practice veterinary medicine; a person whose primary concerns are in an area outside of science; a person who is not affiliated with the institution and who is not a family member of an individual affiliated with the institution; other members as required by institutional needs. One individual may adequately fill more than a single role on the IACUC but the committee must consist of at least five members. The IACUC is authorized to: review, approve, or disapprove protocols and significant changes in ongoing activities; conduct at least twice each year an inspection of the facilities; study areas and review the overall program of animal care and use providing the IO with a written report of the findings; investigate concerns, complaints, or reports of noncompliance; suspend an activity; make recommendations regarding the development and implementation of policies and procedures to facilitate, support, and monitor humane care and use of animals; and perform other functions required by institutional need and applicable laws, regulations, and policies. The oversight role of the IACUC includes consideration, review, and monitoring of many programmatic aspects of animal care and use. IACUC oversight includes but is not limited to: protocols and written operating procedures; handling and transport; restraint; husbandry, housing, and biosecurity; environmental enrichment and social needs of the animals; personnel qualifications; occupational health and safety; hazardous materials; genetically altered animals; animal procurement, quarantine, acclimation, and stabilization; the program of veterinary care; surgery, anesthesia, analgesia, and postsurgical care; residue avoidance for research or teaching animals that may enter the human food chain; recordkeeping; drug storage and control; and euthanasia.

REUSE

Animal Welfare Act/Animal Welfare Regulations

The AWRs and standards make no special provisions for reuse of animals other than to prohibit reuse for major operative procedures unless justified for scientific reasons or as part of routine veterinary care.

Public Health Service Policy and Office of Laboratory Animal Welfare

The PHS Policy and OLAW make no specific recommendations regarding the reuse of animals other than to prohibit reuse for major operative procedures unless justified for scientific reasons or as part of routine veterinary care.

Guide

The *Guide* recommends that procedures, which may induce substantial post-procedural pain or impairment, should be scientifically justified if performed more than once in a single animal. On page 5 the *Guide* states, "*Refinement and reduction goals should be balanced on a case-by-case basis. Principal investigators are strongly discouraged from advocating animal reuse as a reduction strategy, and reduction should not be a rationale for reusing an animal or animals that have already undergone experimental procedures especially if the well-being of the animals would be compromised.*"

Ag Guide

The *Ag Guide* does not address the reuse of animals except with regard to performing more than one major surgery on a single animal. A distinction is made between multiple surgeries for therapeutic reasons such as for cesarean section or displaced abomasum in cattle, and surgeries necessary to complete the scientific aims of the study. In the latter case, scientific justification is required to perform more than one major operative procedure.

SETTING FREE/REHOMING

Animal Welfare Act/Animal Welfare Regulations

The AWRs and standards make no special provisions for setting free/rehoming.

Public Health Service Policy and Office of Laboratory Animal Welfare

The PHS Policy does not specifically address the use of wildlife and has no setting free/rehoming requirements. OLAW guidance on the use of wildlife states that "Investigators are encouraged to consult relevant professional societies, available guidelines, wildlife biologists, and veterinarians, as applicable, in the design of the field studies" and that "proposed studies are in accord with the Guide."[23]

Guide

The *Guide* addresses the issue of rehoming by stating on page 32 that "*When species are removed from the wild, the protocol should include plans for either a return to their habitat or their final disposition, as appropriate.*"

Ag Guide

The *Ag Guide* does not address the topic of setting free/rehoming.

While there is no mandate in the US for setting free/rehoming, "animal adoption" is a practice that institutions may employ when possible. Careful consideration of issues related to transfer of animal ownership, institutional liability, long-term health and welfare of the animals, provision of proper husbandry and the animal's overall needs, such as environment and veterinary care is prudent. Ensuring that animals are rehomed conscientiously with responsible stewardship is essential.

OCCUPATIONAL HEALTH AND SAFETY

Animal Welfare Act/Animal Welfare Regulations

The AWRs and standards make no special provisions for occupational health and safety.

Public Health Service Policy and Office of Laboratory Animal Welfare

The PHS Policy requires that a health program be in place for personnel who work in laboratory animal facilities or have frequent contact with animals. Further the PHS Policy requires that programs of animal care and use meet the requirements of the *Guide*, which describes the essential elements of an OHSP. OLAW provides more specific guidance that "An effective occupational health and safety program must encompass all personnel that have contact with animals." and "Minimally, the program should include: pre-placement medical evaluation; identification of hazards to personnel and safeguards appropriate to the risks associated with the hazards; appropriate testing and vaccinations; training of personnel regarding their duties, any hazards, and necessary safeguards; policies and facilities that promote cleanliness; provisions for treating and documenting job-related injuries and illnesses; facilities, equipment, and procedures should be designed, selected, and developed to reduce the possibility of physical injury or health risk to personnel; good personal hygiene practices, prohibiting eating and drinking, use of tobacco products, and application of cosmetics and/or contact lenses in animal rooms and laboratories; and personal protective equipment (PPE)."[24]

Guide

The *Guide* states on page 17 that "*Each institution must establish and maintain an occupational health and safety program as an essential part of the overall Program of animal care and use…The nature of the OHSP will depend on the facility, research activities, hazards, and animal species involved. The OHSP must be consistent with federal, state, and local regulations and should focus on maintaining a safe and healthy workplace.*" The *Guide* describes many aspects of an OHSP on pages 18–24, and describes key elements of an OHSP. An OHSP should include programs for control and prevention strategies for the protection of personnel, which comprise a hierarchy of importance: (1) appropriate design and operation of facilities as well as the use of safety equipment, (2) the development of standard operating procedures; and (3) the provision of appropriate personnel protective equipment.

Another key feature of an OHSP is hazard identification and risk assessment. The identification of work-related risks and hazards is essential to ensure that risks are reduced to minimal and acceptable levels. The assessment of work related risks such as biologic agents, radiation, physical hazards, chemical agents, animal bites, scratches and kicks, allergy, and zoonoses is important and should be performed on an ongoing basis. Strategies for minimizing risk should be based on these periodic assessments. The *Guide* states on page 18 that, "*The extent and level of participation of personnel in the OHSP should be based on the hazards posed by the animals and materials used (the severity or seriousness of the hazard); the exposure intensity, duration and frequency (prevalence of the hazard); to some extent the susceptibility (e.g., immune status) of the personnel; and the history of occupational illness and injury in the particular workplace.*"

An OHSP program includes appropriate facilities, equipment, and monitoring. The facilities should preferentially employ engineering controls and equipment to minimize personnel exposure and risk. Aspects of personnel hygiene can also be important to minimize exposure (e.g., changing, washing and showering facilities, and supplies). When biologic agents are used the *Guide* recommends following the safety guidelines published in the Center for Disease Control (CDC) and NIH publication, *Biomedical safety in Microbiological and Biomedical Laboratories* (BMBL).[25]

Personnel training in occupational health and safety is another key aspect to an effective OHSP. The *Guide* states on page 20 that, "*Personnel at risk should be provided with clearly defined procedures, practices and protective equipment to safely conduct their duties, understand the hazards involved and be proficient in implementing the required safeguards. They should be trained regarding zoonoses, chemical, biologic, and physical hazards (e.g., radiation and allergies), unusual conditions or agents that might be a part of experimental procedures (e.g., the use of human tissue in immunecompromised animals), handling of waste materials, personal hygiene, the appropriate use of PPE, and other considerations (e.g., precautions to be undertaken during pregnancy, illness, or immunosuppression) as appropriate to the risk imposed by their workplace.*"

Occupational health and safety professionals should be involved in the development and implementation of programs for medical evaluation of personnel and preventive medicine. A preemployment health evaluation is advisable and periodic medical evaluation should occur at appropriate intervals given the nature of risks posed. The *Guide* states, "*Zoonoses surveillance should be part of the OHS. Personnel should be instructed to notify supervisors of potential or know exposures and of suspected health hazards and illness.*" Further the *Guide* states that, "*Clear procedures should be established for reporting all accidents, bites, scratches and allergic reactions and medical care for such incidents should be readily available.*"

Ag Guide

The *Ag Guide* requires that an OHSP be established for those who work with agricultural animals. Key elements of an OHSP include adherence to federal, state, and local regulations; risk assessment by health and safety specialists; consideration of hazards posed by the animals, the materials used, the duration, frequency, and intensity of exposure; the susceptibility of personnel; and the history of occupational injury and illness in the workplace. Health assessments of personnel prior to job placement and periodically thereafter are recommended. Mechanisms for surveillance to ensure protection from health hazards and provision of occupational health care services are also recommended. There should be an educational component of the OHSP to educate personnel about risks associated with: zoonoses; physical hazards; containment of and protection from hazardous agents; personal hygiene; noise; safety procedures for personnel protection; and special precautions for individuals who may be at higher risk (pregnancy, chemical and radiation hazards, etc.). Appropriate immunizations should be available and animal caretakers should receive tetanus immunization at least every 10 years. Considerations of allergies and physical injury are important for those working with agricultural animals. Personnel should be knowledgeable about the risks associated with agricultural animals, the specific hazards that may be involved in the study, their overall participation in the OHSP, and be sufficiently trained so that risks of animal contact are minimized.

Occupational Health and Safety in the Care and Use of Research Animals

In the US there is an additional occupational health and safety reference that is widely used as recommended guidelines for personnel in the animal research environment, "*Occupational Health and Safety in the Care and Use of Research Animals*" (NRC, 1997).[26] These guidelines are important, comprehensive recommendations for the structure and function of OHSPs involving research animals. The *Guide* and *Ag Guide* reference the use of these guidelines when establishing OHSPs. Key sections of the guidelines include: program design and management; physical, chemical, and protocol-related hazards; allergens; zoonoses; principle elements of an OHSP; and occupational health care services. These guidelines are detailed and complete and establish the current standards in the US for OHSPs involving research animals.

EDUCATION, TRAINING, AND COMPETENCE OF PERSONNEL

Animal Welfare Act/Animal Welfare Regulations

It is the responsibility of the research facility to provide training to scientists, animal technicians, and other personnel involved with animal care and treatment. This training should include the use of humane methods of animal maintenance and experimentation; the concept, availability, and use of research or testing methods that limit the use of animals or minimize animal distress; the proper use of anesthetics, analgesics, and tranquilizers for any species of animals used by the facility; and methods whereby deficiencies in animal care and treatment can be reported. In addition the training should include information on the utilization of library services available to provide information on these subjects.

Public Health Service Policy and Office of Laboratory Animal Welfare

The PHS Policy requires that institutions describe the "*training or instruction in the humane practice of animal care and use, as well as training or instruction in research or testing methods that minimize the number of animals required to obtain valid results and minimize animal distress, offered to scientists, animal technicians, and other personnel involved in animal care, treatment, or use,*" and that the IACUC ensure that "*Personnel conducting procedures on the species being maintained or studied will be appropriately qualified and trained in those procedures.*" In addition, there is a specific requirement that the veterinarian with "*direct or delegated program authority and responsibility for activities involving animals at the institution*" has "*training or experience in laboratory animal science and medicine.*"

Guide

The *Guide* further elaborates on the tenets of the PHS Policy stating that "*All personnel involved with the care and use of animals must be adequately educated, trained, and/or qualified in basic principles of laboratory animal science to help ensure high-quality science and animal well-being.*" Specific expectations and recommendations for training and qualifications for veterinary, animal care, and research staff as well as IACUC members are described in this chapter. Training expectations for specific activities, such as surgery, euthanasia, anesthesia, or hazards, such as exposure to hazardous agents or workplace risks, are also described in the *Guide*.

Veterinary staff must have experience or training and expertise necessary to evaluate the health and well-being of the animals in the context of the animal use at the institution. Training or experience in other areas may also be important for veterinarians, such as facility administration and management, facility design and renovation, human resources management, pathology of laboratory animals, comparative genomics, facility and equipment maintenance, diagnostic laboratory operations, and behavioral management. AC personnel should be appropriately trained and the institution should provide formal and/or on-the-job training. Staff should receive specific training, or have experience, to complete their duties. External training programs for technical staff are available and include the American Association of Laboratory Animal Science (AALAS) and Laboratory Animal Welfare and Training Exchange (LAWTE).

All members of the research team should receive training or have the necessary knowledge, experience, and expertise to conduct the specific procedures with the species used. Training should be aimed at the specific needs of the research groups and specifically include the topics of animal care and use legislation; IACUC function; ethics of animal use and concepts of the 3Rs; methods for reporting concerns about animal use; occupational health and safety issues pertaining to animal use and animal handling; aseptic surgical technique; anesthesia and analgesia; and euthanasia.

The institution also has a responsibility to provide the necessary training for the IACUC members so that they understand their role and responsibilities. Training should include formal orientation to the institutional program of animal care and use; relevant legislation, regulations, guidelines, and policies; animal facilities and laboratories in which animal studies are conducted; and the process of animal protocol review. The overall frequency of training should ensure that all personnel who work with animals are properly trained before they begin animal work. The IACUC, veterinary, and AC staffs should be provided ongoing opportunities for training. All personnel training should be documented.

Ag Guide

On page 3, the *Ag Guide* states that, "*It is the responsibility of the institution to ensure that scientists, agricultural animal care staff, students, and other individuals who care for or use agricultural animals are qualified to do so through training or experience.*" Training programs should be tailored to user needs and include information about: husbandry; handling; surgical procedures and postprocedural care; methods for minimizing the numbers of animals used; techniques for minimizing pain and distress; proper use of anesthetics, analgesics, tranquilizers, as well as nonpharmacologic methods of pain relief; methods for reporting deficiencies in the AC program; use of information services; and methods of euthanasia. Records of personnel training should be maintained. Those responsible for animal care should participate regularly in training and education activities relevant to their responsibilities. Both formal and on-the-job training opportunities should be made available. It is desirable for the AC staff to undergo professional training and certification. The AALAS and the American Registry of Professional Animal Scientists (ARPAS) are two certification organizations that provide such training and certification.

TRANSPORT

Animal Welfare Act/Animal Welfare Regulations

The requirements for transportation are covered in the standards sections for the species covered by the AWA. These standards include requirements for consignment to carriers and intermediate handlers, the enclosures used to transport the animals, the cargo space of the primary conveyance, provision of food and water, care in transit, terminal facilities, and handling of the animals to include things such as shelter from sunlight, rain or snow, and cold weather.

Public Health Service Policy and Office of Laboratory Animal Welfare

The PHS Policy does not specifically address the transport of animals; however, the US Government Principles state that the transportation of animals should be in accordance with the AWA and other applicable federal laws, guidelines, and policies. "OLAW expects all parties involved in the transportation of animals to apply due diligence in assuring that animals are shipped under appropriate conditions to prevent morbidity or mortality due to temperature extremes or other adverse events. OLAW expects shipping institutions to report adverse events that occur to animals in transit."[27]

US Government principle number one states, "*The transportation, care and use of animals should be in accordance with the Animal Welfare Act (7 U.S.C. 2131 et. seq.) and other applicable Federal laws, guidelines, and policies.*" This reference includes adherence to the *Guide*.

Guide

The *Guide* describes appropriate transport of animals in detail on pages 107–109 stressing the guidance in the AWRs and the need to comply with various governmental agency requirements involving animal transportation. A number of agencies and the requirements for animal transportation are cited including, the National Research Council publication Guidelines for the Humane Transportation of Research Animals (NRC 2006).[11] Careful planning and coordination of transportation procedures should be employed to ensure animal safety and well-being. Personnel responsible for transportation should be well-trained and transit time should be minimized with delivery times occurring during normal business hours. Animals in transit between institutions should be accompanied by appropriate documentation (health certificates, agency permits, sending and receiving institution's addresses, contacts, emergency procedures, and veterinary contact information). The *Guide* indicates that key aspects of transportation include an appropriate level of animal biosecurity while minimizing zoonotic risks; protecting against environmental extremes; avoiding overcrowding; providing for the animals' physical, physiologic, or behavioral needs and comfort; and protecting the animals and personnel from physical trauma. "Transportation of animals in private vehicles is discouraged because of potential animal biosecurity, safety, health, and liability risks for the animals, personnel, and institution."

Ag Guide

The *Ag Guide* contains detailed recommendations for the handling and transport of agricultural animals and highlights the goal of calm, respectful, stress-free methods for the benefit of the animals and to promote sound research practices. There are specific guidelines for several species with topics that include: flight zone and behavior principles; aids for moving animals; considerations for animal perception, hearing, vision, and visual distractions; flooring; proper maintenance of equipment; sanitation; and principles to prevent behavioral agitation. Specific recommendations are provided for beef and dairy cattle, horses, sheep and goats, pigs, and poultry. There are also specific recommendations for safe animal transport that include: guidelines regarding animals not suitable for transport (near parturition, new born, weak, debilitated, etc.); thermal environment, bedding, and floor surfaces during transport; space provisions during transport (the guidelines published by the National Academy of Sciences are referenced) (ILAR Transportation Guide, 2006)[11]; lairage; adherence to guidelines and federal regulations regarding duration of transportation; and methods for loading and unloading.

HOUSING AND ENRICHMENT

Animal Welfare Act/Animal Welfare Regulations

The requirements for housing are contained within the sections of the standards on facilities and operating standards for the species covered by the AWA. Included in this section are specific requirements for facilities, in general, including that they be structurally sound and maintained in good repair to protect the animals from injury and to keep them secure. In addition, the site must be maintained in a clean and orderly manner. There are requirements for the surfaces of the facility including the need to maintain and replace surfaces on a regular basis and for cleaning on a regular basis. There are specific requirements for the provision of water and electric power, storage, drainage and waste disposal, and washrooms and sinks for the employees. In addition to the general requirements for facilities there are specific requirements for indoor, sheltered, outdoor, and mobile facilities, which include areas such as heating, cooling, temperature control, ventilation, lighting, and interior surfaces. There are also specific requirement for the design and construction of primary enclosures.

Part 3 of the AWRs—Standards, describe the requirements for environmental enhancement to promote the psychological well-being of nonhuman primates. Research facilities must develop, document, and follow an environmental enhancement plan. The enhancement plan must include consideration of social grouping; environmental enrichment; special consideration for infants, animals exhibiting psychological distress, animals with IACUC approved restricted activity, animals individually housed, and great apes weighing more than 110lbs; restraint devices; and animals that are exempt from participating in the plan and the reasons for their exemption. The AWRs also require provisions for the compatible social housing for dogs and cats as well as exercise for dogs. Dogs or cats that are housed in the same primary enclosure must be compatible with special consideration given for animals in estrus, young animals, animals with aggressive dispositions, and animals that have or suspected of having a contagious disease. The exercise requirement for dogs includes provisions for dogs housed individually or in groups, and the amount of space that is made available for exercise as well as the methods and period of providing the opportunity to exercise. Methods shall be determined by the AV and approved by the IACUC. Protocols that involve exercise procedures (swimming, treadmills, etc.) are considered unacceptable for meeting the exercise requirements of the regulations.

Public Health Service Policy and Office of Laboratory Animal Welfare

While the PHS Policy does not specifically address the topics of housing and enrichment, US Government Principles are endorsed by the Policy. Principle number VII states, "*The living conditions of animals should be appropriate for the species and contribute to their health and comfort. Normally, the housing, feeding, and care of all animals used for biomedical purposes must be directed by a veterinarian or other scientist trained and experienced in the proper care, handling, and use of the species being maintained or studied. In any case, veterinary care shall be provided as indicated.*" Also, OLAW provides guidance on the use of performance standards to (1) assess and set enclosure space guidelines for rodents and rabbits and (2) assess environmental enrichment issues. "An institution's environmental enrichment practices must be species-specific and appropriate for the animals."[28] OLAW categorically states that "There is universal agreement among oversight agencies that nonhuman primates should be socially housed."[29] and provides guidance on associated issues such as staff training on nonhuman primate socialization and IACUC and veterinary review of exemptions to the social housing expectation. OLAW does not specifically address aquatics other than to state, when discussing management of heating, ventilation, and air conditioning system (HVAC) failures, that "Institutions are also responsible to ensure the welfare of fishes, amphibians, and other vertebrates whose environment is aquatic, with the emphasis on water temperature and quality, including oxygenation, circulation and filtration."[30]

Guide

The *Guide* provides extensive and specific requirements and recommendations in regard to housing, environmental enrichment, and behavioral management of all vertebrate species, including aquatics. Guidelines and recommendations are described for the provision of adequate temperature and humidity; ventilation and air quality; illumination; noise and vibration; terrestrial housing; microenvironment of the primary enclosure; environmental enrichment and social housing; sheltered or outdoor housing; space requirements; behavioral and social management; husbandry; population management; and water quality and life support systems for aquatic species.

Chapter 3 of the *Guide* provides information regarding a number of key topics related to housing and enrichment. Specifically, the *Guide* offers recommendations regarding the microenvironment or primary enclosures. Animals should be housed under conditions that provide sufficient space and supplementary structures and resources to meet their physical, physiologic, and behavioral needs. The social needs of animals are prominently featured.

The primary enclosure should provide a secure environment and be well-constructed of materials that are safe, durable, and sanitizable. Flooring should be solid, perforated or slatted, and slip resistant. Adequate bedding or substrate should be provided as well as structures for resting or sleeping.

Opportunities for species-typical behaviors such as burrowing and nest building should be considered and accommodated whenever possible. Considerations for animal biosecurity and minimization of airborne particles that may transfer between cages may be important and the potential need for specialized caging systems may be necessary. Appropriate housing strategies should be developed and implemented by animal care and use management in consultation with investigators and the veterinarian and be reviewed by the IACUC. Recommendations for sheltered, outdoor, or naturalistic housing strategies are described.

A series of tables are provided as recommended minimum cage space guidelines for a variety of species including: rodents; rabbits; dogs; cats; poultry; nonhuman primates; sheep; goats; swine; cattle; and horses. Of equal or greater importance are the performance criteria describing adequacy of cage space. The *Guide* acknowledges that determining the appropriate size for primary enclosures includes the consideration of many aspects and not simply body weight/size alone. The performance criteria described in the *Guide* provide recommendations for functional cage space given a variety of factors that should be considered, recognizing that cage dimensions alone represent only one of the factors necessary when determining adequacy of cage space. Some of the performance criteria for adequacy of cage space include space for the animals to rest away from areas soiled by urine and feces, ability to make normal postural adjustments, consideration of ancillary equipment/structures that may otherwise limit floors space, sufficient cage height for arboreal nonhuman primates, opportunities to escape aggression, and increased space for breeding animals.

The *Guide* also provides a detailed discussion of the need for environmental enrichment of animals. The *Guide* states, "*The primary aim of environmental enrichment is to enhance animal well-being by providing animals with sensory and motor stimulation, through structure and resources that facilitate the expression of species-typical behaviors and promote psychological well-being through physical exercise, manipulative activities, and cognitive challenges according to species-specific characteristics*." Examples of enrichment are provided. Well-conceived enrichment provides animal with choices and some degree of control over their environment, which affords the ability to better cope with environmental stressors. Recommendations for the cautious application of enrichment items are provided so as to avoid the unintentional introduction of stressors as some animals may be averse to novel items. Similarly, enrichment structures may act as fomites and increase the potential for disease transmission. The *Guide* specifies that the enrichment program should be reviewed by the IACUC, researchers, and veterinarian on a regular basis to ensure it functions as intended, keeping in mind the benefit to the animals and the scientific goals. Enrichment can be a source of environmental variation if not judiciously applied and the *Guide* recommends the careful implementation of a uniformly applied program that avoids an undesirable impact on scientific outcomes while improving animal well-being.

The need to consider social housing is emphasized in the *Guide*. Appropriate social interactions among conspecifics are considered essential for normal development and well-being. Consideration of behavior such as an animal's being naturally territorial or communal are important when determining whether they are best housed singly, in pairs, or in groups. The *Guides* states that, "*Single housing of social species should be the exception and justified based on experimental requirements, or veterinary-related concerns about animal well-being.*" Not all members of a social species are compatible, and in such cases single-housing may also be justified. The *Guide* further indicates that animals should be housed under conditions that meet their behavioral and social needs and that social animals should be housed in stable pairs or groups of compatible individuals. When animals are socially housed, often structural accommodations within the enclosure are necessary (perches, visual barriers, refuges, and the multiple opportunities for obtaining food and water). Socially housed animals require careful and frequent monitoring to identify and manage antagonistic interactions among conspecifics, should they occur. Whenever animals are single-housed it should be limited to the minimum period necessary and where possible, visual, auditory, olfactory, and tactile contact with compatible conspecifics should be provided. The housing needs of animals, particularly those in single housing, should be reviewed on a regular basis by the IACUC and the veterinarian.

Ag Guide

The *Ag Guide* provides extensive recommendations regarding housing conditions for agricultural animals. With regard to facilities and environment there are guidelines regarding the widespread options for housing (pasture to confinement) and the need to understand the impact of environmental stressors on the animal and on the research activity. Criteria for well-being are detailed. Topics include: macro- and microenvironment; thermal indices/comfort; temperature, vapor pressure, and ventilation; air quality; pasture, feedlot, and confinement housing; social needs; protection from the elements for animals housed outdoors; husbandry; feed and water; bedding; ventilation; lighting; provision for space; standard agricultural practices; waste management; and special considerations. There are wide-ranging recommendations for the enrichment of agricultural animals and enrichment is often considered part of the refinement effort for animals used in research and teaching. Five categories of enrichment are described: social; occupational (enrichment that provides both psychological benefits and physical challenges to encourage exercise); physical enrichment (altering the size and complexity of the enclosure or the addition of accessories); sensory (stimuli that are visual, auditory, tactile, olfactory, or taste); and nutritional enrichment (providing varied or novel food types or the methods of delivery). Methods of enrichment for each species, which employ these five basic strategies, are provided. The specific methods recommended vary with the species of animal but strategies for social enrichment are emphasized except with boars, and the careful application of social housing for males of other agricultural species is also described. The importance of gentle and calm human–animal interaction is emphasized. The benefits of paddock and pasture environments in promoting natural grazing behaviors and allowing for appropriate social interactions, as well as the need to recognize and minimize abnormal behaviors is provided.

HUSBANDRY AND ENVIRONMENT

Animal Welfare Act/Animal Welfare Regulations

The standards contain sections for the covered species on animal health and husbandry. The environmental requirements are contained in the section on facilities and operating standards. The husbandry standards include requirements for feeding, watering, cleaning, sanitization, housekeeping, and pest control. In addition, the standards require that licensed and registered facilities have enough employees to carry out the husbandry and care required by the standards and be supervised by someone with the appropriate knowledge, background, and experience with the species being maintained. The employer must also ascertain that the staff can perform to the prescribed standards. The standards for some species also address issues such as compatibility, exercise, classification and separation, and environmental enrichment.

Public Health Service Policy and Office of Laboratory Animal Welfare

The PHS Policy does not specifically address daily care activities or environmental conditions other than to state in the US Government Principle VII that, "*The living conditions of animals will be appropriate for their species and contribute to their health and comfort. The housing, feeding, and nonmedical care of the animals will be directed by a veterinarian or other scientist trained and experienced in the proper care, handling, and use of the species being maintained or studied.*" OLAW similarly does not go into specifics of daily care activities or environmental conditions other than to reference the *Guide* and specify the need for planning for HVAC failures.

Guide

The *Guide* provides extensive and specific requirements and recommendations in regard to the provision of daily animal care. Key sections include temperature and humidity; ventilation and air quality; illumination; noise and vibration; space provisions; food, water, bedding, and nesting materials; sanitation procedures and frequency of bedding changes; cleaning and disinfection activities; assessing efficacy of sanitation; waste disposal; pest control; weekend, emergency, and holiday care; identification methods; record-keeping; breeding; genetics; and nomenclature. There are detailed descriptions of environmental enrichment and social housing as well as appropriate overall environmental conditions for all vertebrate species including aquatics.

For terrestrial animals, there are species-specific recommended ranges for establishing set points for temperature and humidity with the expectation that day-to-day variation around the set point be minimal so that unnecessary demands are not placed on the animals, which requires that they adapt to changes in environmental temperature and humidity. Similarly, there is a need to minimize temperature and humidity variation because of the potential impact such variation may have on scientific outcomes. Ventilation and air quality are also considered important aspects of the animal's environment. Recommendations for air exchange previously emphasized engineering criteria, and the *Guide* now stresses the importance of performance-based criteria to determine and assess the adequacy of ventilation rates. Factors such as species housed, animal density, management and husbandry practices, and type of housing may impact the requirements for adequate ventilation of animal areas. Likewise the scientific aims and specific animal models employed may effect ventilation and supply and exhaust air requirements. Animals in barrier and containment facilities may require a high degree of air filtration as well as areas for surgery and necropsy. The *Guide* cautions the use of recycled air as supply for animal areas because of risks of biosecurity and the potential for spread of pathogens. The *Guide* provides specific recommendation for illumination, light intensity, consideration of photoperiod, and monitoring the function of lighting systems to ensure they are appropriate for the animals and adequately support scientific needs. The inadvertent exposure of light (during the dark phase) is of particular concern as many physiologic functions can be affected. There are recommendations regarding noise and vibration in the animal environment with specific details provided about the need to identify and dampen sources of vibration and minimize sources of noise (at frequencies important for the species of animal housed).

Recommendations for the overall husbandry program include detailed considerations and requirements for the source and provision of food and water. There are descriptions of dietary formulations and their potential uses for conventional, microbiologically defined, and immune-compromised animals. Recommendations for water quality, monitoring, and methods of drinking water delivery systems are provided. The requirements for bedding and nesting materials are provided as are the considerations of types of bedding and their potential impact on animal well-being and scientific outcomes. Nesting materials can be beneficial for a number of animal species and the *Guide* recommends careful consideration for the provision of nesting materials as a means of providing environmental enrichment (to promote species-typical behaviors) as well as a means to improve the animal's behavioral repertoire for thermoregulation and rearing young.

The *Guide* provides detailed information regarding goals, methods, and monitoring of sanitation practices. This section applies to sanitation of the animal facility, secondary enclosures, as well as primary enclosures. Specific performance outcomes are defined while providing latitude for a particular institution's best approach to achieve them. Specific procedures of cleaning, sanitation, and disinfection are provided. There are guidelines regarding the methods and intervals of sanitation. A regular, microbiologically based method to verify the effectiveness of sanitation is recommended. The *Guide* also contains recommendations for waste disposal and the need to effectively manage waste while ensuring safe and efficient means of personnel and public protection. Practices must be in accordance with all applicable federal, state, and local regulations. Hazardous waste must be handled in a manner that ensures personnel safety. The *Guide* states that, "*Hazardous wastes must be rendered safe by sterilization, containment or other appropriate means before their removal from the facility*." Adherence to the regulatory requirements of local and federal authorities is of particular importance for hazardous wastes.

The *Guide* also contains recommendations for program of pest control, emergency, weekend and holiday care, population management, and record-keeping. Pest control programs are essential in the animal environment and the *Guide* indicates that a "*regularly scheduled and documented program of control and monitoring should be implemented*." The use of pesticides can be effective means to minimize or eliminate arthropod pests, but the *Guide* recommends they be used with caution as the effects of these chemicals may impact animal health and scientific outcomes. These chemicals should be used in the facility only when necessary and researchers should be consulted prior to their use to ensure that there will be no impact on the studies. The *Guide* requires that animals be cared for by qualified personnel every day, including weekends and holidays. Emergency veterinary care must be available at all times. A disaster plan that includes the needs of both animals and personnel must be in place. Population management of research animals includes a proper means of identification and record-keeping. A discussion of the various methods is provided and ranges from procedures to identify individual animals to cage card information and radio frequency identification. Record-keeping is an important aspect of animal facility and program management. The *Guide* states that, "*animal records are essential for genetic management and historical assessments of colonies. Records of rearing and housing histories, mating histories and behavioral profiles are useful for the management of many species, especially nonhuman primates.*" Medical records for individual animals of several species are important and should contain pertinent information regarding clinical and diagnostic procedures, inoculations, history of surgical procedures, and postoperative care. Information regarding research procedures and necropsy findings may also be important.

The *Guide* has recommendations regarding the breeding, genetics, and nomenclature for research animals and underscores the importance of these aspects of the animal care and use program. There are recommendations regarding the use of outbred stocks and genetic management of such colonies. With regard to the use of inbred animals, the *Guide* emphasizes the need for careful monitoring of F1 progeny to ensure appropriate genetic selection. Similarly, the careful monitoring of newly created GMAs is an important aspect of day-to-day animal care and use, which ensures that unexpected phenotypes are appropriately identified and managed so as to minimize potential pain and distress that may result. Careful and detailed records of pedigree and genetic monitoring are essential.

The most recent update of the *Guide* provides recommendations for the care and use of aquatic species. Similar to the terrestrial animal section, the sections for aquatic animals include guidelines for: micro- and macroenvironment; water quality; life support systems; temperature, humidity, and ventilation; illumination; and noise and vibration. There are details provided for: environmental enrichment of aquatic species; sheltered, outdoor, and naturalistic housing; space requirements; behavioral and social management; husbandry; sanitation; pest control; emergency, weekend, and holiday care; population management; and record-keeping. Many of the fundamental tenets of aquatic animal care and use are similar to those for terrestrial animals even though the species requirements for housing and management are quite different and diverse. As in all other sections of the *Guide*, the performance outcomes are defined with the understanding that the means to achieve the outcomes can vary based on the specific needs of a particular species and the scientific aims of the research.

Ag Guide

As described previously there are extensive recommendations in the *Ag Guide* for each species with regard to facilities and husbandry practices, feed and water, common agricultural practices, fencing, predator control, and management techniques to promote health and safety of the animals.

VETERINARY CARE

Animal Welfare Act/Animal Welfare Regulations

The regulations require that each research facility employ an AV with appropriate authority to ensure the provision of adequate veterinary care and to oversee other aspects of animal care use. Each research facility must establish and maintain a program of adequate veterinary care that includes: appropriate facilities, personnel, and equipment; methods to control, diagnose, and treat diseases; daily observation and provision of care; guidance to personnel on the use of anesthetics, analgesics, and euthanasia procedures; and pre- and postprocedural care. The AV must be a graduate of an accredited school of veterinary medicine or have equivalent formal education, have received training and/or experience in the care and management of the animals under his/her care, and have direct or delegated authority for activities involving animals at the facility. The AV shall be voting member of the IACUC.

Public Health Service Policy and Office of Laboratory Animal Welfare

Neither the PHS Policy nor OLAW specifically address veterinary care activities other than to state in section IV.C.e. of the Policy that "Medical care for animals will be available and provided as necessary by a qualified veterinarian."

Guide

The *Guide* provides extensive and specific requirements and recommendations regarding the provision of veterinary care, preventive medicine, animal procurement, health monitoring, and the conduct of surgery. An entire chapter is devoted to the topic of veterinary care. Programs of veterinary care are based on the requirement stated on page 112 that, "All animals should be observed for signs of illness, injury, or abnormal behavior by a person trained to recognize such signs. As a rule, such observation should occur at least daily." The *Guide* further states that "An adequate veterinary care program consists of assessment of animal well-being and effective management of

- animal procurement and transportation
- preventive medicine (including quarantine, animal biosecurity, and surveillance)
- clinical disease, disability, or related health issues
- protocol-associated disease, disability, and other sequelae
- surgery and perioperative care
- pain and distress
- anesthesia and analgesia"

The *Guide* places a significant amount of authority and responsibility on the AV. Throughout Chapters 2 and 4, one finds an overall description of the role of the AV that encompasses many topics and ultimately features the AV prominently as a leader, in collaboration with the IO and IACUC, of the animal care and use program. The *Guide* indicates that the animal care and use program is the collaborative responsibility of the IO, the AV, and the IACUC. The AV is responsible for the well-being and clinical care of the animals throughout all phases of the animal's life. The program of veterinary care must uphold the highest standards of care and ethics and there must be sufficient authority of the AV to provide treatment and relieve pain and distress, including euthanasia. Although the AV is imbued with a significant level of institutional authority, the *Guide* emphasizes the need for a collaborative approach between the AV and investigators to achieve the best possible outcome for the animals and the needs of the science. While the AV's areas of responsibility encompass essentially all aspects of animal care, the *Guide* acknowledges that other individuals in the institution may assume responsibility for some of the duties related to veterinary care but this must be done under the authority and direction of the AV. Regarding the qualifications, the *Guide* stipulates that veterinarians must have experience, training, and expertise necessary to appropriately evaluate the health and well-being of the species used. Veterinarians should remain knowledgeable about the latest practices to ensure the highest quality of veterinary medical care. Further the veterinarian must have access to all animals and the provision of veterinary medical care must be available at all times. The veterinarian must have oversight of programmatic aspects such as preventive medicine, health surveillance, medical treatment, establishment of sedation, anesthetic and analgesic, and handling guidelines and should oversee other aspects of the program such as husbandry and housing. The AV is also tasked with providing guidance for surgery and perioperative care. The *Guide* indicates that the AV or designee has responsibilities related to the IACUC. There should be ongoing, regular communication among the AV, the IACUC, and the IO. The veterinarian has definite responsibilities during protocol review and particularly for the development of study removal criteria and pain and distress management plans. The IACUC and AV together are responsible for assessing the qualifications for those performing surgery and anesthesia as well as the classification of major versus minor surgical procedures.

The American College of Laboratory Animal Medicine (ACLAM) has developed guidelines that describe expectations for adequate veterinary care.[31] The guidelines were developed in 1996 and provide a detailed description of the key aspects related to the role of veterinarians and programs of veterinary care, and these guidelines are cited as a reference document in the *Guide*. Important features of these guidelines include the requirement that institutional veterinarians be qualified through postgraduate training or experience in laboratory animal science and medicine; institutional authority of the veterinarian must be provided to allow the veterinarian to fulfill their duties; a veterinarian must be a full member of the IACUC; and be actively involved in the review of protocols, projects, and institutional programs involving animals, and inspection of facilities. Further, the ACLAM guidelines specify that the provision of adequate veterinary care involves: disease detection and surveillance, prevention, diagnosis, treatment, and resolution; handling and restraint; anesthetics, analgesics, and tranquilizer drugs; methods of euthanasia; surgical and postsurgical care; animal well-being; and the appropriate use of animals in research and testing. The guidelines also describe other aspects related to veterinary care, which include: participating in the development and administration for staff training; assisting institutional health officials to establish and maintain an occupational health program; monitoring for zoonotic disease; advising and monitoring of standards of hygiene; and advising and monitoring of biohazard control policies and procedures.

Ag Guide

The program of veterinary care is the responsibility of the AV and include: having direct or delegated authority to develop and oversee the program of veterinary care and other aspects related to husbandry, nutrition, sanitation, zoonoses, and hazard containment; have access to all animals; establish guidelines for anesthetic and analgesics; involvement in protocol review and study removal criteria; involvement in training for animal users and responsible conduct of research activities; and available to provide guidance and advice to research personnel. The program of veterinary care should consist of elements such as: preventive medicine; recognition of sick or injured animals so that timely veterinary medical care can be provided; maintenance of health records; considerations for multiple, major survival surgeries; oversight of surgery personnel; surgical facilities and aseptic technique; recognizing pain and distress; anesthesia and analgesia; postprocedural care; zoonoses; biosecurity and vermin control; and euthanasia.

CONDUCT OF EXPERIMENTAL PROCEDURES

Animal Welfare Act/Animal Welfare Regulations

The AWA prohibits the promulgation of rules, regulations, or orders related to the design, outlines, or guidelines of actual research or experimentation. That said, the research facility is required to follow professionally acceptable standards for the care, treatment, and use of animals during actual research or experimentation. These standards apply specifically to procedures that minimize pain and distress including adequate veterinary care to include the use of tranquilizers, analgesics, and anesthetics; the provision of pre- and postsurgical care consistent with established veterinary medical and nursing procedures; and the prohibition of the use of paralytics without anesthesia. If withholding of tranquilizers, analgesics, anesthetics, or euthanasia is scientifically necessary, it shall be done for only the necessary period of time. On an annual basis each registered research facility must submit a report that assures that the facility is adhering to the standards and regulations governing the care and use of animals including the appropriate use of anesthetics, analgesics, and tranquilizers, and that investigators considered alternatives to painful procedures. They must also report the number of animals that were held and not used; used for procedures that caused no more than momentary pain and distress; involved pain and distress that was relieved by the use of anesthetics, analgesics, or tranquilizers; and/or involved pain and distress that could not be relieved by the use of anesthetics, analgesics, or tranquilizers because such use would have adversely affected the data being generated.

Public Health Service Policy and Office of Laboratory Animal Welfare

The HREA, which provides the statutory mandate for the PHS Policy, describes some fundamental aspects regarding the conduct of animal studies. The proper use of tranquilizers, analgesics, anesthetics, paralytics, and euthanasia agents/procedures is required. Also described are the requirements for appropriate pre- and postsurgical care as well as veterinary medical and nursing care. There is also a requirement that scientists, animal technicians, and other personnel involved with animal care, treatment, and use have available to them instruction or training in the humane practice of animal maintenance and experimentation as well as the testing methods that employ concepts that limit use of animals or animal distress. Also described in the PHS Policy are the US Government Principles, all of which relate to the conduct of animal studies. The PHS Policy lists the fundamental requirements in the conduct of animal studies: that procedures avoid or minimize animal discomfort, distress, and pain; procedures that cause more than momentary or slight pain or distress will employ appropriate sedation, analgesia, or anesthesia unless withholding of such agents is scientifically justified; description of procedures designed to assure that discomfort and injury to animals will be limited to that which is unavoidable in the conduct of scientifically valuable research; animals that would otherwise experience severe or chronic pain or distress will be humanely euthanatized with a description of the methods used; the living conditions of the animals will be appropriate and contribute to their health and comfort; rationale for involving animals and for the appropriateness of the species and numbers used; a complete description of the proposed use of the animals.

Further the IACUC is tasked with conducting continuing review of each previously approved ongoing activity at appropriate intervals. Finally, the PHS Policy requires that conduct of PHS funded animal studies must be in accordance with the institution's assurance statement, the AWA, and the *Guide*. The PHS Policy and OLAW do not specify classification of animal activities based on severity, either prospective or retrospective. Refinement was discussed above in the subchapter on "Principles" in association with the 3Rs.

Guide

The *Guide* discusses appropriate conduct of experimental procedures extensively and endorses the following principles (excerpted from a larger list): consideration of alternatives (in vitro systems, computer simulations, and/or mathematical models) to reduce or replace the use of animals; design and performance of procedures on the basis of relevance to human or animal health, advancement of knowledge, or the good of society; avoidance or minimization of discomfort, distress, and pain; and establishment of humane endpoints. Essentially, all of Chapter 2 of the *Guide* relates to the proper conduct of animal studies. The *Guide* places particular emphasis on the need to develop detailed, well-written, comprehensive animal study protocols, which describe all animal procedures and their anticipated impact on the animals. A thorough review of the proposed protocols must be provided by the IACUC. Once protocols are approved, the IACUC must provide ongoing PAM to ensure that the studies conducted are in accordance with the approved protocol. Should a need arise to revise animal procedures once a protocol has been approved, the proposed changes must be described in writing by the PI and approved by the IACUC before the revised procedures are conducted. The IACUC has the authority to investigate concerns related to animals use and suspend animal activities.

With regard to pain and distress during the conduct of experiments, the *Guide* states that "*the proper use of anesthetics and analgesics in research animals is an ethical and scientific imperative*," based on the concept that, unless the contrary is known or established, it should be considered that procedures that cause pain in humans may also cause pain in other animals.

Ag Guide

A central feature of the recommendations in the *Ag Guide* is the IACUC and its breadth and scope of responsibility. In essence, all research and teaching procedures conducted on animals must be approved by the IACUC. Procedures must be described, in detail, in either the protocol or in written operating procedures. The IACUC must approve the procedures (protocols or written operating procedures) before the initiation of the research or teaching activity. The IACUC should perform a complete review of ongoing studies at least once every 3 years, or more often at an interval deemed necessary by the IACUC. The IACUC is authorized to investigate concerns, complaints, or reports of noncompliance (e.g., the conduct of procedures that are not described in an approved protocol). The IACUC is further authorized to suspend an activity when it is not in compliance with the approved protocol or written operating procedures.

EUTHANASIA

Animal Welfare Act/Animal Welfare Regulations

Methods of euthanasia must be consistent with the definition contained in the regulations. That definition requires the method used produce rapid unconsciousness and subsequent death without evidence of pain or distress or a method that utilizes anesthesia produced by an agent that causes painless loss of consciousness and subsequent death. The USDA's Animal Care Resources Guide, which is intended to clarify the regulations and standards, indicates that the methods used should be consistent with the current Guidelines on Euthanasia of the AVMA[32] (http://www.aphis.usda.gov/animal_welfare/downloads/policy).

American Veterinary Medical Association Guidelines on Euthanasia

The AVMA published its most recent guidelines in 2013 titled, *AVMA Guidelines for the Euthanasia of Animals: 2013 Edition.*[7] The Guidelines are a comprehensive set of recommendations that are widely used by veterinarians and IACUCs as the standard for euthanasia procedures and techniques. There are very detailed recommendations for species used in research settings with guidance on those methods that are considered "acceptable," "acceptable with conditions," "adjunctive," and "unacceptable." With regard to the agents used there are recommendations for inhaled and noninhaled as well as physical methods. There are considerations for dangerous or fractious animals, emergencies, age or life stage, and animals in free-range conditions. Traditional laboratory animal species included are ruminants, equids, fish, amphibians, reptiles, birds, embryos, fetuses, and neonates, as well as some wild species. Apart from the extensive recommendation for animals in research, there are sections detailing consideration for companion animals, free-ranging wildlife, captive and free-ranging marine mammals. The guidelines also contain recommendations regarding euthanasia as a humane technique and the professional judgment and decision-making process when considering the euthanasia of animals. Veterinary medical ethics, animal and human behavior, stress and distress, as well as pain and its perception are all topics that warrant appropriate consideration.

Species specific sections have been added and expanded to include terrestrial and aquatic species in a variety of environments. The three main parts to these guidelines include Part I—Introduction and General Comments, which details: the historical context of euthanasia; a precise definition of the euthanasia; veterinary medical ethics; and discussion regarding many aspects of the methods of euthanasia such as consciousness and unconsciousness, pain and distress, and human behavior, among other topics. Part II—Methods of Euthanasia, which details aspects related to inhaled agents, noninhaled agents, and physical methods, providing an array of options for consideration depending on species, environment, safety, efficacy, etc. Part II—Methods of Euthanasia by Species and Environment, which includes a thorough discussion related to companion animals, laboratory animals, animals farmed for food and fiber (including finfish and free-ranging nondomestic animals), marine mammals and embryos, fetuses, and neonates. Each section provides detailed information regarding acceptable methods, methods that are acceptable with conditions, adjunctive methods, and unacceptable methods. In many cases there are also descriptions of specific technical aspects of proper technique for various methods, anatomy, and physiology of a variety of species, restraint, and final disposition of animal carcasses. Extensive references are provided.

Overall, the guidelines stress the importance of proper, respectful animal handling for calm, stress-free methods. Key criteria for euthanasia methods include: (1) "ability to induce loss of consciousness and death with a minimum of pain and distress"; (2) "time required to induce loss of consciousness"; (3) "reliability"; (4) "safety of personnel"; (5) "irreversibility"; (6) "compatibility with intended animal use and purpose"; (7) "documented emotional effect on observers or operators"; (8) "compatibility with subsequent evaluation, examination, or use of tissue"; (9) "drug availability and human abuse potential"; (10) "compatibility with species, age, and health status"; (11) "ability to maintain equipment in proper working order"; (12) "safety for predators or scavengers should the animal's remains be consumed"; (13) "legal requirements"; and (14) "environmental impacts of the method or disposition of the animal's remains."

Most IACUCs in the US require that euthanasia be conducted in a manner that is consistent with recommendations of these important Guidelines.

Public Health Service Policy and Office of Laboratory Animal Welfare

In section IV.C.1.g the PHS Policy states that "Methods of euthanasia used will be consistent with the recommendations of the AVMA Panel on Euthanasia [https://www.avma.org/KB/Policies/Documents/euthanasia.pdf], unless a deviation is justified for scientific reasons in writing by the investigator." OLAW provides further guidance on the procedural methods for use of carbon dioxide as a euthanasia agent[33] (http://grants.nih.gov/grants/olaw/faqs.htm#useandmgmt_1 and http://grants.nih.gov/grants/guide/notice-files/NOT-OD-02-062.html).

Guide

The *Guide* discusses euthanasia criteria and methods in detail but does not recommend specific agents other than to state that "*The selection of specific agents and methods for euthanasia will depend on the species involved, the animal's age, and the objectives of the protocol. Generally, chemical agents (e.g., barbiturates, nonexplosive inhalant anesthetics) are preferable to physical methods (e.g., cervical dislocation, decapitation, use of a penetrating captive bolt)*". The *Guide* also endorses the AVMA Guidelines on Euthanasia. Methods used should induce rapid unconsciousness and death without pain, distress, or anxiety. Methods should also be reliable and irreversible and appropriate for the species, compatible with research objectives, and safe for personnel. The *Guide* also acknowledges the importance of considering the emotional effects on humans. When euthanasia is a planned procedure at the end of a protocol, criteria for euthanasia should be in accordance with protocol-specific endpoints that have been carefully developed and approved by the IACUC. Humane endpoints described in protocols must include four key criteria: (1) a precise description/definition of the endpoint; (2) frequency of animal observations; (3) the training and expertise of personnel to ensure recognition of the endpoint; and (4) the appropriate action to be taken once the endpoint is reached. Special considerations are required for fetuses and neonates, and the specific selection of agents and methods will be determined in part, by the age of the animal. The *Guide* acknowledges the controversy of carbon dioxide as an agent for euthanasia of rodents and recommends careful consideration of its use with the general requirement that specific methods be compatible with the *AVMA Guidelines for the Euthanasia of Animals*.

Ag Guide

The *Ag Guide* specifies that protocols for euthanasia should follow the recommendations of the AVMA Guidelines on Euthanasia. There is emphasis on personnel training, experience, skill, and familiarity with the normal behavior of agricultural animals and how handling and restraint affect their behavior.

Further, the AV or qualified scientist should ensure that all personnel performing euthanasia have demonstrated proficiency. Acceptable methods are those that initially depress the central nervous system to ensure insensitivity to pain and techniques should minimize stress and anxiety of the animal and result in rapid unconsciousness followed by cardiac or respiratory arrest and the ultimate loss of brain function. Equipment and devices must be properly maintained and personnel adequately trained in their use. Agents that result in tissue residues cannot be used in animals intended for food unless approved by the Federal Food and Drug Administration. The slaughter of animals entering the human food chain must be accomplished in compliance with regulations of the federal Humane Methods of Slaughter Act (9 CFR. 313.1-90; CFR, 1987),[8] which provide regulations for the humane treatment of livestock before and during slaughter.

EQUIPMENT AND FACILITIES

Animal Welfare Act/Animal Welfare Regulations

The requirements for facilities are contained within the sections of the standards on facilities and operating standards for the species covered by the AWA. Included in this section are general requirements for housing facilities, as well as specific requirements for indoor, sheltered, outdoor, and mobile housing facilities. The specific standards address construction of the structure, surfaces, water and electric power, storage, and drainage and waste removal. For indoor and sheltered housing facilities there are requirements for heating, cooling, ventilation, temperature controls, lighting, and interior surfaces. There are requirements for shelter from the elements for sheltered and outdoor facilities. There are specific space requirements for the primary enclosures of the species covered by the regulations. There are no specific specifications for equipment than the space requirements for primary enclosures.

Public Health Service Policy and Office of Laboratory Animal Welfare

The PHS Policy defines an animal facility as "Any and all buildings, rooms, areas, enclosures, or vehicles, including satellite facilities, used for animal confinement, transport, maintenance, breeding, or experiments inclusive of surgical manipulation. In section III.B of the Policy, a satellite facility is any containment outside of a core facility or centrally designated or managed area in which animals are housed for more than 24 hours." The PHS Policy and OLAW do not specifically address equipment or facility characteristics.

Guide

Chapter 5 of the *Guide* discusses equipment and facilities in detail providing guidance and expectations on construction guidelines for animal housing (corridors, floors, drainage, walls and ceilings, HVAC, power and lighting, storage areas, noise control, vibration control, facilities for sanitizing materials, and environmental monitoring) and support space such as surgical facilities; security and access control; hazardous agent containment; facilities for imaging and whole body irradiation, barrier housing, behavioral studies, and aquatic species housing.

The *Guide* defines the major functional areas of an animal facility as: animal housing and care; receipt, quarantine, separation, and/or rederivation of animals; separation of species or isolation of individual projects when necessary; storage; and specialized laboratories/areas for surgery, intensive care, necropsy, irradiation, diet preparation, experimental procedures, behavioral testing, imaging, clinical treatment, and diagnostic laboratory procedures. Other areas often required include containment facilities and equipment; barrier facilities; receiving and storage of equipment and supplies; areas for washing and sterilizing equipment; waste storage; cold storage; space for personnel areas such as offices, training areas, showers, lockers, toilets, and break rooms; and areas for maintenance and repair of specialized equipment. Personnel spaces should be separate from animal areas.

Corridors should be sufficiently wide to facilitate the movement of personnel and equipment and should be designed to be durable and withstand regular cleaning and disinfection. Floor wall junctions should also be designed to facilitate cleaning. Corridors should provide sound attenuation by the location of double-door vestibules, which isolate the noisy areas of the facility. Animal room doors should be of appropriate size to allow easy passage of equipment and doors should fit tightly within their frames and be self-closing. Doors should open into animal rooms. Door security is important at points of ingress to the animal facility and holding areas in which access should be limited. Windows in animal holding areas may provide benefit but concerns related to control of photoperiod, temperature variation, and security should be considered. Floors in the animal facility should be "*moisture resistant, nonabsorbent, impact resistant, relatively smooth, although textured surfaces may be required in some high moisture areas and for some species (e.g., farm animals).*" Floors should be durable and able to withstand regular cleaning with chemical agents, resist cracking, gouging, or pitting to promote proper cleaning.

A key feature of the animal facility is the HVAC. The *Guide* states that, "*A properly designed and functioning HVAC system is essential to provide environmental and space pressurization control.*" Proper temperature and humidity control are critical to maintain a suitable environment for animals and minimizing environmental variability for the conduct of quality science. Adequate ventilation is essential to remove waste gases and heat associated with animal housing areas. Guidelines for ventilation rates and temperature and humidity settings are provided for several species. Directional airflow is critical in several areas of the animal facility for the protection of both animals and personnel (e.g., containment areas, surgery, necropsy, barriers, isolation, wash rooms, and quarantine). Specific design features of modern HVAC system are described with recommendations for variable air volume systems, safeguards such as reheat coils that fail in the closed position, and automated alarm systems to notify appropriate personnel, which malfunctions or environmental parameters are out of normal ranges. The HVAC system should be designed with sufficient redundancy or back-up systems so that minimal function is maintained, should an overall malfunction occur. An automated system to monitor temperature and humidity at the room level, and which notifies personnel of abnormalities/malfunction, should be in place. Similar systems for monitoring the lighting and photoperiod are also recommended. A source of emergency power should be available to provide function to critical areas of the facility during a power failure.

Overall facilities should be designed and operated to minimize vibration and noise. Facilities for sanitation of cages and equipment should be properly designed and operated with consideration of the following criteria:

- Location with respect to animal rooms and waste disposal and storage areas
- Ease of access, including doors of sufficient width to facilitate movement of equipment
- Sufficient space for staging and maneuvering of equipment
- Soiled waste disposal and prewashing activities
- Ease of cleaning and disinfection of the area
- Traffic flow that separates animals and equipment moving between clean and soiled areas
- Air pressurization between partitioned spaces to reduce the potential of cross contamination between soiled and clean equipment
- Insulation of walls and ceilings where necessary
- Sound attenuation
- Utilities, such as hot and cold water, steam, floor drains, and electric power
- Ventilation, including installation of vents or canopies and provisions for dissipation of steam and fumes from sanitizing processes
- Vibration, especially if animals are housed directly above, below, or adjacent to the washing facility
- Personnel safety, by ensuring that safety showers, eyewash stations, and other equipment are provided as required by code; exposed hot water and steam lines are properly insulated; procedures with a propensity to generate aerosols are appropriately contained; and equipment, such as cage/rack washers, and bulk sterilizers, which personnel enter, are equipped with functioning safety devices that prevent staff from becoming trapped inside

The *Guide* also provides specific recommendations for specialized areas such as surgery facilities. Surgical facilities should contain at least five separate areas: surgical support; animal preparation; surgeon scrub; the operating room; and postoperative recovery. Also specific recommendations are provided for the design and construction of: barrier facilities; imagining areas; whole body irradiation; hazardous agent containment; behavioral studies; and aquatic species housing. Recommendations for animal facility security and access control are also provided.

Ag Guide

Guidelines regarding equipment and facilities are described in section Housing and Enrichment.

MISCELLANEOUS

Animal Welfare Act/Animal Welfare Regulations

The AWR contains a section entitled miscellaneous, which includes requirements for furnishing business information, access to and inspection of records and property, publication of the names of licensed and registered facilities, inspection for missing animals, confiscation and destruction of animals, handling, identification of dogs and cats, health certificates, holding of animals, and holding period and compliance with standards and prohibitions. In addition there are subparts of the regulations on the scope and applicability of the rules of practices including summary actions and stipulations.

REFERENCES

1. AWA. *Animal Welfare Act of 1966 (Pub L. 89-544) and subsequent amendments*. 1966. U.S. Code. vol. 7, Secs. 2131-2157 et seq.
2. PHS Policy. *Public health service policy on humane care and use of laboratory animals*. Office of Laboratory Animal Welfare, National Institutes of Health, Public Health Service; 2002.
3. Health Research Extension Act of 1985 (Pub L. 99-158). *Animals in research*. November 20, 1985.
4. U.S. Government Principles. OER Home Page – Grants Web, Office of Laboratory Animal Welfare: PHS Policy on Humane Care and Use of Laboratory Animals. Available from: http://grants.nih.gov/grants/olaw/references/phspol.htm#USGovPrinciples.
5. National Academy of Sciences. *Guide for the care and use of laboratory animals*. Washington DC: National Academy Press; 2011.
6. Federation of Animal Science Societies (FASS). *Guide for the care and use of agricultural animals in research and teaching*. 3rd ed. January 2010.
7. AVMA. *American veterinary medical association guidelines for the euthanasia of animals: 2013 edition*. 2013. Available from: https://www.avma.org/KB/Policies/Documents/euthanasia.pdf.
8. *Humane Methods of Livestock Slaughter Act of 1958 (Pub L. 85-765) and subsequent amendments*. 1978. 7 U.S. C. 1901 et seq.
9. *Horse Protection Act of 1970 (Pub L. 91-540) and subsequent amendments*. 1976. 15 U.S.C. 1821 et seq.
10. *Twenty-Eight Hour Law. 1873 (49 USC, Section 80502) and later amendments*. 1994.
11. National Academy of Sciences. *Guidelines for the humane transportation of research animals*. Washington DC: National Academy Press; 2006.

12. Sikes RS, Gannon WL, The Animal Care and Use Committee of the American Society of Mammalogists. Guidelines of the American society of mammalogists for the use of wild mammals in research. *J Mammal* 2011;**92**(1):235–53.
13. Gaunt AS, Oring LW. In: Fair JM, Paul E, Jones J, editors. *Guidelines to the use of wild birds in research*. 3rd ed. The Ornithological Council; 2010.
14. National Academy of Sciences. *The psychological well-being of nonhuman primates*. Washington, DC: National Academy Press; 1998.
15. Workman P, Aboagye EO, Balkwill F, et al. Guidelines for the welfare and use of animals in cancer research. *Brit J Cancer* **102**(11):1555–77.
16. National Academy of Sciences. *Guidelines for the care and use of mammals in neuroscience and behavioral research*. Washington, DC: National Academy Press; 2003.
17. American Society of Ichthyologists, Herpetologists. *Guidelines for the use of live Amphibians and reptiles in field research*. American Society of Ichthyologists and Herpetologists; 1987. Available from: http://www.asih.org/files/hacc-final.pdf.
18. United States Department of Agriculture, National Agricultural Library, Animal Welfare Information Center. *Environmental enrichment for nonhuman primate resource guide*. Series No. 32 2006. Available from: http://www.nal.usda.gov/awic/pubs/Primates2009/primates.shtml#ack.
19. United States Department of Agriculture, Animal Plant Health Inspection Service. *Animal welfare regulations. Code of federal regulations, title 9*, vol. 1. 2008. Available from: http://www.aphis.usda.gov/animal_welfare/awr.shtml.
20. Council of Europe. 2006. Appendix A of the European Convention for the Protection of Vertebrate Animals Used for Experimental and other Scientific Purposes (ETS No. 123). Guidelines For Accommodation And Care Of Animals (Article 5 Of The Convention). Approved By The Multilateral Consultation. Cons 123 2006;3.
21. NRC. *Animal biotechnology: science-based concerns*. Washington, DC: National Academies Press; 2002. Available from: http://www.nap.edu/openbook.php?isbn=0309084393&page=R1.
22. FDA. *Guidance for industry 187 regulation of genetically engineered animals containing heritable recombinant DNA constructs*. 2009. Federal Register Volume 74, Number 11.
23. OLAW. 2006-2013. National Institutes of Health, Office of Extramural Research. Frequently asked questions. A6 – Does the PHS policy apply to animal research that is conducted in the field? Available from: http://grants.nih.gov/grants/olaw/faqs.htm#App_6.
24. OLAW. 2006-2013. National Institutes of Health, Office of Extramural Research. Frequently asked questions. G2 – What is required for an occupational health and safety program? Available from: http://grants.nih.gov/grants/olaw/faqs.htm#instresp_2.
25. DHHS. In: Chosewood LC, Wilson DE, editors. *Department of health and human services, public health service center for disease control and prevention, National Institutes of Health. Biosafety in microbiological and biomedical laboratories*. 5th ed. 2009. CDC 21-1112. Available from: http://www.cdc.gov/biosafety/publications/bmbl5/BMBL.pdf.
26. NRC. *Occupational health and safety in the care and use of research animals*. National Academies Press; 1997.
27. OLAW. 2006-2013. National Institutes of Health, Office of Extramural Research. Frequently asked questions. F2 – What are the institution's responsibilities in ensuring that animals are shipped safely and in reporting adverse events that occur in shipment of animals to or from the institution? Available from: http://grants.nih.gov/grants/olaw/faqs.htm#useandmgmt_12.
28. OLAW. 2006-2013. National Institutes of Health, Office of Extramural Research. Frequently asked questions. F17-May performance standards determine environmental enrichment issues? Available from: http://grants.nih.gov/grants/olaw/faqs.htm#useandmgmt_12.

29. OLAW. 2006-2013. National Institutes of Health, Office of Extramural Research. Frequently asked questions. F14 – Is social housing required for nonhuman primates when housed in a research setting? Available from: http://grants.nih.gov/grants/olaw/faqs.htm#useandmgmt_12.
30. OLAW. 2006-2013. National Institutes of Health, Office of Extramural Research. Frequently asked questions. F6 – How can institutions assure animal welfare when HVAC systems malfunction or fail? Available from: http://grants.nih.gov/grants/olaw/faqs.htm#useandmgmt_12.
31. ACLAM. 2013. American College of Laboratory Animal Medicine Position Statement on Adequate Veterinary Care. Available from: http://www.aclam.org/Content/files/files/Public/Active/position_adeqvetcare.pdf.
32. USDA. 2000. United States Department of Agriculture, Agricultural Research Service, National Agricultural Library animal Welfare Information Center. Information Resources for Institutional Animal Care and Use Committees 1985-1999. AWIC Resources Series No.7. Available from: http://www.nal.usda.gov/awic/pubs/IACUC/iacuc.htm.
33. OLAW. 2006-2013. National Institutes of Health, Office of Extramural Research. Frequently asked questions. F1 – Is the use of carbon dioxide an acceptable euthanasia agent? Available from: http://grants.nih.gov/grants/olaw/faqs.htm#useandmgmt_12.

Chapter 3

Canada's Oversight of Animal Ethics and Care in Science

Michael Baar, Julie Dale, Gilly Griffin
Canadian Council on Animal Care, Ottawa, ON, Canada

GENERAL FRAMEWORK

The oversight of animal ethics and care in science has evolved in a somewhat different manner in Canada than in other countries. In 1961, a committee of the Canadian Federation of Biological Societies published the first Canadian standards as "*Guiding Principles on the Care of Experimental Animals*."[1] A few years later, in 1966, the then Medical Research Council (MRC) and the National Research Council (NRC) undertook a study to determine how the implementation of national standards should be overseen in Canada. The subsequent report recommended the formation of a Canadian Council on Animal Care (CCAC), to provide standards and a quality assurance program for all aspects of the ethics and care of animals in science.

The CCAC was officially founded in 1968 as a standing committee of the Association of Universities and Colleges of Canada (now Universities Canada), the national organization representing Canadian universities. In line with recommendations made by the MRC/NRC in the report, the Canadian program developed as a peer review system, drawing on the strengths of many individuals and organizations to reach its goals.

The CCAC continues to operate through the involvement of over 2000 volunteers and 22 member organizations from federal government departments/agencies involved in animal-based science or supporting animal-based research; national charitable organizations funding animal-based research; institutionally based national academic associations; national scientific and academic associations; national organizations representing pharmaceutical companies; and national organizations representing animal welfare and animal care. The Canadian Federation of Humane Societies (CFHS) was included from the outset as a representative of the animal welfare movement in Canada.

Laboratory Animals. http://dx.doi.org/10.1016/B978-0-12-849880-4.00003-9
Copyright © 2018 Elsevier Inc. All rights reserved.

The Board of Directors, elected by representatives of the member organizations, is responsible for guiding the direction of the organization, in line with CCAC's mandate. Four standing committees: Standards, Assessment and Certification, Public Affairs and Communications, and Governance and Nominations provide advice to the Board of Directors. The Assessment and Certification Committee (AACC) oversees and advises on the CCAC's Assessment and Certification Program. In doing so, the committee develops and maintains the policies and practices underpinning the assessment and certification of institutional animal ethics and care programs. The committee also reviews reports resulting from the assessment of institutions and makes decisions regarding certification of institutions that meet CCAC's national standards for animal ethics and care. The Standards Committee oversees the CCAC's Standards Setting and Maintenance Program, providing advice on the development and renewal of CCAC guidelines, policies, and practices, which support the welfare and ethical treatment of animals in Canadian science.

Since 1968, the CCAC has brought about high standards for experimental animal ethics and care through the development of relevant standards and a code of ethics, and through training and the certification of institutional programs. CCAC policies establish the ground rules and basic requirements for each institutional program within the Canadian system of oversight of animal ethics and care.

Prior to 1997, CCAC guidance for the ethics and care of experimental animals consisted of two volumes of the *CCAC Guide to the Care and Use of Experimental Animals*.[2,3] Subsequently, and in particular with the advent of electronic publication, general and specific guideline documents have been written, revised, and expanded in response to changes in scientific and societal attitudes to animals in research, teaching, and testing (http://www.ccac.ca). The full list of policy and guideline documents, which support the implementation of the Canadian system of oversight, can be found on the Standards and Policies section of the CCAC website (http://www.ccac.ca).

The Canadian system relies heavily on oversight at the local institutional level, as further described below and on the CCAC website (http://www.ccac.ca). Each institution wishing to participate in the CCAC programs must have at least one active and effective animal care committee (ACC), responsible for animal research (or teaching or testing) ethics and for ensuring high levels of animal care. ACCs typically report to and derive their authority from the senior administrator responsible for animal ethics and care at the institution (i.e., Vice-President of Research for most academic institutions). The CCAC oversees the work of over 190 local ACCs in over 200 institutions across Canada.

Since 1998, the CCAC has been awarding its Certificate of GAP (Good Animal Practice) to institutions that meet its national standards for animal ethics and care. The certificate is a prerequisite to receiving research funding from

Canadian and American granting agencies, from charitable organizations that fund animal-based research, and is a requirement for institutions wishing to import certain species of research animals into Canada.[4]

CANADA'S LEGAL FRAMEWORK

Under Canada's Constitution Act, 1867, the federal government cannot legislate with respect to experiments involving animals, as research is a provincial responsibility.[5–7] However, there are three areas in which the federal government has influence over animal-based science. Firstly, Section 446 and 447 of the Criminal Code of Canada[8] protects animals (in general) from cruelty, abuse, and neglect. This section of the Criminal Code has been under review for several years. Secondly, the Health of Animals Act (1990)[9] and its regulations aim primarily to protect Canadian livestock from a variety of infectious diseases that would threaten both the health of the animals and people, and Canadian trade in livestock with other countries. This act is used both to deal with named disease outbreaks in Canada, and to prevent the entry of unacceptable diseases that do not exist in Canada. Lastly, the other mechanism through which the federal government supports the humane treatment of animals, while not strictly speaking legislative in nature, is one of the most powerful instruments available to the federal government in establishing national standards. In Canada, to obtain research grants through the two main federal granting agencies requires the host institution to hold a CCAC Certificate of GAP. Where the government itself awards a contract to an academic or nonacademic institution, clause A9015C of Public Works Standard Acquisition Clauses and Conditions Manual[10] imposes conditions related to the ethics and care of experimental animals in public works and government services.

All Canadian provinces and the Yukon territory have legislated in the area of animal welfare through animal protection acts and regulations. Several provinces have specifically occupied the field of animals acquired and involved in research, teaching, and testing, and five of these (Alberta, Manitoba, Newfoundland and Labrador, Nova Scotia, and Prince Edward Island) refer directly to CCAC standards. Ontario is the only province that has its own *Animals for Research Act*[11] and inspectorate, and is unique in Canada in that it creates a system of control based on the registration of research facilities and the issuance of licenses for supply facilities. Further details relating to the legal oversight system can be found in Ontario section.

One of the general concerns about the seemingly "voluntary" participation by institutions as participants in the CCAC programs has been that some private entities and provincial government departments may choose not to participate, and thus may not have third party assessment of the ethics and care of the animals in their charge. However, over the years, many private companies and some provincial government departments have become participants in the CCAC programs, recognizing the scientific and public relations benefits that external monitoring of their animal ethics and care program brings. Although we believe it to be unlikely, there is the possibility that a private company, or a provincial government laboratory operating within a province or territory where there is no requirement for adherence to CCAC standards could carry out research or testing without any recourse if national standards were not met. However, there would still be the potential intervention by provincially appointed officers (e.g., from Societies for Prevention of Cruelty to Animals (SPCAs), or agriculture ministries, depending on the province), or from the federal police force, implementing the Cruelty to Animals section of the Criminal Code,[8] if there was thought to be grounds to suspect cruel treatment of animals.

THE PRINCIPLES

The overarching principles regarding the ethics of animal experimentation in Canada are contained in the *CCAC policy statement on: the ethics of animal investigation*,[12] which is foundational for Canadian animal-based science. It states, "*The use of animals in research, teaching, and testing is acceptable ONLY if it promises to contribute to understanding of fundamental biological principles, or to the development of knowledge that can reasonably be expected to benefit humans or animals.*"[12] Like most national systems of oversight of animal experimentation, the CCAC incorporates the Three Rs (3Rs) tenet[13] in this overarching policy, which also states, "*Animals should be used only if the researcher's best efforts to find an alternative have failed. A continuing sharing of knowledge, review of the literature, and adherence to the Russell-Burch "3R" tenet of "Replacement, Reduction, and Refinement" are also requisites. Those using animals should employ the most humane methods on the smallest number of appropriate animals required to obtain valid information.*"[12] The policy sets requirements for meeting animals' physical and psychological requirements, and points to restrictions for various types of procedures (burns, freezing, fractures, trauma, staged predator/prey encounters). It does not permit certain types of procedures; for example, the use of neuromuscular blockers without anesthetics, and painful experiments or serial invasive procedures conducted solely for the purpose of instruction of students. For studies where death has typically been the endpoint, the policy strongly encourages establishing alternative endpoints that do not result in irreversible pain and distress for the animals, but still meet the requirements of the study.

In 2009, the CCAC established a Three Rs Program,[14] to (1) promote the principles of the 3Rs and (2) support the implementation of the 3Rs. This includes development of strategic initiatives that contribute to: optimizing replacement of animals or refinement of procedures for animals in science, or, when animals are necessary in studies, ensuring appropriate numbers, through sound experimental design.

SCOPE/APPLICABILITY

The CCAC's services extend to any institution involved in animal-based science. This includes academic institutions where institutions sign an Agreement on the Administration of Agency Grants and Awards by Research Institutions, requiring an institution conducting animal-based research to hold a CCAC Certificate of GAP to be eligible for Canadian granting agency funds (http://science.gc.ca/default.asp?lang=En&n=56B87BE5-1). This also includes government laboratories, both federal and provincial, and private companies. In the case of private companies, there is no legal or policy requirement for a CCAC Certificate of GAP, and as indicated earlier these institutions (except for those in Ontario) could, in theory, operate without any third party oversight. In the main, private companies recognize the importance of public accountability and trust for animal-based science, choose to participate in the CCAC programs, and hold the CCAC Certificate of GAP as a tangible confirmation of their commitment to high standards for animal ethics and care. Institutions holding a CCAC Certificate of GAP may choose to have their names listed on the CCAC website (http://www.ccac.ca).

The types of animal-based activities are categorized into six main purposes, described in Appendix D of the CCAC Interpretation Bulletin—Purpose of Animal Use (PAU)[15] and include as follows:

PAU 0—Animals held in breeding colonies or on holding protocols (e.g., rodents, farm animals) that have not been assigned to a particular research, teaching, or testing protocol;

PAU 1—Studies of a fundamental nature in sciences relating to essential structure or function (e.g., biology, psychology, biochemistry, pharmacology, physiology, etc.), including basic biomedical, biological, or agricultural research;

PAU 2—Studies for medical purposes, including veterinary medicine, that relate to human or animal diseases or disorders, including applied research to develop therapies;

PAU 3—Studies for regulatory testing of products for the protection of humans, animals, or the environment, including vaccine efficacy trials and testing of new therapeutic compounds for new drug submissions;

PAU 4—Studies for the development of products or appliances for human or veterinary medicine; and

PAU 5—Training of individuals in postsecondary institutions or facilities.

Protocol review is the responsibility of the institutional ACC(s), but the CCAC provides guidance on the level of invasiveness of protocols, with a view to indicating which activities require a review by an ACC, and which activities should be reported to the CCAC as part of the annual submission of animal data.[16] Categories of invasiveness are assigned prospectively, taking a precautionary approach; i.e., considering the potential greatest degree of pain and distress likely to be experienced by the animals. There are five categories of invasiveness (A–E), with A being the least severe and level E the most severe:

A. Most invertebrates or live isolates
B. Little or no discomfort or stress
C. Minor stress or pain of short duration
D. Moderate to severe distress or discomfort
E. Severe pain near, at, or above the pain tolerance threshold of unanesthetized conscious animals

Category A protocols involve experiments on most invertebrates or on live isolates or observation of wildlife. Possible examples include: the use of tissue culture and tissues obtained at necropsy or from the slaughterhouse; the use of eggs, protozoa, or other single-celled organisms; experiments involving containment, incision, or other invasive procedures on metazoan; or observation of migratory routes. The number of animals involved in these types of protocols do not have to be recorded, or reported to the CCAC; however, review of these protocols does permit ACCs to reflect on relative replacement—i.e., whether investigators are moving toward replacing animals of greater sentiency such as vertebrates, with animals that current scientific evidence indicates have a significantly lower potential for pain perception, such as some invertebrates.[17] Conversely, permitting Category E protocols to be considered ensures that ACCs discuss, reject, or approve protocol procedures, deemed to be at the limits of ethical acceptability, based on scientific necessity.

AUTHORIZATION OF USER-BREEDING INSTITUTIONS/ INSPECTION/PENALTIES BY COMPETENT AUTHORITIES

The CCAC works at the institutional level, as it is the institution itself, which is awarded the CCAC Certificate of GAP to recognize its compliance with CCAC guidelines and policies. The elements that need to be in place for an institution wanting to obtain a CCAC Certificate are listed on the CCAC website (http://www.ccac.ca).

CCAC assessment panels (composed of at least one scientist and one veterinarian, with experience in animal experimentation and care relevant to the institution to be visited, a community representative and a CCAC associate director of assessments) visit institutions at least once every 3 years to ensure that:

- the animal ethics and care program is well structured, effective, and well supported, and is tailored to the size of the institution and the complexity of animal-based studies carried out there;
- the ACC is fulfilling its responsibilities as described in the *CCAC policy statement on: terms of reference for animal care committees*,[18] which includes meeting at least twice every year to review and approve animal-based protocols, visiting all animal facilities at least once every year and carrying out postapproval monitoring of animal-based studies;

- the veterinary and animal care services are well suited for the institution's needs, and meet the Canadian Association for Laboratory Animal Medicine (CALAM/ACLAM)[19] and CCAC standards;
- the training, occupational health and safety, and crisis management programs meet CCAC standards; and
- any animal facilities are suitable for the work carried out at the institution and meet CCAC standards.

Assessments are based on CCAC guidelines, policies, and associated documents. Following any assessment visit to an institution, the CCAC prepares an assessment report, which generally includes recommendations that must be answered satisfactorily in order for the institution to be certified. Further details relating to the assessment report are described on the assessment and certification pages of the CCAC website http://www.ccac.ca. The Certificate of GAP is awarded to an institution once a satisfactory response to the CCAC recommendations has been given. Unsatisfactory institutional responses or the absence of a response can lead to probationary certification, and eventually to certificate removal. For academic institutions this could have the effect of rendering them ineligible for granting agency funding. Details of the certification process can be found in the *CCAC policy statement on: the certification of animal care and use programs*.[20]

Participating institutions in the CCAC program are required to submit animal data to the CCAC on an annual basis. The numbers of animals are categorized according to species, purpose of use (six categories), and level of invasiveness (five levels, A–E), as described earlier. This permits the CCAC to publish reasonably comprehensive pooled data on an annual basis. Annual statistics are made available on the CCAC website (http://www.ccac.ca). Complete data have been published since 1996, with a more limited data set published in the previous years beginning in 1976.

NONHUMAN PRIMATES: SPECIAL CONSIDERATIONS AND RESTRICTIONS

Nonhuman primates continue to be involved in research and testing in Canada. Because of their high level of sentiency and the difficulties in meeting the psychological and behavioral needs of these complex animals, ACCs are expected to take particular care in providing an ethics review of any protocol in which they are involved. No great apes are currently held in Canada for scientific purposes, and while there is no official moratorium, there would have to be considerable scientific justification for their involvement in any laboratory-based study. As is the case for all animals used for scientific purposes in Canada, the CCAC guidelines for their care focus on meeting the physical and psychological needs of the animals. Currently, CCAC guidance for nonhuman primates is under review.

GENETICALLY ALTERED ANIMALS: SPECIAL CONSIDERATIONS

In 1997, the CCAC published *CCAC guidelines on: transgenic animals*,[21] recognizing that there was a rapid increase in the creation of and work with these animals, and that the public was beginning to express concern about genetic modification of animals.[21] The *CCAC guidelines on: transgenic animals*[21] requires that protocols involving generation of genetically modified animals be assigned to category of invasiveness level D, in recognition of the potential for pain and distress both from the procedures and from unanticipated welfare impacts. ACCs are encouraged to revisit the categorization once the actual impact of the procedures on the animals has been evaluated, and reduce the category of invasiveness where warranted.[22] This appears to be a challenge for institutional ACCs, and so the numbers of animals reported associated with level D of invasiveness in the annual CCAC census have risen since 1997, giving a false impression of the numbers of animals actually subject to pain and distress.

While the CCAC does not currently enumerate genetically modified animals separately from other animals, it would appear from the increase in Category D level of invasiveness protocols that the creation of and work with genetically modified animals continues to expand. This was substantiated by a bibliometric study of animal studies in the scientific literature.[23] The 1997 guidelines recognized that "*the creation and use of genetically modified animals is a rapidly evolving field of research.*"[21] As CCAC guidelines for each type of animals are revised, relevant information concerning genetically modified lines of the particular types of animals will be included. This is of particular relevance for mice and for zebrafish, which are currently the most commonly modified species groups.

INSTITUTIONAL AND DESIGNATED PERSONNEL RESPONSIBILITIES

Within each institution, the senior administrator holds the ultimate responsibility for the operation of the institution's animal ethics and care program. The responsibilities of the senior administrator are laid out in the *CCAC policy statement for: senior administrators responsible for animal care and use programs*.[24]

The CCAC oversees a wide variety of institutions that fall into three main categories: academic, government, and private. Despite the variety of type and size of institutions, each is required to have the following main elements.

1. At least one functional ACC, the membership of which includes scientists/lecturers experienced in animal-based science; at least one veterinarian trained and/or experienced with the relevant species and the types of work carried out at the institution; a member who works at the institution, but who does not carry out animal-based research, teaching, or testing; at least one community representative who has no association with the institution and no links to animal-based science (most institutions' workloads require more

than one community representative); the director/manager of the animal facilities; a representative from the technical staff; for academic institutions a student representative; and a paid committee coordinator.
2. Sufficient veterinary and animal care service staff with the relevant expertise needed to care for the animals involved in animal-based science at the institution.
3. A process whereby investigators describe any proposed animal-based studies in a protocol, and cooperate with the ACC to ensure that the proposed work is in line with CCAC guidelines and policies and with the requirements of the institution.

OVERSIGHT AND ETHICS REVIEW PROCESS

Oversight

The CCAC provides oversight of institutional animal ethics and care programs through its Assessment and Certification Program, which carries out regular visits to institutions. At the institutional level, considerable reliance is placed on local ACCs to oversee animal-based research, teaching, and testing on an ongoing basis. This includes the ethics review of protocols (as detailed later). ACCs are also expected to monitor the progress of approved protocols through a postapproval monitoring program, designed to meet the specific needs of the institution. Postapproval monitoring includes ensuring that:

1. the veterinarian(s) and animal care staff are helping researchers, study directors, and/or course instructors to remain in compliance with the approved protocols as well as relevant CCAC and institutional standards, in particular (a) ensuring that individual researchers are competent in handling animals and carrying out procedures successfully and (b) that any unnecessary distress is avoided through the use of endpoints as agreed with the ACC;
2. the ACC coordinator or other ACC members are available to answer any questions from the researcher, and assist in revising a protocol (for example, if the original protocol is found to present some technical or logistical difficulties);
3. site visits and discussion of protocols with research/study teams are held with ACC members or other colleagues, to address any concerns with the study; and
4. careful assistance and follow-up is in place for new procedures and where procedures may result in animal pain or distress.

Ethics Review

As indicated in section Institutional and Designated Personnel Responsibilities, all animal-based work must be included on an animal protocol and submitted to an institutional ACC for ethics review, prior to commencement of the work.

Before an ACC can review a protocol, there must be evidence of scientific, pedagogical, or regulatory merit. In the case of research studies[25] evidence of scientific merit can be given through the peer review process of the main granting agencies; however, where evidence of good peer review is absent, the institutional ACC must use a peer review process that has appropriate independence and expertise. The *CCAC policy statement on: scientific merit and ethical review of animal based research* provides further guidance to institutions on how to achieve suitable review of scientific merit.[26] For teaching protocols, the pedagogical merit should be established prior to ethics review. Guidance on providing independent peer review for teaching protocols is given in the *CCAC policy statement on: pedagogical merit of live animal-based teaching and training*.[27] For animal-based tests, required by government regulations, the study director needs to demonstrate that: (1) the study meets current regulatory requirements, acceptable to the Canadian regulatory agency(ies); (2) the study meets the requirements of the *CCAC policy statement on: ethics of animal investigation*[12]; and (3) the number of animals required does not exceed the number of animals specified in regulatory authority test guidelines.

The elements that must be included in an animal protocol are listed in the *CCAC guidelines on: animal use protocol review*[25] and the *CCAC policy statement on: terms of reference for animal care committees.*[18] ACCs base their decisions concerning the acceptability of protocols on CCAC standards (guidelines and policies) as well as on professional judgment and common sense.

Once a protocol has been approved by the ACC, the investigator is permitted to proceed with the planned studies. However, he/she is expected to provide a progress report to the ACC on an annual basis. This may also necessitate requesting amendments to the protocol.

INVOLVEMENT OF ANIMALS IN ADDITIONAL STUDIES (REUSE)

Involvement of animals in additional studies (reuse) is permitted within the Canadian system. Specifically, guidelines for involvement in additional studies are described in the *CCAC guidelines on: procurement of animals used in science.*[28] When considering the procurement of animals from another scientific institution or from another researcher/study director/course instructor within the same institution, a request should be made for documentation detailing the original source of the animal and the history of the animal while in captivity (e.g., conditioning, housing, nutrition, previous studies). For transfers between institutions, a health certificate for the animals is required, and the animal care services at the receiving institution should be notified of the transfer and be given a copy of the health certificate. These animals should only be procured if they are suited to the intended study and to the conditions under which they will be housed, and they have not been subjected to procedures that would preclude their involvement in any further procedures (typically previous invasive procedures).

In general, additional studies are more frequently carried out in the case of larger animals, and so specific guidance is included within guidelines documents relevant to these types of animals. For example, from the *CCAC guidelines on: the care and use of farm animals in research, teaching and testing*[29] Guideline 16 states: "*Animals subjected to invasive surgery must not be used in additional studies, without explicit approval of the animal care committee.*" The text goes on to explain that "*Occasionally, animals that have been used for a study and have not been subjected to invasive procedures may be used for a further scientific study. As well, a second major surgery may be performed on an animal if it is a nonsurvival procedure. Minor procedures such as biopsies may be performed more than once, but only if they can be done with effective anesthesia and analgesia and do not significantly impact the well-being of the animal.*"

Larger animals are often involved in more than one teaching protocol,[29] so guidelines referring to these animals also make recommendations for restricting the number of times that this may occur. For example, Section 8.2 of the *CCAC guidelines on: the care and use of farm animals*,[29] Guideline 33: Frequency-of-Use, states that "*When planning student exercises with animals, the instructor must carefully weigh the pedagogical merit of the procedure against the invasiveness of the procedure and how often it will be carried out on each animal,*" and the supporting text explains that: "*Even relatively innocuous procedures, when done repeatedly, can be harmful to animals. Special precautions need to be taken when potentially painful or distressing procedures* (e.g., *rectal palpation, tail or jugular bleeding, etc.) are taught….Animals should not be maintained indefinitely for teaching purposes. There should be an established length of time and/or number of training procedures that an animal is involved in before disposition of the animal and replacement.*"

SETTING FREE/REHOMING

Investigators are required to indicate the intention for the disposition of the animals at the end of any study. In general, rehoming of laboratory animals is not encouraged; however, some institutions have put in place programs, which aim to rehome animals. Normally, these are companion animals (dogs and cats), which have been socialized and are able to make the transition to a home environment.

Rehoming is considered more specifically in the guidelines for various species' groups. For example, the *CCAC guidelines on: the care and use of fish in research, teaching and testing*[30] details expectations if fish are to be rehomed: "*Some institutions release healthy research fishes (not genetically modified fishes) that are commonly accepted pet or companion species to individuals with the knowledge and ability to provide adequate care. No genetically modified fish may be removed from research facilities to private premises. If fishes are to be released to the care of an individual as companion animals, the institution should develop an appropriate policy describing the conditions that need to be fulfilled before their release.*"

OCCUPATIONAL HEALTH AND SAFETY

The CCAC takes occupational health and safety seriously, as those working with experimental animals risk exposure to physical hazards (e.g., heat, noise, radiation), chemical hazards (e.g., disinfectants, cleaning solutions), as well as intestinal parasites, enteric bacteria, pathogenic organisms, and animal bites.[31] Zoonotic diseases are a particular concern depending on the species. Guidelines for working with biohazards (e.g., bacteria, viruses, parasites, fungi, and other infectious agents) are provided in the Canadian Biosafety Standards and Guidelines.[32] These guidelines include such items as biohazard containment, laboratory design, personal hygiene, and safety facilities, and can be used to provide training for employees as mandated by Canada's Workplace Hazardous Materials Information System (WHMIS; https://www.canada.ca/en/health-canada/services/environmental-workplace-health/occupational-health-safety/workplace-hazardous-materials-information-system.html), and support the implementation of the Human Pathogens and Toxins Act.[33] In addition, the Canadian Food Inspection Agency (CFIA) Containment Standards for Facilities Handling Aquatic Animal Pathogens[34] applies to aquatic facilities involved in research and testing.

Additional guidance related to occupational health and safety is included in each of the CCAC species-specific guidelines; for example, see the *CCAC guidelines on: the care and use of farm animals in research, teaching and testing.*[29]

TRAINING AND COMPETENCE OF PERSONNEL

In Canada, all personnel from institutions participating in CCAC programs and involved in animal-based research, teaching, or testing must be competent and adequately trained in the principles of animal ethics and care. When assessing institutions, the CCAC ensures that appropriate institutional training programs are implemented as an essential component of any institutional animal ethics and care program.

Those working with animals in science (including investigators, graduate students, postdoctoral fellows and research staff, study directors), animal health professionals, ACC members, and institutional officials (senior administrators) directly responsible for animal ethics and care programs have different training needs. The CCAC develops a variety of training and reference materials to support the needs of these audiences, or makes materials available through external links. These resources and events include a training syllabus, training modules, and other materials, workshops, and web-based seminars.

The *CCAC guidelines on: training of personnel working with animals in science*[35] requires all investigators, graduate students, postdoctoral fellows and research staff, study directors, etc. to have an appreciation of the ethics of animal-based studies in Canada; to understand the principles of experimental animal science relevant to their area of work; and to demonstrate the technical ability to carry out any procedure they are required to perform. While the CCAC provides access to training modules to support the *CCAC Recommended syllabus for an institutional animal user training program*,[35] institutions provide the practical hands-on training as appropriate and are expected to ensure that those working with animals are competent to carry out any relevant animal-based procedures.

Institutions are also expected to provide continuing education opportunities for their animal health professionals as required in the *CCAC policy statement for: senior administrators responsible for animal care and use programs*,[23] the *CALAM/ACMAL standards of veterinary care*,[19] and the *CCAC policy statement on: terms of reference for animal care committees*.[18]

TRANSPORT

Under the *Health of Animals Act*[9] and its regulations, the Canadian Food Inspection Agency (CFIA) is responsible for the humane transportation of animals in Canada. The regulations include requirements for such elements as the provision of food, water, rest, protection from adverse weather, use of proper containers and transport vehicles, and segregation of incompatible animals. This Act and its regulations also specify that livestock, poultry, animal embryos, and animal semen exported from Canada must be accompanied by a health certificate issued or endorsed by a CFIA veterinary inspector, and that CFIA is responsible for testing, inspection, permit issuing, and quarantine activities for live animals imported to Canada. Guidelines specific to the transportation of animals for research purposes are included in the *CCAC guidelines on: procurement of animals used in science*.[28] This guidelines document is based on recommendations made by the US NRC *Guidelines for the Humane Transportation of Research Animals*[36] and the UK Laboratory Animal Science Association *Guidance on the Transport of Laboratory Animals*.[37] The CCAC guidelines recommend that these two documents be consulted for more detailed information on suitable practices for the transportation of animals. The guidelines refer not only to transportation of animals between institutions, but also within an institution, recognizing that transportation can be an acutely stressful event for laboratory animals. Further recommendations regarding transportation are also included in the *CCAC guidelines on the care and use of farm animals in research, teaching and testing*[29] and the *CCAC guidelines on: the care and use of fish in research, teaching and testing*,[30] as the transportation of these animals poses unique challenges.

HOUSING AND ENRICHMENT

The CCAC guidance for housing many of the types of animals involved in science is under revision, which will include consideration of enclosure dimensions. Emphasis is placed on grounding requirements in expert peer advice and current interpretation of scientific evidence. For species groups where guidelines have already been revised, namely, farm animals and fish,[29,30] guidance is provided on enclosure dimensions, but more specifically on the needs for the particular animals in question. So, for example, where dairy cattle are to be kept in tie-stalls the *CCAC guidelines on: the care and use of farm animals*[29] states in Guideline 59: "Dairy *cows kept in tie-stalls should be allowed a period of exercise every day,*" and the supporting text states that "*Cows kept in tie-stalls should be allowed a period of exercise every day unless experimental procedures or inclement weather preclude it. This also allows assessment of mobility and other health problems. Moving cows in and out of tie-stalls should be done with care and slowly so that they do not fall and injure themselves.*"

The CCAC has always had a strong focus on meeting the social and behavioral needs of animals. Prior to the revision of the *CCAC Guide to the Care and Use of Experimental Animals* in 1993,[3] the CCAC implemented a policy statement on: social and behavioral requirements of experimental animals.[38] Social housing is expected to be the norm for animals housed within Canadian institutions. Protocols that involve single housing must describe proposed measures for meeting the social requirements of the isolated animal (e.g., where appropriate, increased positive human contact).

As the guidelines on the care of the various species groups are revised, emphasis will be placed on meeting the basic needs of animals as well as on providing enrichment of the environment to improve the quality of life. However, while providing positive experiences for the animals is a laudable goal; it is important to ensure that the enrichment strategies that are used do not interfere with scientific goals (see, for example, Refs. 39–42).

HUSBANDRY AND ENVIRONMENT

General guidelines in relation to daily care of animals in a laboratory animal facility are currently under development, and detailed information is being included in the guidelines for various types of animals as they are revised. This has already been completed for farm animal species[29] and for fish.[30] However, in particular in the case of fish the guidelines acknowledge upfront that the greatest challenge in providing *guidelines on: the care and use of fish* is the wide variety of fish used in Canada, and the diversity of their habitats, behavior, life history, and environmental and husbandry requirements. It is also acknowledged that the scientific information required to define the preferred conditions for fish welfare is limited.

In general, the CCAC approach to developing guidelines is to provide the framework for the implementation of good animal practices, rather than stating the details of those practices, to allow for the evolution of refinements to animal-based procedures, etc. This approach also allows for coverage of a large range of types of animals, even in situations where scientific information on preferred conditions (such as for fish welfare) is limited. Within the guidelines, efforts are made to define the most appropriate conditions, based on expert peer advice and current interpretation of the scientific evidence available. The CCAC provides links to additional information concerning improved practices as it becomes available, and encourages institutions to share good practice information through its assessment process and through publication on the CCAC website.

For general environmental parameters such as room temperature, humidity, and ventilation, basic recommendations are included in the *CCAC guidelines on: laboratory animal facilities—characteristics, design and development.*[43] More specific information relevant to individual animal types is included in the relevant guidelines, which can be accessed on the CCAC website (http://www.ccac.ca).

VETERINARY CARE

In general, the CCAC refers to the *CALAM/ACMAL Standards of Veterinary Care*[19] as the national standard in relation to veterinary care for animals in science. The CALAM/ACMAL standards include the CALAM Position on Standards of Veterinary Care, which states that a "*veterinarian with authority and responsibility for supporting an institutional animal care and use program must be involved in all issues and activities that relate in any way to animal care and use.*" The types of institutions and facilities vary widely, and are recognized in the position statement, which also states that "*The extent of the program of veterinary care will depend on several factors, such as the number and species of animals used and the nature of the activities that involve animal use. The structure of the veterinary program, including the number of licensed veterinarians and their background and training must be appropriate to fulfill the programs' requirements and to ensure that the CALAM/ACMAL Standards of Veterinary Care are met. This will vary by institution.*" The importance of having veterinary expertise available at all times is also recognized, "*In all cases, formal arrangements must be made by the senior administration of the institution to ensure that veterinary services are readily available at all times to meet both routine and emergency needs.*"

The CCAC uses the CALAM/ACMAL standards[19] during its assessment visits as the basis for recommendations made in CCAC assessment reports to institutions. In addition, throughout CCAC guidelines documents there are recommendations relating to areas where the services of a veterinarian should be used. For example, the *CCAC guidelines on: the procurement of animals used in science*[28] indicates in Guideline 5 that "*the institutional veterinarian should have ultimate responsibility for ensuring procurement of healthy animals.*"

This includes identifying potential sources and suppliers, routine inspection of suppliers, developing in-house quarantine and conditioning programs, providing assistance in the selection of an appropriate animal model, and generally developing a good relationship with suppliers to ensure appropriate holding times, delivery and transportation methods, and thus minimize any stress on the animals.

While the responsibilities relating to veterinary care for "traditional" laboratory animals is well defined in the *CALAM/ACMAL Standards for Veterinary Care*,[19] for some of the other species groups, the CCAC guidelines make particular reference to the responsibilities of the veterinarian. In addition, the guidelines reinforce the importance for veterinarians to receive adequate training in the relevant types of animals. For example, from the *CCAC guidelines on: the care and use of farm animals in research, teaching and testing*,[29] "*Veterinarians attending farm animals should have special training in farm animal health management in research, teaching or testing environments*" and the *CCAC guidelines on: the care and use of fish in research, teaching and testing*[30] "*Veterinarians working at institutions with large populations of fishes are encouraged to have special training in fish health management in research, teaching or testing environments*." Animal protocols involving wildlife represent another special case where oftentimes, the researchers are the most knowledgeable about the particular species in question. The *CCAC guidelines on: the care and use of wildlife*[44] encourages consultation and/or participation of veterinarians having experience with wildlife in projects where there is the potential for animal health concerns, for example, in translocation of animals or for medical or surgical procedures including procedures requiring chemical immobilization. The document also emphasizes that veterinarians are liable for the use of any pharmaceuticals they dispense and for veterinary care, so they should be an integral member of the research team when medical or surgical procedures are part of the study.

CONDUCT OF EXPERIMENTAL PROCEDURES

As described earlier, before the start of any animal-based study an animal protocol must be approved by the institutional ACC.[25] The *CCAC policy statement on: the ethics of animal investigation*[12] describes fundamental principles for the conduct of animal-based science, including minimizing any potential pain and distress for the animals, and ensuring that the numbers of animals are appropriate for the study.[12] The expectation is that investigators will carry out a 3Rs search prior to drafting the protocol to ensure that the possibility for using Replacement, Reduction, or Refinement alternatives has been considered.[18] To assist in this regard, the CCAC has developed a Three Rs Search Guide, which investigators can use to help develop a search strategy and collect the relevant information (http://3rs.ccac.ca/en/searches-and-animal-index/guide/). Investigators are also expected to work with the institutional laboratory animal veterinarian, to ensure that the animals will receive appropriate housing and husbandry, and that any potential pain and distress is minimized. At the time of approval, animal protocols are also assigned a category of invasiveness based on the potential pain and distress that could be experienced by the animals.[16] Guidance on the conduct of experimental procedures is further elaborated in many of the CCAC guidelines documents, listed in the Standards section of the CCAC website (http://www.ccac.ca).

EUTHANASIA

The *CCAC guidelines on: euthanasia of animals used in science*[45] is based on recommendations made by the International Council for Laboratory Animal Science (ICLAS) Working Group on Harmonization[46] and the two international reference documents on euthanasia recommended by ICLAS: the American Veterinary Medical Association (AVMA) *Guidelines on Euthanasia*[47] and *Recommendations for Euthanasia of Experimental Animals*, Part 1[48] and Part 2[49]; with some modification to fit the Canadian context. An overview of acceptable methods of euthanasia for common species groups involved in research, teaching, and testing is based primarily on these two major reference documents. Information is included on other methods that may be acceptable for specific purposes providing they comply with the general guiding principles and receive the approval of the ACC reviewing the application.

Summary charts of the acceptable methods of euthanasia, and of the conditionally acceptable methods (i.e., methods where there is greater potential for operator error or safety hazards, they might not consistently produce humane death, or they are not well documented in the scientific literature), are published in the *CCAC guidelines: on euthanasia*.[45] Further details for each of the methods are provided within the text of the guidelines. In addition, the CCAC has published additional information on the potential impact of each euthanasia method on research data. This is available on the CCAC website in association with the guidelines document (http://www.ccac.ca).

EQUIPMENT AND FACILITIES

The CCAC requires that facilities for the care and conduct of procedures for all animals in research, teaching, and testing provide for the well-being and safety of the animals, provide an appropriately appointed and safe workplace for personnel, and afford a stable environment for research studies. The *CCAC guidelines on: laboratory animal facilities—characteristics, design and development*[43] is intended to assist the users and designers of laboratory animal facilities to achieve optimal levels of animal care and facilitate good research without curtailing new and innovative facility design ideas.

Any renovation or new construction of laboratory animal facilities must meet certain basic functional criteria to be in compliance with the spirit and intent of the CCAC guidelines. These basic criteria are outlined in the text of this guidelines document.

The *CCAC guidelines on: laboratory animal facilities—characteristics, design and development*[43] applies to laboratory animals such as rats, mice, rabbits, dogs, and cats held in controlled environments, but not to animals held in field settings. Facilities for farm animals, fish, and short-term holding of captive wildlife are described in other CCAC guidelines (*CCAC guidelines on: the care and use of farm animals in research, teaching and testing*[29]; *CCAC guidelines on: the care and use of fish in research, teaching and testing*[30]; and *CCAC guidelines on: the care and use of wildlife*[44]), although many of the general principles described within the *CCAC guidelines on: laboratory animal facilities*[43] are generally applicable to any types of animals maintained in captive environments for the purposes of research, teaching, and testing. The guidelines do not attempt to address building codes or safety codes and standards. It is the responsibility of consultant architects and engineers to address these issues in consultation with the responsible institutional officials.

If institutions are planning and designing biosafety containment facilities, the *CCAC guidelines on: laboratory animal facilities*[43] should be used in conjunction with the relevant biosafety guidelines, such as the *Canadian Biosafety Standards and Guidelines*[32] or the *CFIA Containment Standards for Facilities Handling Aquatic Animal Pathogens.*[34] These biosafety guidelines must be implemented whenever facilities will be used to house animals that are experimentally infected with human and/or animal pathogens.

The *CCAC guidelines on: laboratory animal facilities—characteristics, design and development*[43] refers to barrier systems for reducing or minimizing cross-contamination since these are important concepts in all animal facilities. Barriers are commonly used to separate animals of different or unknown disease statuses, such as dogs, cats, mice, specific pathogen-free animals, and genetically modified animals.

ONTARIO

Ontario is the one province that has enacted legislation directed solely at animal-based science. The Ontario Animals for Research Act[11] extends to all nonhuman vertebrate animals. Bare minimum requirements for research facilities are included in regulation 24 to the Ontario Animals for Research Act "Research Facilities and Supply Facilities."[50]

According to the Animals for Research Act, operators of research facilities and animal suppliers are required to register the facility with the Ontario Ministry for Agriculture, Food and Rural Affairs (OMAFRA). The operator of the facility is responsible for ensuring that there are an appropriate number of individuals competent in the care of animals in every research facility. He/she is also required to maintain a record of all animals in the research facility, and to keep those records within the research facility for at least 2 years in the case of dogs and cats and 1 year for all other animals.

The OMAFRA ensures that institutions comply with the legislation through unannounced inspections of research and supply facilities. Currently, there is only one inspector to cover the entire province of Ontario.

Regulations to the Ontario Animals for Research Act[11] requires each registered facility to establish an ACC, one of the members of which must be a veterinarian.[51] The ACC responsibilities include the coordination and review of "*(a) the activities and procedures relating to the care of animals; (b) the standards of care and facilities for animals; (c) the training and qualifications of personnel that are engaged in the care of animals; and (d) procedures for the prevention of unnecessary pain including the use of anesthetics and analgesics.*"[11]

Research facilities in Ontario are required to submit an annual report to the OMAFRA, which includes "(1) the total number of every species of animal used for research in the research facility in the year; (2) the total number of dogs and the total number of cats purchased or otherwise acquired from (a) other research facilities, (b) pounds, (c) supply facilities, and (d) other sources; and (3) the total number of dogs and the total number of cats that in any experiment or surgical procedure did not recover from anesthesia."[51]

There is a special mention of nonhuman primate requirements within the Animals for Research Act.[11] These requirements relate to ensuring that the animals are tested for tuberculosis (TB) upon arrival in the research facility and either isolated or euthanized if found to be carrying TB, unless it is a necessary element of the research study. In addition, no one with active TB may be employed to take care of nonhuman primates. There are also requirements for isolation of nonhuman primates when they arrive at the facility to give a period of acclimatization to the research environment. No more than 25 nonhuman primates are permitted to be housed together in the same cage or pen. The regulations to the Act also make restrictions on the use of restraint chairs, requiring that they only be used to the extent required by the research protocol.[50]

The OMAFRA has made specific regulations concerning transportation under the Ontario Animals for Research Act. These regulations include requirements for the vehicles, the shipping containers, feed, water, and duration of the transportation, as well as for personnel, both to accompany the animals and to be present to receive the animals at their destination.

While the Act includes basic requirements for housing, no cage sizes, etc. are given. Rather the regulations to the Act make generalized statements regarding the behaviors that the animals should be able to perform, e.g., being able to extend their legs, sit, stand, and lie in an extended position. The Act requires that enclosures are designed in such a manner that the animals can be readily observed and cannot escape, and that the enclosures can be readily sanitized and are designed to minimize the transfer of pathogens as far as practicable.[50] General requirements for husbandry and the environment are also included in the regulations to the Act.

As far as veterinary care is concerned, the Act states that "*The Lieutenant Governor in Council may make regulations...classifying research facilities, requiring the operators of any class of research facility to provide for the services of a veterinarian in connection with the care of animals in the research facility and prescribing the terms and conditions on which such services shall be provided in respect of any such class*"[11]; however, regulations in respect of veterinary services have yet to be made.

Specific statements regarding euthanasia are included in the regulations "*Where euthanasia is carried out with respect to any animal in a research facility or supply facility, it shall be carried out, (a) by a person or persons properly trained in the euthanasia procedure to be used; (b) in such manner that the death of the animal occurs without unnecessary pain, delay or discomfort; and (c) in a manner that does not endanger or disturb other animals in the research facility or supply facility.*" and "*No person shall use a euthanasia procedure with respect to any animal in a research facility or supply facility unless it is a procedure that is permitted under section 29, 30, 31 or 32.*"[50] The regulations list specific permitted methods for mammals as well as for fish and other cold-blooded mammals.

Some of the methods of euthanasia listed are no longer considered acceptable by the CCAC. For the institutions in Ontario that participate in the CCAC program, more humane practices, as defined in the *CCAC guidelines on: euthanasia of animals used in science*,[45] are expected to be used. In general, the Ontario Ministry inspector keeps apprised of the new CCAC guidelines, which then help inform inspections carried out in the province.

SUMMARY

The Canadian system of oversight for animals in science has evolved in a unique manner. The oversight is not reliant on legislative instruments, but uses a peer-based approach. This means that the system is able to adapt more readily to changes in animal-based science, and thus provide for high standards of animal ethics and care.

REFERENCES

1. Canadian Federation of Biological Societies—CFBS. *Guiding principles on the care of experimental animals*. Ottawa, ON: CFBS; 1961. p. 1.
2. Canadian Council on Animal Care—CCAC. *Guide to the care and use of experimental animals*, vol. 2. Ottawa, ON: CCAC; 1984. p. 372.
3. Canadian Council on Animal Care—CCAC. *Guide to the care and use of experimental animals*, vol. 1. Ottawa, ON: CCAC; 1993. p. 212.
4. Canadian Food Inspection Agency. *Requirements for non-human primates imported into Canada*. 2009. http://www.inspection.gc.ca/animals/terrestrial-animals/imports/policies/live-animals/2009-1/eng/1320891194183/1320892143537.
5. Wilson P. *Legislative jurisdiction over animals used in research, testing and testing. Canadian Council on Animal Care commissioned legal opinion*. Ottawa, ON: Osler, Hoskin & Harcourt; 1998.
6. Létourneau L. *The protection of animals used for the purpose of xenotransplantation in Canada*. 2000. [unpublished, archived at Health Canada].

7. Canada. *Constitution act/Lois constitutionnelle. 30 & 31 Victoria, c 3 (UK)*. Ottawa, ON: Department of Justice Canada; 1867.
8. Canada. *Criminal code/code criminal RSC, 1985, c C-46*. Ottawa, ON: Department of Justice Canada; 1865.
9. Canada. *Health of Animals Act/Loi sur la santé des animaux. SC 1990, c 21*. Ottawa, ON: Department of Justice Canada; 1990.
10. Canada. *Standard acquisition clauses and conditions manual (SACC)*. Ottawa, ON: Public Works and Government Services Canada; 2010.
11. Ontario, Canada. *Animals for Research Act/Loi sur les animaux destinés à la recherché RSO 1990, Chapter A22*. Ottawa, ON: The Ontario Ministry of Agriculture Food and Rural Affairs; 1990.
12. Canadian Council on Animal Care—CCAC. *CCAC policy statement on: ethics of animal investigation*. Ottawa, ON: CCAC; 1989. p. 2.
13. Russell WMS, Burch RL. *The principles of humane experimental technique*. London, UK: Universities Federation for Animal Welfare—UFAW; 1959/1992. Special edition.
14. Griffin G. Establishing a three Rs programme at the Canadian Council on Animal Care. *Altern Lab Anim* 2009;**37**(S2):63–7.
15. Canadian Council on Animal Care—CCAC. *Interpretation bulletin no. 1-1 animal use data form*. Ottawa, ON: CCAC; 2006. p. 20.
16. Canadian Council on Animal Care—CCAC. *CCAC policy statement on: categories of invasiveness in animal experiments*. Ottawa, ON: CCAC; 1991. p. 2.
17. Fenwick N, Ormandy E, Gauthier C, Griffin G. Classifying the severity of scientific animal use: a review of international systems. *Anim Welf* 2011;**20**:281–301.
18. Canadian Council on Animal Care—CCAC. *CCAC policy statement on: terms of reference for animal care committees*. Ottawa, ON: CCAC; 2006. p. 12.
19. Canadian Association for Laboratory Animal Medicine—CALAM. *CALAM/ACMAL standards for veterinary care*. CALAM; 2007. p. 15.
20. Canadian Council on Animal Care—CCAC. *CCAC policy statement on: the certification of animal ethics and care programs*. Ottawa, ON: CCAC; 2016. p. 10.
21. Canadian Council on Animal Care—CCAC. *CCAC guidelines on: transgenic animals*. Ottawa, ON: CCAC; 1997. p. 9.
22. Griffin G, Dansereau M, Gauthier C. Categories of invasiveness – a precautionary approach. *Altern Anim Test Exp* 2007;**14**(special issue):715–20.
23. Ormandy EH, Schuppli CA, Weary DM. Worldwide trends in the use of animals in research: the contribution of genetically-modified animal models. *Altern Lab Anim* 2009;**37**(1):63–8.
24. Canadian Council on Animal Care—CCAC. *CCAC policy statement for: senior administrators responsible for animal care and use programs*. Ottawa, ON: CCAC; 2008. p. 30.
25. Canadian Council on Animal Care—CCAC. *CCAC guidelines on: animal use protocol review*. Ottawa, ON: CCAC; 1997. p. 12.
26. Canadian Council on Animal Care—CCAC. *CCAC policy statement on: scientific merit and ethical review of animal-based research*. Ottawa, ON: CCAC; 2013. p. 2.
27. Canadian Council on Animal Care – CCAC. *CCAC policy statement on: pedagogical merit of live animal-based teaching and training*. Ottawa, ON: CCAC; 2016. p. 2.
28. Canadian Council on Animal Care—CCAC. *CCAC guidelines on: the procurement of animals used in science*. Ottawa, ON: CCAC; 2007. p. 27.
29. Canadian Council on Animal Care—CCAC. *CCAC guidelines on: the care and use of farm animals in research, teaching and testing*. Ottawa, ON: CCAC; 2009. p. 135.
30. Canadian Council on Animal Care—CCAC. *CCAC guidelines on: the care and use of fish in research, teaching and testing*. Ottawa, ON: CCAC; 2005. p. 87.

31. Soave O, Brand CD. Employer responsibility for employee health in the animal environment. *Lab Anim* 1991;**20**(2):41–6.
32. Public Health Agency of Canada, Canadian Food Inspection Agency–PHAC/CFIA. *Canadian biosafety standards and guidelines*. Ottawa ON: Government of Canada; 2013. p. 343.
33. Canada. *Human Pathogens and Toxins Act/Loi sur les agents pathogènes humains et les toxines*. Ottawa, ON: Department of Justice Canada; 2009.
34. Biohazard Containment, Safety Science Branch. *Containment standards for facilities handling aquatic animal pathogens*. 3rd ed. Canadian Food Inspection Agency—CFIA; 2010. p. 65.
35. Canadian Council on Animal Care—CCAC. *CCAC guidelines on: training for personnel working with animals in science*. Ottawa, ON: CCAC; 2015. p. 26.
36. Institute for Laboratory Animal Research—ILAR. *Guidelines for the humane transportation of research animals*. Washington, DC: National Academies Press; 2006. p. 164.
37. Swallow J, Anderson D, Buckwell AC, Harris T, Hawkins P, Kirkwood J, et al. Guidance on the transport of laboratory animals. *Lab Anim* 2005;**39**(1):1–39.
38. Canadian Council on Animal Care—CCAC. *CCAC policy statement on: social and behavioural requirements of experimental animals*. Ottawa, ON: CCAC; 1990. p. 1.
39. Griffin G. Evaluating environmental enrichment is essential. *Enrich Rec* 2012:29–33.
40. Cao L, Liu X, Lin E-JD, Wang C, Choi EY, Riban V, et al. Environmental and genetic activation of a brain-adipocyte BDNF/leptin axis causes cancer remission and inhibition. *Cell* 2010;**142**(1):52–64.
41. Hamm RJ, Temple MD, O'Dell DM, Pike BR, Lyeth BG. Exposure to environmental complexity promotes recovery of cognitive function after traumatic brain injury. *J Neurotrauma* 1996;**13**(1):41–7.
42. Passineau MJ, Green EJ, Dietrich WD. Therapeutic effects of environmental enrichment on cognitive function and tissue integrity following severe traumatic brain injury in rats. *Exp Neurol* 2001;**168**:373–84.
43. Canadian Council on Animal Care—CCAC. *CCAC guidelines on: laboratory animal facilities—characteristics, design, and development*. Ottawa, ON: CCAC; 2003. p. 108.
44. Canadian Council on Animal Care—CCAC. *CCAC guidelines on: the care and use of wildlife*. Ottawa, ON: CCAC; 2003. p. 66.
45. Canadian Council on Animal Care—CCAC. *CCAC guidelines on: euthanasia of animals used in science*. Ottawa, ON: CCAC; 2010. p. 32.
46. Demers G, Griffin G, De Vroey G, Haywood JR, Zurlo J, Bedard M. Harmonization of animal care and use guidance. *Science* 2006;**312**(5774):700–1.
47. American Veterinary Medical Association—AVMA. *AVMA guidelines on euthanasia*. Schaumburg, IL: AVMA; 2007. p. 39.
48. Close B, Banister K, Baumans V, Bernoth EM, Bromage N, Bunyan J, et al. Recommendations for euthanasia of experimental animals: Part 1. *Lab Anim* 1996;**30**(4):293–316.
49. Close B, Banister K, Baumans V, Bernoth EM, Bromage N, Bunyan J, et al. Recommendations for euthanasia of experimental animals: Part 2. *Lab Anim* 1997;**31**(1):1–32.
50. Ontario, Canada. *Research Facilities and Supply Facilities/Services de Recherche et Animaleries R.R.O. 1990, Regulation 24 of The Animals for Research Act*. Ottawa, ON: Department of Justice Canada; 1990.
51. Ontario, Canada. *General/dispositions générales R.R.O. 1990 regulation 22 of the animals for research act*. Ottawa, ON: Department of Justice Canada; 1990.

Chapter 4

Laboratory Animal Legislation in Latin America

Ekaterina A. Botovchenco Rivera[1], Rafael Hernández-González[2], Cecilia Carbone[3], Juan M. Baamonde[4], Tamara A. Botovchenco Rivera[5], André S. Carissimi[6]

[1]*Federal University of Goiás, Goiania, Brazil;* [2]*Universidad Nacional Autónoma de México, Mexico City, Mexico;* [3]*Universidad de La Plata, La Plata, Argentina;* [4]*Centro de Estudios Científicos, Valdivia, Chile;* [5]*State of Goiás Public Ministry, Goiania, Brazil;* [6]*Federal University of Rio Grande do Sul, Porto Alegre, Brazil*

INTRODUCTION

Latin America is a multicultural region due to its people's different ethnic features, colonial background, religion, language, and level of educational and financial development. Therefore, there are many different ways of thinking. Almost every country has Spanish as their common language, with the exception of Brazil, but these countries greatly differ in the development of their technological, educational, health, and economic systems. Of course, this reflects the development of science too, and we may state that we have centers of excellence in some places and in other places science is just starting to bloom. This discrepancy may be seen in some countries like Argentina, which as early as 1891, was concerned to regulate the protection of animals, and a national law against animal cruelty called "Law Sarmiento" No. 2786[1] (http://www:animanaturalis.org/p/1387) was approved by the government. Also in Brazil, in 1934, a Federal Decree 24.645[2] was issued to assure the protection of experimental animals (http://www.planalto.gov.br/ccivil_03/decreto/1930-1949/D24645.htm), but this Decree did not come into force. On the other hand, these same countries have fairly new cities (for example, Brazil's capital, Brasilia is only 60 years old) where universities and technical institutions are also very young, and as a consequence, science is too. Thus, it is not difficult to imagine that regulations on the scientific use of experimental animals is a new subject in Latin America, and in a universe of 23 countries, only three (Brazil, Mexico, and Uruguay) have a national law specific to the use and care of experimental animals. However, substantial effort has been placed in developing regulations, knowing that this is a very important aspect not only in developing a successful laboratory animal science, but also

Laboratory Animals. http://dx.doi.org/10.1016/B978-0-12-849880-4.00004-0
Copyright © 2018 Elsevier Inc. All rights reserved.

for it to be internationally harmonized. To better understand the evolution of laboratory animal science in Latin American countries and the current scenario, a little bit of history might be useful.

Brazil had no regulatory or other normative issues related to the use of animals in research, teaching, or testing until 2008, when the Law known as the "Arouca Law" was issued.[3] Historically, the first attempt to call the attention of the scientific community on how research should be conducted that is based on ethical principles was in 1983, when a veterinarian called Dr. Fernando Sogorb invited some scientists, mainly veterinarians who worked with laboratory animals, to create an association called the National College on Animal Experiments (COBEA; www.cobea.org.br). One of the first activities of this association was to issue the COBEA Ethical Principles, based on the CIOMS (Council for International Organizations of Medical Science) principles.[4] In 2008, members of Brazilian College of Animal Experiments (COBEA) decided to change its name to Brazilian Laboratory Animal Science Association (SBCAL), because it was in fact an association and not a college, and also because people thought COBEA performed animal experiments giving rise to many protests from activists. Brazilian scientists knew about laws in other countries and felt uneasy, as they were aware of the importance of having some type of regulations if animals were to be used in their scientific projects. Since there were no laws, COBEA/SBCAL and scientists from different universities thought something had to be done and organized a task force to implement ethics committees in their institutions, using the Canadian model as a reference (http://www.ccac.ca). These were easier to implement than having a law passed in Congress. No efforts were spared to achieve these goals, and thus, with lectures, seminars, and congresses, the word was spread. And the first ethics revision committee in the use of experimental animals was in place at the Federal University of Goiás, in 1992 (http://www.hc.ufg.br/pages/24568).

We can say, without any doubt, that this was a very important first step toward better science and better animal care in Brazil. It completely changed the approach scientists had toward laboratory animals. As a consequence of such drastic changes, the Brazilian Scientific Council (CNPq; www.cnpq.br), the most important scientific body in the country, decided that no grant was to be given to researchers if their project had not passed by an ethics revision committee. The main scientific publications also started to accept only articles in which the research had been approved by an ethics committee. Although Brazil had no specific law on the use of laboratory animals, there were some tentative federal law projects against the cruelty to animals in experiments. These projects were the Federal Decree No. 24,645, which assured protection to experimental animals (not enacted); and the Law No. 6638/79[5] on the protection of experimental animals. However, it was not enacted. In 1995, a physician called Dr. Sergio Arouca, also a federal representative, coordinated a group of scientists who issued the first draft of a law project concerning the scientific use of animals. Although one might be tempted to shake one's head in disbelief, the law was approved 13 years later,[3] with lots of amendments, and without the supervision of Dr. Arouca, who died in 2003. As this Brazilian law does not cover all the important topics related to the use and care of experimental animals, the National Council on the Control of Animal Experiments (CONCEA) and its group of *ad hocs* have issued the *Brazilian Guide on the Care and use of Laboratory Animals*[6] (http://www.mct.gov.br/upd_blob/0238/238057.pdf) covering the following items: ethics and animal welfare, environment, education, training and competence of personnel, transport, housing, enrichment, husbandry and environment, veterinary care, conduct of experimental procedures, equipment, and facilities. CONCEA is also issuing normative regulations that must be followed by animal users on compulsory points related to the law. Meanwhile, CONCEA accepts international guides as references, such as Australian Guide for the Care and Use of Laboratory Animals,[7] National Institutes of Health Guide for the Care and Use of Laboratory Animals,[8] CCAC Guide for the Care and Use of Experimental Animals.[9a,9b]

Mexico also has a specific law on the care and use of experimental animals.[10] This country differs greatly from other countries in the area. The Mexican economy is 1 of the 10 largest world economies and is a member of the Organization for Economic Development (OECD; www.oecd.org), while other countries of the region have a totally emerging economy. For example, Cuba has a socialist economic model and Puerto Rico has a political and economic system under the jurisdiction of the United States (US). Since Puerto Rico applies US's legislation, its case will not be considered herein. Additionally, Mexico has the largest North American university, the greatest technological development among this group of Hispanic countries, and has the most demanding population of health and education services. Mexico, Cuba, and Costa Rica produce the majority of PhD graduates in different knowledge areas and have health coverage for almost 100% of their people. The three countries have developed a scientific research system recognized in the region and have centers of reference for laboratory animals. Mexican official norm specific for laboratory animals is NOM 062/1999.[10] An important aspect of this law is that it derived from the initiative of veterinarians who worked with laboratory animals, scientists interested in regulating the use of these animals, and officials of the Ministry of Agriculture, Livestock, Rural Development, Fishery, and Food (SAGARPA), who were all interested in improving the use and welfare of laboratory animals in Mexico. The norm has its roots in three documents internationally accepted as reference: Guide for the Care and Use of Laboratory Animals, National Academy of Sciences of the United States of America,[8] Guide for the Care of Laboratory Animals, Canadian Council on Animal Care,[9a,9b] and the American Veterinary Medical Association (AVMA) Guidelines for the Euthanasia of Animals.[11]

Uruguay, like Brazil and Mexico, also has a law specific to experimental animals, Law No. 18.611/2009.[12] Despite the small number of scientists,

Uruguay has an important background in laboratory animal matters. In 2001, in the University of the Republic (UdelaR) a group of professionals identified the need for a framework at a national level that would regulate the care and use of experimental animals. And they called it "Regulations: Use of Animals in Testing, Research and Teaching." This working group was named The Honorary Committee for Experimental Animals (CHEA), which worked until 2008 under the Secretary of Science and Technology from the Universidad de la República. The CHEA promoted the creation of the University Ethics Committees and is responsible, in this institution, for: the training of the animal facility staff and users, the personnel accreditation system, and the creation of research and teaching protocols to be evaluated by the Ethics Committee.

Other Latin American countries are trying to develop their own specific laboratory animal legislations such as **Argentina** and **Colombia**. In 2011 the Argentinean Association for Laboratory Animal Science and Technology (AACyTAL; http://aacytal.wixsite.com/aacytal/sitio-oficial-de-aacytal) presented a new project for a national law taking as examples the European Directive 2010/63/EU[13] and the Guide for the Care and Use of Laboratory Animals, eighth edition.[8] It was discussed and agreed by the Argentine biomedical community in 2012 and now it is being revised by a national committee. All private or national institutions that use animals for scientific purposes in the national territory will be covered by this law. The Three Rs (3Rs) (Replacement, Reduction, and Refinement) are addressed as general principles that support the document. Occupational health and safety, housing, enrichment, husbandry and environment, education and training of personnel and veterinary care are included in the law project. Regarding **Colombia,** there were two Law Project drafts, from 2009 to 2011 to modify the general animal protection law: Law No. 84, December 27, 1989 on the Protection of Animals.[14] But, there is still no specific law on the use of laboratory animals in research and teaching, not even a draft of a project.

Other countries in Latin America do not have a specific law on the care and use of laboratory animals, but have laws of general scope, which include articles related to them.[15–17]

Belize is a singular case, in 2000 the Law Revision Commissioner amended the Animal (Control of Experiment) Act from 1958, Chapter 148.[15] It is a very short document stating that: "The Minister may grant a licence to any person to perform any experiment for any purpose specified in such licence...," and "No person except a licensee shall perform any experiments." For this law "experiment" means an experiment performed on an animal and calculated to cause pain thereto and "animal" means a living vertebrate animal. Prohibition of performance of experiments includes teaching and the purpose of attaining manual skills. However, the Director of Health Services may grant to licensee under this Act a teaching permit to perform any experiment specified in such license when it is absolutely necessary for the due instructions of persons attending any course of lectures for the purpose of acquiring physiological knowledge or any knowledge, which will be used for saving or prolonging life, or alleviating suffering, or combating disease whether of human beings, animals, or plants.

In Colombia the Federal Law No. 84, December 27, 1989 on the Protection of Animals[14] states that "all animals in the national territory must be protected from pain and suffering caused by man." *Animal* means any domesticated or wild animal, no matter its habitat. In Chapter II, Article 6, item "s" states that "vivisection is banned unless it is scientifically sounded and performed in authorized places."

In **Chile**, Law No. 20.380/2009 on the Protection of Animals[18] is a generic Law with a wide range of application, starting from Article 1, which states that "the Act is intended to establish rules to know, protect and respect animals as living beings and part of nature, with the purpose to give proper treatment and to avoid unnecessary suffering." There is no clear definition if this applies to vertebrates, invertebrates, nonhuman primates, and if it includes fetal forms in their final stages, among other omissions.

In 1994, the Act of Animal Welfare (Ley No. 7451) was published in **Costa Rica**, as a law of general applicability, which, in its Chapter III, Articles 3, 10, 11, 12, and 13, refers to the use of experimental animals and covers ethical considerations for their use.[19] Also the Minister of Science and Technology edited the rules for Articles 3, 10, 11, 12, and 13 of the Ley No.7451 (Decreto No. 266668-MICIT) in which became compulsory for all researchers to obey when laboratory animals are used for experimental research. Hence, the University of Costa Rica has a specific regulation for the use of Laboratory Animals, the "Reglamento para el Cuidado y Uso de Animales de Laboratorio en la Universidad de Costa Rica."[20] It describes general procedures and researcher's responsibilities for the proper management of laboratory animals.

In **Peru,** there are two main official documents that address the care and use of laboratory animals: one is the national Law No. 27265/2000 for the "Protection of Domestic and Wild Animals Maintained in Captivity"[21]; the other official document is an institutional one, the Ethics Code for the Care and Use of Experimental Animals approved by the National Institute of Health in Lima in April 2012.[22] It covers all the studies, diagnostic assays, and experiments with animals that are developed in this institution. It is The Public Health Service Policy on Humane Care and Use of Laboratory Animals and describes the Institutional Animal Care and Use Committee (IACUC) functions. Similar to the previously mentioned countries, **Venezuela** has no specific law related to experimental animals, but the Law for the Protection of the Fauna: Domestic, Wild, and in Captivity Animals, No. 39.338, January 4, 2010—Asamblea Nacional de la República Bolivariana de Venezuela—has some specific articles covering experimental animals. In Venezuela, this law is applied when needed.[23]

We have seen that Brazil and Mexico have specific laws on the use of laboratory animals, but it is also worth mentioning other nonspecific laws where articles on laboratory animals can be found.

In **Brazil,** there is another law called Law of Environmental Crimes, also known as The Law of Nature (IBAMA-Brazilian Institute of the Environment) No. 9605/98 regulated by the Decree No. 3179/99,[24] which is not specific for laboratory animals, but it has one article (Article No. 32) that states: "anyone who treats animals in experiments with cruelty is subjected to the penalties of this law." It is very vague and difficult to interpret. However, it was the first Law to address the care and use of experimental animals.

In **Mexico,** there are also laws of general and unspecified scope, which cover, among others, the use of experimental animals:

1. Act of General Health and Its Regulation.[25]
2. NOM-042-SSA-2006 or Mexican Official Norm for the Disease Prevention and Control. Health Specifications for Canine Control Centres.[26]

There is also an Animal Welfare Act under discussion, but it refers for the case of laboratory animals to the NOM 062 already presented.[10]

The NRC Guide for the Care and Use of Laboratory Animals[8] is used as a reference in almost all Latin American countries, mainly in the countries where there is no regulation on the use of experimental animals. Ecuador also follows the OIE (World Organisation for Animal Health) Principles from the Terrestrial Code.[27]

El Salvador since May 4, 2016 has an Animal Welfare Act: Decreto No. 330. *Ley de protección y promoción del bienestar animal de animals de compañía.*[28] The Law refers a specific chapter (Article 23) for situations where pets are used as research animals. There is a clear indication that any experimental research must be performed under a scientific protocol considering pain and distress reduction. Protocol approved by the Agricultural and Livestock Ministry is required.

Guatemala has no specific legislation; only the University of San Carlos has an Internal Regulation of their main Laboratory Animal Facility, the "Reglamento Interno del Bioterio Central de la Universidad de San Carlos." It describes general procedures for the proper management of laboratory animals issued by some scientists working in research.

Honduras has also recently promulgated an Animal Welfare Act: Decreto No. 115–2015. *Ley de protección y bienestar animal*, April 2016.[29] Although the Law is for general there is a specific chapter (Article 12) for animal experimentation: "Animal experiments should be supervised by the Secretary of Health, National Service for Food Quality and Safety, Honduras Veterinary Medical Association and the Veterinary Schools."

Nicaragua has legislation for both domestic animals and wild life under captivity since 2011: Ley No. 747. Ley para la protección y el bienestar de los animales domésticos y animales silvestres domesticados.[30] There is a specific chapter for animal research and experimentation (Section VII), where is pointed out that any particular research using animals must be supervised by the Ministry of Agricultural, Livestock, and Forest. Moreover, the law sets down 10 specific principles similar to the well-known five freedom proclaimed by the Royal Society for the Prevention of Cruelty to Animals[31] as well as definitions for animal welfare, slaughter, euthanasia, vivisection, and animal cruelty.

The laws in Honduras and Nicaragua have a section for word definitions where *vivisection* means: "dissection of live animals with scientific purposes"; although is well known that scientists never use this word since good science practice does not allow performing surgery on animals without anesthesia or desensitization. On the other hand, there is also a misunderstanding about the word *euthanasia* since both laws are not considering in the explanation, the use of terms such as painless dead, absence of suffer or anger. *Euthanasia* is considered only as synonymous of killing of animals with clinical signs of terminal disease.

There is no information available on regulations/laws/guidelines on the care and use of experimental animals from the following countries: *Paraguay, Bolivia, Guiana, Suriname, French Guiana*, and *Panamá*.

COUNTRIES WITH A SPECIFIC LAW IN THE CARE AND USE OF EXPERIMENTAL ANIMALS

Brazil, **Mexico,** and **Uruguay** are the three Latin American countries with specific laws on the care and use of experimental animals and they all have the same core principles with some slight differences due to regional demands and different legal systems. All of them are compulsory.

In **Mexico**, in addition to some domestic regulations, which are subdivided in other laws of general applicability, that relate, among many other aspects, to the use of laboratory animals, the specific document addressing the laboratory animals is the NOM-062-ZOO-1999 or Mexican Official Norm for The Production, Care and Use of Laboratory Animals.[10] This document will be discussed in this chapter.

The **Brazilian** Law concerning the scientific use of animals is Law No. 11.794 from October 8, 2008, called Arouca Law.[3] The law is administered by the Ministry of Science, Technology, Innovations and Communications. This law creates, in its Article 4, the CONCEA (http://www.mct.gov.br/index.php/content/view/310553.html). This is an advisory, deliberative, and appellate body. CONCEA has various competences, among others:

- To issue regulations and guidelines on the ethical and humane use of animals in research, teaching, and testing
- To see if these regulations are being followed
- To approve and register institutions that breed or use animals for experimental purposes

- To monitor and evaluate the introduction of alternatives to replace the use of animals
- To keep an updated data bank of the institutes, scientists, and researchers
- To give advice to the Government, when asked, about the activities treated in this law

CONCEA is composed of members and substitutes from 12 institutions, including representatives, among others, of the Ministry of Education, Ministry of Health, and the Brazilian Academy of Sciences, and who are nominated by the Ministry of Science, Technology, and Innovation. The members must be Brazilian citizens with a PhD degree or recognized knowledge in research and teaching.

In **Uruguay,** the National Law is Law No. 18.611, called "Use of Animals in Experiments, Teaching and Research Activities," was published in October, 2009.[12] Like in Brazil, the Uruguayan Law created (Articles 4, 5) a statutory national body responsible for the control of the activities in which animals are used named the National Commission for Animal Experimentation (CNEA). The CNEA is responsible for the revision of guidelines on the accommodations and care of animals used for experimental and other scientific purposes. Animal breeders, facilities, and research institutions where animals are used for scientific purposes or teaching should be registered in a National Register (Article 8).

THE PRINCIPLES

The principles of 3Rs[32] are the core of the laws regulating the use of animals in all the countries having a specific law on experimental animals and also in those general laws on the protection of animals in other countries. The overarching policy on the scientific use of animals in Mexico is reflected in the following NOM 062[10] policy statement: "The use of animals in research, teaching, and testing is acceptable only if it promises to contribute to development of knowledge that can be expected to benefit humans or animals." It incorporates the principles of the 3Rs when stating, "Animals should be used only if the researcher cannot find an alternative method." A continuing sharing of knowledge, review of the literature, and adherence to the principles of 3Rs are also requisites. Those using animals should employ "the best existing methods according to animal welfare principles on the smallest number of appropriate animals required to obtain valid information."

In the Brazilian Law,[3] the 3Rs are clearly mentioned as noted in the following chapters:

- Chapter II, Article 5, Item III: CONCEA has to evaluate the introduction of alternatives that will replace the use of animals in teaching and research (Replacement).
- Chapter IV, Article 14, paragraph 3: use of alternatives instead of animals in teaching procedures (Replacement).

- Chapter IV, Article 14, paragraph 4: use the minimal possible number of animals to achieve the expected results (Reduction).
- Chapter IV, Article 14, paragraph 5: If an experiment may cause pain and distress sedatives, analgesia or anaesthesia must be used (Refinement).

The principles of 3Rs are not included in detail in the Uruguayan Law.[12] Only the principle of reducing the number of animals used in research and testing is mentioned in Article 17. Alternative methods and their validation are not included.

The Costa Rica' Animal Welfare Act (Ley No. 7451) refers in Chapter III, Articles 3, 10, 11, 12, and 13, the use of experimental animals and covers ethical considerations for their use. The 3Rs are the basis of this law when referring to the use of laboratory animals. Article 10 paragraph *b* refers to the Reduction on the number of animals used, paragraphs *c*, *d*, and *e* of the same Article refer to Refinement and Article 11 treats specifically on alternatives.[19]

In Peru, the national Law No. 27265/2000 for the "Protection of Domestic and Wild Animals Maintained in Captivity" in its Chapter IV contains regulations on the use of animals for scientific purposes mainly emphasizing the use of alternative methods (Article 10).[21]

In Venezuela, the law for the Protection of the Fauna: Domestic, Wild, and in Captivity Animals, No. 39.338, January 4, 2010—Asamblea Nacional de la República Bolivariana de Venezuela—states: "Laboratory Practices. The use of domestic animals is permitted only in technical schools, universities, for the development of laboratory practices included in the curricula approved by the competent authorities when no other methods or techniques are available to obtain the same results."[23]

SCOPE/APPLICABILITY

In the Brazilian Law[3] Chapter 1, Article 1, paragraph 1 indicates that the use of animals for teaching is restricted to academic institutions and technical schools of the biomedical area. And paragraph 2 of the same article considers scientific research as all the activities related to basic science, applied science, technological development, drugs production and control, food, immune procedures, instruments, or any other procedure using animals. Paragraph 3 emphasizes that all zootechnical practices are not considered as research. So, they do not need approval of an ethics committee. Article 2 of the previously mentioned chapter specifies the animal species covered by the law: the law regulates the use of animals (Phylum: Chordata, Subphylum: Vertebrata) in research, teaching, and testing. In a similar manner, in Uruguay the Law[12] regulates the use of animals in research, teaching, and testing, and the animals included in the scope of its law in its Article 3 are animals of the Phylum Chordata, Subphylum Vertebrata.

The Mexican NOM 062 only covers species directly referred to in the document: laboratory rodents (mice, rat, gerbil, hamster, and guinea pig), rabbits, dogs, cats, pigs, and nonhuman primates. All these animals used for research, teaching, or testing, or for display purposes or eventual use in research, teaching, or testing must be the subject of a written animal use protocol to be approved by IACUC.[10]

AUTHORIZATION OF USER/BREEDING INSTITUTIONS

In Mexico, the NOM 062[10] applies to all institutions using animals for scientific purposes. This includes academic institutions, government laboratories, and private companies. The Mexican Agriculture Department has a list of institutions with a commitment to practice NOM 062 standards for the ethical use of animals. Both the Animal Welfare Act and the NOM 062 do not allow the use of animals in primary or secondary school science classes.

Only institutions licensed by CONCEA can breed or use experimental animals in Brazil and to be accredited, they must have an active Ethics Committee, called the Commission on the Ethical Use of Animals (CEUA) (Chapter III, Article 8). The requirements for licensing institutions are specified in CONCEA's resolutions available on the web page (http://www.mct.gov.br/index.php/content/view/313178/Resolucoes_Normativas.html).

In Uruguay, universities, research institutes, and technical schools can use animals under the conditions of Law No. 18.611/2009 (Article 2).[12]

Law No. 20.380 on the Protection of Animals in Chile[18] also states that experiments should be performed in suitable facilities according to the species and status of the animal.

In Colombia, the Federal Law No. 84, December 27, 1989 on the Protection of Animals[14] refers to authorized places in its Chapter II, Article 6, item "s": "vivisection is banned unless it is scientifically sounded and performed in authorized places."

The Law on the Protection of Animals from 1994 in Costa Rica in its Article 12 indicates that all animal experiments must be registered in the Ministry of Science and Technology, except in the case of wild animals, which have to follow Law No. 7317, October 30, 1992.[19] The experiments not registered will be subjected to the penalties covered by this law.

Some places are allowed to breed laboratory animals if there is no alternative. In the draft project from Argentina the 3Rs are addressed as general principles that support the document.

In Venezuela, the Law for the Protection of the Fauna: Domestic, Wild, and in Captivity Animals, No. 39.338, January 4, 2010—Asamblea Nacional de la República Bolivariana de Venezuela—in its Chapter III, Article 52 states that:

> *"It is permitted to use domesticated animals in research only in licensed places, legally authorized by the competent authorities, if these animals are necessary for the advancement of science in areas like diagnosis, prevention and treatment of diseases which affect man and other living beings."*[23]

INSPECTIONS

The Mexican Agriculture Department provides oversight of animal care and use programs through its Verification–Certification program, which carries out regular visits to institutions, typically on a 3-year basis. Assessment visits are usually conducted by a representative of the Mexican Agriculture Department. When conducting a visit, the representative assesses the structure and resources of the animal care and use program; the composition, functioning, and effectiveness of the Institutional Animal Care and Use Committee (IACUC); and the appropriateness of animal care and use practices, procedures, and facilities. Assessments are based on NOM 062 guidelines and associated documents.[10]

In Uruguay, the statutory national body responsible for the control of the activities in which animals are used is the Comisión Nacional de Experimentación Animal (CNEA) (Articles 4, 5 of the Law).[12] Animal breeders, facilities, and research institutions where animals are used for scientific purposes or teaching should be registered in a National Register (Article 8). In Brazil Chapter V, Article 21 states: "The inspection of all the activities regulated by this law is under the responsibility of the Ministry of Agriculture, Ministry of Education, Ministry of the Environment, and Ministry of Science."[3] At moment, specific standards for inspection are currently under discussion in CONCEA.

PENALTIES

Following Chapter V, Article 17 of the Brazilian Law,[3] all Brazilian institutions that develop activities regulated by this law are subject, if they transgress its requirements, to several potential penalties. They include, in increasing range of importance: advertence, fines, temporary suspension of activities, and also of governmental financial support, and at last, definitive suspension of activities or closing of the facilities. Penalties for violations of the law 11794 are applied by CONCEA after administrative procedure provided in normative resolutions (normative resolutions No. 11 and No. 24: http://www.mct.gov.br/index.php/content/view/313178/Resolucoes_Normativas.html).

The observance of the norm's implementation in Mexico[10] relies on the following bodies:

- The laboratory's Institutional Animal Care and Use Committee (IACUC).
- Ministry of Agriculture, Livestock, Rural Development, Fishery and Food (SAGARPA).
- State and Mexico City, Federal District's Governments. Provisions declared under NOM-062.[10]

In **Uruguay**, Articles 18–22 of the Law[12] correspond to the penalties and punitive measures that the *law* imposes for the performance of an act that is proscribed.

In **Chile**, infringements related to this subject will be sanctioned by the corresponding Ministerial Regional Secretary of Education.

Costa Rica has recently reviewed and enforced penalties against animal cruelty.[33,34]

NONHUMAN PRIMATES

Specification for the care and use of nonhuman primates are described in Mexico in the NOM 062.[10] Animal health monitoring must ensure that the animals are tested for tuberculosis upon arrival in the research facility and either isolated or euthanized if found to be carrying tuberculosis, unless it is a necessary element of the research study. In addition, no person who is known to have active tuberculosis may be employed in the care of nonhuman primates. There are also requirements for isolation upon arrival to permit the animal to become used to the research environment. IACUCs are expected to take particular care in providing an ethical review of any protocol involving their use. No great apes are currently used in Mexico.

The Brazilian Law[3] has no special considerations or restrictions on the use of nonhuman primates. However, there are a large number of nonhuman primates being used in research in this country and ethical considerations are of concern in the use and care of these animals. So, CONCEA published the "Brazilian policies for use and care for scientific and teaching use of animals" (called DBCA[35]) where states considerations on the use of nonhuman primates. Also, the normative resolution No. 28[36] establishes conditions for breeding, husbandry, and use for nonhuman primates (http://www.mct.gov.br/upd_blob/0238/238242.pdf). Furthermore, the role of regulating their use also stays with the Ethics Committees (CEUAs) of every institution.

Nonhuman primates are not specifically mentioned in the Law in Uruguay.[12]

GENETICALLY ALTERED ANIMALS

In 2005, the Mexican government published the Biosafety Act for genetically modified organisms, recognizing that there was a rapid increase in the creation and use of these animals, and that the public was beginning to express concern about the genetic modification of animals. In 2008, the Mexican government published a policy within the Biosafety Act for Genetically Modified Organisms and in 2009, a new edition was published with some amendments. The Act does not separately enumerate genetically engineered bacteria, virus, plants, or animals. The 2009 Biosafety Act creates the National Register of Genetically Modified Organisms in Mexico.[37]

In Brazil, Law No. 11794/08[3] does not cover genetically altered animals because there was already a specific law in existence. This is referred to as Law No. 8974, January 5, 1995, and it created the National Technical Commission on Biosafety (CTNBIO; http://ctnbio.mcti.gov.br/). In 2005, the Law No. 8974 was repealed and a new legislation was enacted (Lei 11105, March 24, 2005).[38] CTNBIO is the organization that proposes regulations, guidelines, the Ethics Code on the use of genetically altered animals, and all other procedures related to these animals.

Genetically modified animals are not mentioned in the Uruguayan Law.[12]

INSTITUTIONAL AND DESIGNATED PERSONNEL RESPONSIBILITIES

In Mexico and Brazil, they are almost the same, with minor changes but the same core. Within each institution, the senior director holds the ultimate responsibility for the operation of the respective institutional committees, IACUC (Mexico) or the CEUA (Brazil). The responsibilities of the senior director are laid out in the NOM 062[10] and in Law 11 794/08,[3] respectively. Despite the variety of type and size of animal facilities, all institutions are required to have the following main elements:

1. At least one functional institutional animal care and use committee (IACUC–CEUA) composed of scientists/teachers experienced in animal use; at least one veterinarian with experience/training with regard to the species being used and the types of work; an institutional member who does not use animals in his/her work; the director/manager of the animal facility and the IACUC–CEUA coordinator. Also, one person paid by the institution who assists the IACUC–CEUA in its work.
2. Competent veterinary and animal care service providers whose numbers and expertise match the nature and scope of the institutional program and an appropriate number of laboratory animal care technicians. The NOM 062 demands the registration of the facility with the Mexican Agriculture Department and in the case of Brazil with CIUCA, Ministry of Science, Technology, Innovations and Communications. The operator of every research facility is also required to maintain a record of all animals in the research facility preserving the record within the research facility for at least 1 to 5 years from the date that the animals entered the research facility.

The quality and professional competence of the personnel supervising procedures in Uruguay, as well as of those performing procedures or supervising those taking care of the animals on a daily basis is not described in the Law.[12]

In 2010 CNEA was consolidated, chaired by the Ministry of Education and Culture and from 2010 until 2014 its operation rules were drawn and subcommissions to work on different themes were formed. The National Register of Institutions and Personnel Accreditation National System were set. From 2014, CNEA started to work in the implementation of the National Control System: setting minimum standard guidelines for institutions about the correct state and operation of facilities, staff, and experimental procedures.

ETHICAL REVIEW PROCESS

In Mexico, Brazil, and Uruguay the Ethical Review Process is mandatory by law as aforementioned and they all have the same requirements.

Each research institution using laboratory animals must have at least one active and effective Institutional Animal Care and Use Committee (called IACUC in Mexico, or CEUA in Brazil and Uruguay), responsible for ensuring ethical animal use and optimal levels of animal care. These committees are composed of researchers, veterinarians, animal care technicians, and animal facility managers. Each committee typically reports to and derives their authority from the senior administrator of the institution, and has a number of important responsibilities including reviewing animal use protocols, defining and reviewing standard operating procedures, policy development, as well as postapproval monitoring and facility inspection.

Protocol review and approval are based on the principles of 3Rs. Protocols are reviewed for its ethical merit. The committees base their decisions on the existing law and also guidelines, professional judgment, and common sense. In addition, each committee has to submit an annual report to the Mexican Agriculture Department in Mexico, to CONCEA in Brazil, and to CNEA in Uruguay, respectively.

In the Brazilian Law,[3] the composition, authority, responsibilities, and functioning of the Ethical Committees (CEUAs) are defined in written terms of reference in Chapter III, Articles 9 and 10 and also reinforced in Chapter IV Article 13, paragraphs 1 and 2. The CEUAs are to be composed of at least one veterinarian, biologists, researchers, and teachers of specific areas, and one representative of an animal protection society. The CEUAs have the responsibility of assuring the compliance of this law when animals are used in teaching and research. They have to make an ethical review of all the research protocols of the institution, keep an updated register of all the research conducted and that of researchers, and send a copy to CONCEA. Also, if needed, the CEUAs are to issue certificates to financial organizations or scientific journals and to immediately notify CONCEA and sanitary authorities if any accident occurs with research animals and what actions are to be taken. If there is a problem with a CEUA decision, the question can be sent to CONCEA.

In Mexico, as well as in Brazil and Uruguay, there must be an IACUC/CEUA in every institution using animals in research, to which must be submitted, for ethical review, all the protocols prior to the commencement of the work.

Before the IACUC/CEUA can review a protocol, there must be evidence of scientific, pedagogical, or regulatory merit. In the case of research studies, information provided within the protocol review form should provide the IACUC/CEUA with a clear sense of the need for the experimental project, and of the relationship between the proposed experiment and the overall scientific objective. For the use of animals in teaching, the pedagogical merit for the use of animals should be established, and for testing projects, an indication that the testing has been planned according to the most current regulatory requirements. Animal use protocols should include: the project title and descriptive procedures; name(s) of the primary investigator and all personnel who will be handling the animals; department affiliation; proposed start and end dates; funding sources; an indication of scientific or pedagogical merit, or requirement for regulatory purposes; summary and indication on the use of any hazardous materials; evidence of addressing the 3Rs; a description of why sentient animals must be used; justification for the numbers and species of animals selected; anesthesia and analgesia; medical treatments; housing and husbandry; improvements to procedures; improvements to the duration of the experiment and holding times; a description and time course of procedures (flow diagram where possible); a description of humane endpoints; a description of capture, etc., for field studies; method of euthanasia; ultimate fate of the animals; and any other relevant information, such as relevant information from studies previously carried out. Decisions concerning the acceptability of the protocol are generally taken by consensus of the IACUC/CEUA members.

Once a protocol has been approved by the IACUC/CEUA, the investigator is allowed to initiate the research, teaching, or testing studies. Researchers should report back to the IACUC/CEUA on an annual basis with a progress report and requesting any amendments to the protocol. IACUCs/CEUAs are also required to implement an effective postapproval monitoring program, designed to meet the specific needs of the institution, including actions such as:

- seeing if individual animal users are comfortable handling animals and successfully carrying out procedures, and that they are able to do so in appropriate conditions;
- assessing if endpoints are applied as approved by the IACUC/CEUA to avoid unnecessary distress to the animals;
- conducting site visits and discussing the protocols with animal users by IACUC/CEUA members; and
- addressing concerns through good communication.

The Uruguayan Law[12] states that a CEUA should be created in all the institutions in Uruguay where animals are used for experiments or teaching (Article 9). The CEUAs are of central importance to the application of the law to animal research in Uruguay. They review all projects involving animals to ensure that they are justified by their benefits and minimize any animal pain or suffering that might occur. This includes research, teaching, and testing. The CEUA regularly inspects all projects using animals and all projects housing animals. The CEUA is comprised of at least one veterinarian, one scientist or teacher, users of laboratory animals, and one representative of the community (Articles 10, 11).

In more recent years, several institutions in other countries that do not have specific laws on the use and care of experimental animals have decided to create Ethics Committees, also called CICUAL or IACUC to oversee and evaluate all aspects of the institution's animal care and use program. This is a way to have some kind of control on the use of laboratory animals.

In Bolivia and Ecuador, few institutions working with laboratory animals have an Ethics Committee. In Ecuador, they are called IACUCs and follow the norms of the IACUCs as in the United States.

In Chile, IACUCs have been established by institutions to compensate for the gaps in the existing animal protection law, and to comply with international recommendations and the pressure exercised by national agencies such as the National Fund for Scientific and Technological Development. This is a public program administered by the National Commission for Scientific and Technological Research. These IACUCs often have problems in their compositions and in some institutions, they only serve as a committee to review scientific projects that involve the use of laboratory animals, and not to implement all activities of an IACUC as in the United States.

An important fact in Argentina is the voluntary establishment of IACUCs, which are of central importance to the application of general principles to animal research. Many institutions that use animals for scientific purposes have formed an IACUC. As a consequence, an IACUC network was created in May 2011 and it has coordinated the member activities ever since.

In Venezuela, since 2008, there have been two active ethics committees, which analyze animal protocols: one in Caracas (Hospital Vargas) and the other one in the Universidad de los Andes (Mérida).

In Colombia, there is an ethics committee (CICUAL) in almost every university, but there are no special regulations on the composition and functioning of these committees. The CICUALES is the Colombian net of ethics committees and is trying to define criteria on how to establish ethics committees at the institutional level, thus promoting the correct use of experimental animals.

In Costa Rica, any experimental research at the University of Costa Rica must be CICUAL approved and their judgments take as a reference the University regulations and the "Guide for the Care and Use of Laboratory Animal".[20,39]

Since 1992, the "Code of Ethics for Science Workers" has existed in Cuba and is mandatory setting forth the workers' responsibility to prevent any unnecessary injury to the animals. The Vadi Resolution 4/00, of 2000 establishes the incorporation of ethics committees in the Cuban health institutions working on basic, clinical, and social investigation. Chapter V and VIII specifically mention the use of laboratory animals. Regulation 64/12, of 2004[40] requires and defines the creation of the Laboratory's Institutional Animal Care and Use Committee (IACUC or CICUAL for its acronym in Spanish), which revises the animal protocols used in every institution.

Peruvian Law No. 27265/2000 for the "Protection of Domestic and Wild Animals Maintained in Captivity"[21] in Article 10 contemplates the functioning and composition of the ethical committees, called IACUC's.

In Chile the Law No. 20.380/2009 on the Protection of Animals[18] says that within 30 days after the publication of the Law, the Council of Deans from the Chilean Universities will have the obligation to designate the members of an Animal Bioethics Committee comprising the following members:

a. Two academics appointed by the Council of Deans from the Chilean Universities.
b. A scientist designated by the Director of the Public Health Institute of Chile.
c. A researcher nominated by the President of the Institute of Agricultural Research.
d. A scientist nominated by the President of the National Commission of Scientific and Technological Research.
e. A representative of the Guild Association of Veterinarians, the oldest association in the country.
f. A representative from nongovernmental organizations on the protection of animals, which has legal personality and is nationally represented.

This committee would define the guidelines under which the experiments on live animals may be developed in accordance with the norms of this Law. These members would serve ad honorem for a period of 3 years and may be eligible for reappointments for successive periods.

REUSE

There is no specific reference on the reuse of animals in the Mexican NOM 062.[10] However, reuse occurs more frequently in the case of large farm animals, rather than rodents, rabbits, or dogs and cats. It occurs in pigs, goats, sheep, and cattle mainly for teaching purposes.

The Brazilian Law[3] indicates in its Chapter IV, Article 14, paragraph 8, that an animal cannot be reused after the main objective of the research has been achieved. And paragraph 9 states that in teaching programs many traumatic procedures may be carried out in only one animal if all the procedures are performed under anesthesia, and that the animal be euthanized before recovering consciousness.

In Uruguay, Article 15 of the Law states that reuse is not permitted only in very well-based exceptions.[12]

SETTING FREE/REHOMING

In Mexico, scientists are required to indicate how the animals will be disposed of at the end of the study. In general, rehoming of laboratory animals does not occur, although it is known that companion animals (dogs and cats), which have been socialized, are able to make the transition to a home environment. There are no programs on rehoming animals that have been used.

In Chapter IV, Article 14, paragraph 2, the Brazilian Law[3] clearly indicates that, exceptionally, animals used for demonstration purposes or research, and if not submitted to euthanasia, can be rehomed or set free after approval by the Ethics Committee (CEUA) and under the responsibility of an animal protection society or a responsible person.

This option is not mentioned in the Uruguayan Law.[12]

OCCUPATIONAL HEALTH AND SAFETY

In Mexico, the NOM 062[10] has a specific section for occupational health and safety for those working with experimental animals. Section 10 of the NOM 062, "Biosafety and Health and Safety in the Workplace for Personnel Working with Laboratory Animals" provides information about regulatory requirements, dealing with biological hazards, zoonotic diseases, working with nonhuman primates safely, allergies, physical injuries, chemical hazards and radiation, and ultraviolet light.

In Brazil, Law 11794/08[3] also addresses this topic and Chapter IV, Article 14, paragraph 10 states that anyone working with animals in a closed system must follow international guidelines on safety and occupational health.

The Uruguayan Law[12] does not address occupational health and safety, but this is regulated by the CNEA.

EDUCATION, TRAINING, AND COMPETENCE OF PERSONNEL

In Mexico, NOM 062[10] considers that appropriate institutional training programs are to be implemented as an essential component of any institutional animal care and use program. Animal users (including scientists, graduate students, postdoctoral fellows, other research staff, and study directors), animal health professionals, animal care committee members, and institutional officials directly responsible for animal care and use programs should receive specific education and training according to their needs.

Law No. 20.380 on the Protection of Animals, Article 7, from Chile[18] states that only qualified personnel may use laboratory animals to prevent their suffering. It defines qualified personnel as anyone who has studied veterinary, medical, or other related science, and is certified and recognized by a state academic institution.

The Brazilian Law[3] has no articles on this topic. However, education and training is one of the main concerns in this country. The new Brazilian guide, which is being prepared, emphasizes this subject. Although not required by law, there are a number of courses for technicians and researchers provided by various universities. Also, on site courses are given to personnel working at laboratory animal houses.

In Uruguay, the CNEA created the Personnel Accreditation National System, which ensures the training and gives licenses to those who use animals in experimentation, teaching, and scientific research.

The available courses in Latin America can be found at: http://iclas.org/committees/america-regional-committee/activities.

TRANSPORT

There is no specific regulation in Mexico for the transport or mobilization of laboratory animals. However, the NOM 051 ZOO 1995 from the Mexican Agriculture Department[41] addresses the humane transportation of animals. The NOM 051 details requirements for the provision of food, water, rest, protection from adverse weather, use of proper containers and transport vehicles, and segregation of incompatible animals. This NOM also specifies that livestock and poultry must be accompanied by a health certificate issued or endorsed by a veterinary inspector certified by the Mexican Agriculture Department.

The Brazilian Law[3] does not mention transportation of laboratory animals, but the Brazilian Ministry of Agriculture is responsible for the humane transportation of animals. It requires the following OIE Principles for the Humane Transportation of Animals[27] and also internal rules from the Ministry of Agriculture.

Article 12 of the Uruguayan Law states that the CNEA guidelines must be followed for the transport laboratory animals.[12]

HOUSING AND ENRICHMENT

In Mexico, NOM 062[10] gives specific guidelines on housing animals. Enclosure dimensions recommended by the NOM are only used as a bare minimum. The dimensions used are based on the NRC Guide for the Care and Use of Laboratory Animals.[8] Protocols that involve single housing must describe proposed measures for meeting the social requirements of the isolated animal.

In Brazil, the CONCEA resolution No. 15 (http://www.mct.gov.br/index.php/content/view/313178/Resolucoes_Normativas.html) establishes the conditions of housing for rodents and lagomorphs. However, there is no mention of environmental enrichment and normally the Ethics Committees (CEUAs) are very demanding on this topic. Every project has to describe the housing and enrichment of the animals, which will be used, and they have to follow the recommendations of the NRC Guide.[8]

Housing and enrichment are not mentioned in the Uruguayan Law.[12]

HUSBANDRY AND ENVIRONMENT

In Mexico, NOM 062[10] considers general guidelines in relation to the daily care of animals in a laboratory animal facility. Moreover, it provides specific information on regulated species. This includes detailed requirements for the nutritional, minimum space requirements, and environmental conditions including some behavior recommendations.

NOM 062 requires that facilities for the care and use of all animals in research, teaching, and testing must be conducive to the well-being and safety of the animals, provide an appropriately appointed and safe workplace for personnel, and establish a stable research environment. It is intended to assist the users and designers of laboratory animal facilities in achieving these objectives. The goal is to promote optimal levels of animal care and facilitate good research without curtailing new and innovative ideas for facility design. The NOM 062 includes guidelines on spaces and areas, construction materials, sanitary requirements, localization and distribution areas, caging, bedding, nesting materials, etc. The NOM 062 "guidelines on: spaces and areas" applies to animals such as mice, rats, gerbils, hamsters, guinea pigs, rabbits, dogs, cats, pigs, and nonhuman primates. The general principles described within the "guidelines on laboratory animal facilities" are applicable to most species maintained in captive environments for the purposes of research, teaching, and testing.

In Brazil, CONCEA states conditions to rodents and lagomorph's facilities by normative resolution No. 15 (http://www.mct.gov.br/index.php/content/view/313178/Resolucoes_Normativas.html), covering subjects' localization and distribution areas, construction materials, environmental and housing requirements. The Ethics committees (CEUAs) also require the laboratory animal breeders and scientists to follow the NRC Guide principles on husbandry and environment.[8] This is already a solid integrated routine for almost all animal users in Brazil, and the ones who have not incorporated it yet are trying to do so.

In Uruguay Law 18611/09, in its Article 12 states that CNEA guidelines must be followed in the case of husbandry and environment.[12]

VETERINARY CARE

In Mexico, NOM 062[10] refers that a veterinarian with knowledge of laboratory animal medicine should be responsible for the daily health care of the animals. Although the types of institutions and facilities widely vary, at any institution the senior director must make formal arrangements to ensure that veterinary services are readily available at all times to meet both routine and emergency needs. Also, the veterinary must have authority to ensure animal welfare. Animal caregivers and users must be able to report an animal health or welfare concern (e.g., injury, ill health, or death) at any time, and a veterinarian must be available to respond to the concern. While the responsibilities relating to veterinary care for "traditional" laboratory animals are well defined in the NOM 062, for some of the other species groups, there is no regulation at all when they are used as laboratory animals.

The Brazilian Law[3] has a normative resolution (NR6, Article 9, paragraph II) stating that any place having laboratory animals must have a veterinarian in charge of the animals. The veterinarian must have a specialization in laboratory animal medicine. As a result, this is now obliged by law.

The Uruguayan Law does not address veterinary care.[12]

CONDUCT OF EXPERIMENTAL PROCEDURES

The Brazilian Law[3] in its Chapter IV emphasizes the need to prevent suffering in animals (Article 4). If experiments will entail suffering or pain, analgesia or anesthesia must be used (Article 5) and projects that cause pain or suffering have to have special authorization by the Ethics Committee (CEUA) (Article 6). The CEUAs are very strict on this point.

As stated before, in Mexico the IACUCs analyze the protocols and one of the main objectives is to thoroughly describe the suffering and pain caused to the animal and how to avoid it.

In Uruguay the Law states in its Article 15 that suffering must be prevented, and like in the Brazilian Law, if experiments will entail suffering or pain, analgesia or anesthesia must be used, and projects that cause pain or suffering have to have special authorization by the Ethics Committee (CEUA).[12]

At Belize, a specific License granted by Minister is required to perform experimental procedures in animals.[15]

In Chile, the Law[18] emphasizes the use of anesthesia for surgical interventions, and that these must be carried out by a veterinarian or another qualified professional.

In the Venezuelan Law,[23] Article 54 states: "Domesticated animals which have been subjected to an experiment affecting its physical conditions in an irreversible manner or if death is inevitable, due to the research protocol, must be culled without pain."

In Cuba, there are guides for the determination of the humane endpoints in the Regulation 64/12,[40] and guidelines set forth to define the critical point or suspension of an experiment in case the animals present severe pain, suffering, are dying or have other conditions seriously affecting their welfare.

In Mexico, the Ethics Committees (IACUCs) use the Canadian Guide[9a,9b] and the NRC Guide[8] as references and require scientists to follow these guidelines. Brazil has now its own Directive (DBCA), which contains the policies for the use and care of laboratory animals, and it has to be followed for anyone using laboratory animals. International guides may be used if necessary.

EUTHANASIA

In Mexico, the NOM 062[10] "guidelines on: euthanasia of laboratory animals" is based on recommendations made by the American Veterinary Medical Association (AVMA) Guidelines on Euthanasia 2007 with some modification to fit the Mexican context.[11] An overview of acceptable methods of euthanasia for common species used for research, teaching, and testing primarily based on the reference documents is provided, and information is included on other methods of euthanasia that are not considered best practice, but that may be acceptable for specific purposes providing they comply with the general guiding principles and receive the approval of the IACUC reviewing the application.

In the Brazilian Law,[3] Chapter IV, Article 14, paragraph 1 requires that an animal will be euthanized when technically recommended after the end of the experiment or in any of its phases if the animal is suffering. The euthanasia must follow strict technical procedures according to the guidelines of the Ministry of Science, Technology, Innovations, and Communications. The Brazilian Euthanasia guidelines (CONCEA normative resolution No. 13: http://www.mct.gov.br/index.php/content/view/313178/Resolucoes_Normativas.html) is based on the AVMA Guidelines on Euthanasia 2013.[11]

In Uruguay Article 15 states that euthanasia must be performed following CNEA strict directions.

In Venezuela, the Law[23] states in its Article 6 that: "Death with no pain must be performed by a veterinarian or by a person under its supervision who must have the necessary experience, to guarantee that the death will not entail cruelty or pain to the animal."

NATIONAL LABORATORY ANIMAL SCIENCE ASSOCIATIONS

It is worth mentioning that national Laboratory Animal Science Associations in Latin America have been paramount in changing the way animals were used in experiments, testing, or teaching. Generally composed of a small number of scientists and minimal funding, they have been doing a wonderful job, organizing seminars, congresses, giving short courses, and mainly trying to call authorities' attention to the need of a law on the care and use of laboratory animals. In some places, they had success such as the Association for Laboratory Animal Science and Technology from Uruguay (AAUCyTAL; www.aucytal.org), which has had a strong role in the development of a regulatory framework for the ethical use of laboratory animals in this country.

The Brazilian Association, former COBEA, now called SBCAL (www.cobea.org.br/) also played a very important role in the development of the law and, ever since, it has been a reference for other Latin American countries.

Some are in the process of organizing and achieving their legal status, such as the Asociación Boliviana de la Ciencia del Animal de Laboratorio (ABOCAL) in Bolivia or the "Asociación Panameña de la Ciencia de los Animales de Laboratorio" (APCAL) in Panama.

An important step in Latin America has been the activation of two Federations: The Federation of South American Societies and Associations of Laboratory Animal Science (**FeSSACAL**), and the Federation of Hispanic Societies and Associations of North America, Central America and Caribe for Laboratory

Animal Science (**FeSAHANCCCAL**) (http://www.fesahancccal.com/), both to be the umbrellas under which all the associations will work together in their regions. Some of them are also members of the International Council for Laboratory Animal Science (http://iclas.org/), including those in Brazil, Chile, Uruguay, Mexico, Guatemala, Costa Rica, and Argentina.

The current Laboratory Animal Science Associations in the area are:

ASOCHICAL (https://www.asochical.org/). Chilean Association of Laboratory Animal Science.

AACyTAL (http://aacytal.wixsite.com/aacytal/sitio-oficial-de-aacytal). Asociación Argentina de la Ciencia y Tecnología de los Animales de Laboratorio.

AVECAL (http://avecal.wix.com/pag). Asociación Venezolana de la Ciencia de los Animales de Laboratorio.

ACCBAL Asociación Colombiana de la Ciencia y el Bienestar de los Animales

de Laboratorio. CICUALES is a network of Ethics Review Committees very active in Colombia.

SCCAL Sociedad Cubana de la Ciencia del Animal de Laboratorio.

AMCAL (http://amcal-ac.blogspot.com.br). Asociación Mexicana de la Ciencia de Animales de Laboratorio.

ACCMAL Asociación Científica Centroamericana y Mexicana de Animales de Laboratorio.

AGCAL Asociación Guatemalteca de la Ciencia de los Animales de Laboratorio.

ARDCAL Asociación de la República Dominicana de la Ciencia de los Animales de Laboratorio.

APCAL Asociación Panameña de la Ciencia de los Animales de Laboratorio.

FeSAHANCCCAL (http://www.fesahancccal.com/). The Federation membership includes: Costa Rica, Cuba, Dominican Republic, Guatemala, Mexico, and Panama Laboratory Animal Science Associations.

REFERENCES

1. República de Argentina. *Ley Sarmiento no 2786; Prohibición de Malos Tratos a los Animales*. 1891. Argentina. Available from: http://www:animanaturalis.org/p/1387.
2. Presidência da República. *Decreto no 24.645, de 10 de julho de, 1934*. 1934. Brasil. Available from: http://www.planalto.gov.br/ccivil_03/decreto/1930-1949/D24645.htm.
3. Presidência da República, Casa Civil, Subchefia para Assuntos Jurídicos. Lei no 11.794 de 8 de outubro, 2008 (Arouca Law) Brazil.
4. CIOMS. *International guiding principles for biomedical research involving animals*. 1985. Available from: http://www.cioms.ch/publications/guidelines/1985_texts_of_guidelines.htm.
5. Presidência da República. Lei no 6.638, de 8 de maio de 1979. Brasil. Available from: http://www.planalto.gov.br/ccivil_03/leis/1970-1979/L6638.htm.
6. Conselho Nacional de Controle da Experimentação animal – CONCEA. Guia brasileiro de produção, manutenção ou utilização de animais em atividades de ensino ou pesquisa científica. Brasil. Available from: http://www.mct.gov.br/upd_blob/0238/238057.pdf.

7. *Australian code of practice for the care and use of animals for scientific purposes*. 8th ed. 2013. Australia. Available from: https://www.nhmrc.gov.au/guidelines-publications/ea28.
8. National Research Council. *Guide for the care and use of laboratory animals*. 8th ed. National Academy Press; 2011.
9. a. Canadian Council on Animal Care. 2nd ed. *Guide to the care and use of experimental animals*, vol. 1. 1993. Canada. Available from: http://ccac.ca/Documents/Standards/Guidelines/Experimental_Animals_Vol1.pdf.
   b. Canadian Council on Animal Care. *Guide to the care and use of experimental animals*, vol. 2. 1984 (Adopted May 1999). Canada. Available from: http://www.ccac.ca/en_/standards/guidelines/additional/vol2_rats.
10. Secretaría de Agricultura, Ganadería, Desarrollo Rural, Pesca y Alimentación. NOM-062-ZOO-1999. Especificaciones Técnicas para la Producción, Cuidado y Uso de Animales de Laboratorio. Mexico *Diario Oficial de la Federación* 22 de agosto de 2001. Available from: http://www.fmvz.unam.mx/fmvz/principal/archivos/062ZOO.PDF.
11. American Veterinary Medical Association (AVMA). *AVMA guidelines for the euthanasia of animals*. 2013. Available from: https://www.avma.org/KB/Policies/Documents/euthanasia.pdf.
12. República del Uruguay. *Ley no 18.611. Utilización de Animales en Actividades de Experimentación, Docencia e Investigación Científica*. 2009. Uruguay. Available from: http://www.iibce.edu.uy/ETICA/ley-18611-oct-2-2009.pdf.
13. The European Parliament and the Council of the European Union. Directive 2010/63/EU of the European Parliament and of the Council of 22 September 2010 on the protection of animals used for scientific purposes. *Off J Eur Union* 2010;**L276**:33–79.
14. Estatuto Nacional de Protección Animal. *Ley 84 de, 1989. Por la cual se adopta el Estatuto Nacional de Protección de los Animales y se crean unas contravenciones y se regula lo referente a su procedimiento y competencia*. 1989. Colombia. Available from: http://spac-05.tripod.com/id24.html.
15. León GM. Bienestar animal en las legislaciones de América Latina. U.N.C.R. *Rev Ciencias Veterinarias* 2006;**24**:185–221. Available from: http://www.droit-aliments-terre.eu/documents/sources_lascaux/articles/Marlen_Leon_Guzman_Bienestar_Animal_2006_ES.pdf.
16. Rojas H, Stuardo L, Benavides D. Políticas y prácticas de bienestar animal en los países de América: estudio preliminar. *Rev Sci Tech Off Int Epiz* 2005;**24**(2):549–65. Available from: http://www.oie.int/doc/ged/D2234.PDF.
17. Aguilar NME. *Bioética y normatividad en el uso de animales en investigación en América Latina*. [Tesis Maestro en Ciencias]. Mexico: Escuela Superior de Medicina. IPN; 2008.
18. Ministerio de Salud. *Subsecretaría de Salud Pública. Ley no 20.380/2009 sobre Protección de Animales*. 2009. Chile Available from: http://www.leychile.cl/Navegar?idNorma=1006858.
19. Asamblea Legislativa de la República de Costa Rica. *Ley de Bienestar de los Animales no 7451*. 1994. Available from: http://www.animallaw.info/nonus/statutes/stat_pdf/stcrlaw7451_welfare.pdf.
20. Rectoría Universidad de Costa Rica. Reglamento para el cuidado y uso de los animales de laboratorio en la Universidad de Costa Rica. *La Gaceta Universitaria* 2006;**XXX**(10):9–11.
21. República del Peru. *Ley no 27265/2000 de Protección a los Animales Domésticos y a los Animales Silvestres Mantenidos en Cautiverio*. 2000. Perú. Available from: http://www.ins.gob.pe/repositorioaps/0/2/jer/cia/Peru%20Ley%20protecc%20animales%20cautiverio%2027265%20del%202000.pdf.
22. Instituto Nacional de Salud del Perú. *Procedimiento para el Uso de Animales de Laboratorio en el Instituto Nacional de Salud*. 2012. Available from: http://www.ins.gob.pe/repositorioaps/0/2/jer/normatividad_01/MANUAL%20PROCEDIMIENTOS%20CIEA.pdf.

23. Asamblea Nacional de la República Bolivariana de Venezuela. Ley para la Protección de la Fauna Doméstica Libre y en Cautiverio. *Gaceta Oficial* del 4 de enero de, 2010;**39.338**. Available from: http://es.scribd.com/doc/47545158/LEY-PARA-LA-PROTECCIONDE-LA-FAUNA-DOMESTICA-LIBRE-Y-EN-CAUTIVERIO/.
24. Presidência da República. *Casa Civil, Subchefia para Assuntos Jurídicos. Lei no 9605/98, regulated by the Decreto no 3179/99*. 1999. Available from: http://www.planalto.gov.br/ccivil_03/leis/L9605.htm.
25. Secretaria de Salud. *Ley General de Salud México*. 2007. Available from: http://www.salud.gob.mx/unidades/cdi/legis/lgs/index-indice.htm.
26. Secretaria de Salud. *NOM-042-SSA2-2006. Mexican Official norm for the Disease Prevention and Control. Health Specifications for Canine Control Centres. Diario oficial de la Federación 06/11/2008*. 2008. Mexico. Available from: http://www.spps.gob.mx/images/stories/SPPS/Docs/nom/NOM-042-SSA2-2006.pdf.
27. OIE. *Terrestrial animal health code. Section 7, animal welfare*. 2012. Available from: http://www.oie.int/en/international-standard-setting/terrestrial-code/access-online/?htmfile=titre_1.7.htm.
28. Asamblea Legislativa República de el Salvador. *Decreto No. 330. Ley de protección y promoción del bienestar de animales de compañía*. 2016. Available from: http://www.asamblea.gob.sv/eparlamento/indice-legislativo/buscador-de-documentos-legislativos/exonerase-del-pago-de-impuestos-la-introduccion-al-pais-de-un-vehiculo-a-favor-de-la-alcaldia-municipal-de-suchitoto-departamento-de-cuscatlan/archivo_documento_legislativo/.
29. República de Honduras. Ley de protección y bienestar animal. Decreto No. 115-2015. Honduras. *La Gaceta* 5 de abril de 2016;**34,000**(Sección A. Acuerdos y Leyes): A23–A32. Available from: http://www.poderjudicial.gob.hn/CEDIJ/Leyes/Documents/Ley%20de%20Proteccion%20y%20Bienestar%20Animal.pdf.
30. Asamblea Nacional República de Nicaragua. Ley No. 747. Ley para la protección y bienestar de los animales domésticos y animales silvestres. *La Gaceta Diario Oficial de Nicaragua* 2011;**96**:2920–30. Available from: http://observatorio.uraccan.edu.ni/sites/default/files/documentos/Ley%20757.%20Ley%20de%20trato%20digno%20y%20equitativo%20a%20Pueblos%20Indigenas%20y%20Afrodescendientes.pdf.
31. *RSPCA. Shelter design and management guidelines. UK*. 2006. Available from: http://www:shelterdesignandmanagementguidelines.pdf. 2008.
32. Russell WMS, Burch RL. *The principles of humane experimental technique*. 1959. Methuen, London.
33. Madrigal RJP. *Legislación de bienestar animal en Costa Rica: Un análisis jurídico formal sobre la protección animal. Tesis UCR. Costa Rica*. 2015.
34. Asamblea legislativa de la República de Costa Rica. *Proyecto de Ley: Reformas al Código Penal, Ley No. 4573, de 4 de mayo de 1970 y Reformas de la Ley de bienestar de los animales, Ley No. 7451, del 17 de noviembre de 1994. Expediente No. 18.298*. 2016. Available from: http://www.asamblea.go.cr/sil_access/ver_texto_base.aspx?Numero_Proyecto=18298.
35. Conselho Nacional de Controle da Experimentação animal – CONCEA. Diretriz Brasileira para o cuidado e a utilização de animais em atividades de ensino ou de pesquisa científica – DBCA. Brasil. Available from: http://www.mct.gov.br/upd_blob/0238/238683.pdf.
36. Conselho Nacional de Controle da Experimentação animal – CONCEA. Resolução normativa n. 28-Primatas não humanos mantidos em instalações de instituições de ensino ou pesquisa científica. Brasil. Available from: http://www.mct.gov.br/upd_blob/0238/238242.pdf.
37. Cámara de Diputados del H. Congreso de la Unión. *Ley de Bioseguridad de Organismos Genéticamente Modificados. Diario Oficial de la Federación 18 de marzo, 2005*. 2005. Mexico.

38. Presidência da República, Casa Civil, Subchefia para Assuntos Jurídicos. *Lei 11105/2005*. 2005. Available from: http://www.planalto.gov.br/ccivil_03/_Ato2004-2006/2005/Lei/L11105.htm.
39. Granados ZJ. Uso de animales de experimentación en la investigación biomédica en Costa Rica. *Acta méd Costaric* 2010;**52**(3):134–6. Available from: http://www.redalyc.org/pdf/434/43415399002.pdf.
40. Ministerio de Salud Pública. *Cuba. Regulación 64/12, Lineamiento para la Constitución y Funcionamiento de los Comités Institucionales para el Cuidado y Uso de los Animales de Laboratorio. Cuba.* 2013.
41. Dirección General Jurídica de la Secretaría de Agricultura, Ganadería y Desarrollo Rural. *NOM 051 ZOO*. 1995. Mexico. Available from: http://www.fmvz.unam.mx/fmvz/p_estudios/apuntes_bioet/051zoo_movilizacion.pdf.

Chapter 5

The European Framework on Research Animal Welfare Regulations and Guidelines

Javier Guillén[1], Jan-Bas Prins[2], Bryan Howard[3], Anne-Dominique Degryse[4], Marcel Gyger[5]

[1]AAALAC International, Pamplona, Spain/Frederick, MD, United States; [2]Leiden University Medical Centre and Leiden University, Leiden, The Netherlands; [3]Private Consultant, Preston, United Kingdom; [4]ESLAV&ECLAM (Past President), Puylaurens, France; [5]Private Consultant, Lausanne, Switzerland

GENERAL FRAMEWORK

It would be an impossible task to describe a general framework for a continent of 49 countries without the existence of two political organizations that group them up in certain ways. These organizations are the Council of Europe (CoE) and the European Union (EU). Both organizations have issued the documents that constitute the European framework on the protection, care, and use of animals in research, education, and testing that will be described in this chapter. The CoE, through the European Convention for the Protection of Vertebrate Animals Used for Experimental and Other Scientific Purposes,[1] published in the European Treaty Series 123, which is better known as ETS 123, and the EU through the Directive 2010/63/EU.[2]

The Council of Europe

The CoE (http://www.coe.int) is an intergovernmental organization that virtually covers the entire European continent, with its 47 member countries including all of the 28 EU Member States. It seeks to develop throughout Europe common and democratic principles based on the European Convention on Human Rights (http://www.echr.coe.int/Documents/Convention_ENG.pdf) and other reference texts on the protection of individuals. The CoE has no legislative power and seeks voluntary cooperation within the member countries through recommendations, agreements, and conventions. Member countries can sign and ratify the conventions. If they do this, they are bound to implement the conventions into their national legislation. One of these conventions is the ETS 123.

Laboratory Animals. http://dx.doi.org/10.1016/B978-0-12-849880-4.00005-2
Copyright © 2018 Elsevier Inc. All rights reserved.

European Treaty Series 123

The importance of ETS 123 relies on the fact that it was the first document trying to establish a common European framework on the protection, care, and use of laboratory animals, in 1986.

ETS 123 has 38 articles distributed in 12 Parts and 2 Appendices (A and B). The first Part of ETS 123 (Articles 1–4) relates to the General Principles, where the scope and definitions are included. Part II (Article 5) focuses on General Care and Accommodation and refers to Appendix A that contains the Guidelines for Accommodation and Care of Animals. Appendix A, revised in 2006,[3] is one of the most important documents of the European framework because of the impact of its recommendations (for all common species) on housing and on environmental enrichment and care, some of which (cage size especially) have been incorporated in the legislation at the EU level. The minimum cage sizes recommended in the revised Appendix A are significantly larger than those in the original version, and although this was at first a controversial issue, they have been gradually accepted and implemented. Appendix A has short general recommendations on physical facilities, environment and its control, and care, and a much longer species-specific section where recommendations on environment, health, housing, enrichment, and care are offered for all the commonly used species. Many of these recommendations are performance based.

ETS 123 has been signed and ratified by a number of member countries, which means that its provisions should have entered into force by means of national legislation. The list of countries that have done so and the dates are available at http://www.coe.int/en/web/conventions/full-list/-/conventions/treaty/123/signatures.

Many of the dates of entry into force of national laws adopting ETS 123 are previous to the revision of Appendix A. The final adoption of some of the Guidelines for Accommodation and Care of Appendix A (i.e., minimum cage sizes) in the EU has occurred thanks to the transposition of Directive 2010/63/EU, which includes these housing standards as a mandatory requirement. Previously, the EU had ratified the ETS 123 in 1998.

Part III of ETS 123 (Articles 6–12) discusses the conduct of procedures, stressing (implicitly) the implementation of the principles of replacement, reduction, and refinement (3Rs). The articles in Part III focus on the use and promotion of alternative methods, choice of species, minimization of pain and distress, reuse, and euthanasia.

The authorization of procedures and persons carrying out procedures is required in Part IV (Article 13).

Administrative measures for breeding or supplying establishments are defined in Part V (Articles 14–17), which includes registration of establishments and records and identification of animals.

Part VI (Articles 18–24) deals with the requirements for user establishments, including those related to animal needs and personnel (responsible persons, veterinarian, and training). A list of animals required to be purpose-bred (originated from breeding establishments) is included. In this original list, the nonhuman primates are not included.

The education and training of personnel is specifically addressed in Part VII (Articles 25–26), and the statistical information in Part VIII (Articles 27–28), which refers to Appendix B containing the statistical tables. Parts VIII–XII (Articles 30–38) relate to legal aspects on the implementation of the convention.

The principles and some of the content in ETS 123 were the basis for the subsequent legislation at the EU level, Directive 86/609/EEC[4] that established the minimum legal requirements applicable in the EU for 25 years. This old Directive included the care and accommodation standards (i.e., cage sizes) of the original Appendix A of ETS 123; however, only as a recommendation. Also, definitions and scope closely resembled those in ETS 123. However, this Directive failed in harmonizing the standards across the EU, as it allowed Member States to go beyond its provisions. Some Member States issued stricter rules (e.g., the United Kingdom (UK)), while others just transposed the Directive literally. The revision of ETS 123 Appendix A occurred in parallel to the revision of the old Directive, and as noted above, the new Directive incorporated the housing standards of the revised Appendix A.

The European Union

The EU is a supranational economic and political partnership between 28 European countries (at the time of writing this chapter, the UK was still a Member State). Member States delegate sovereignty to the EU, so this can develop its activities on the treaties that are approved voluntarily and democratically by all of the Member States. The European Commission (EC) is the EU body responsible to propose new legislation to the European Parliament and the Council of the European Union (which is not the CoE described earlier) under the EU treaties. The EC is also responsible for ensuring that the adopted legislation is implemented correctly by the Member States. The legislation can be in the form of regulations, which are rules that must be complied directly by the Member States without being transposed into national legislation, and directives, which give Member States some time to transpose and implement them.

Directive 2010/63/EU

During the revision process of Appendix A of ETS 123, the EC started the revision of Directive 86/609/EEC with the main objectives of harmonizing the legislative framework within the EU (leveling the playing field), increasing significantly the animal welfare, and implementing explicitly the principles of 3Rs. During this process, the EC had already issued a recommendation to follow the guidelines of the revised Appendix A of ETS 123 with regard to the care and accommodation standards, as a first step of incorporating it in the future legislation.[5] The revision of the old 1986 Directive finalized with the publication of Directive 2010/63/EU. The Directive gave 24 months to Member States for the transposition of the provisions into national legislation. This period ended in November 2012, and the different transpositions at national level had to enter into force in January 2013. However, only a few of the Member States had finished the implementation by then, and many transpositions took some time longer.

One of the most important features of the new Directive, in contrast to the old one, is that it does not allow Member States to establish stricter measures. This provision applies one of the main objectives of the EC, which is to level the playing field in the EU. Therefore, in theory, all Member States should be applying very similar national laws. However, it will be discussed how this is not entirely true with regard to several of the provisions, as they can be interpreted and implemented differently in practice. Member States were allowed to communicate and maintain stricter measures that were in place before the publication of the Directive. The other main objectives of the Directive are pursued by the explicit requirement of the implementation of the 3Rs principles across the whole document.

Directive 2010/63/EU has an introductory part or preamble with 56 recitals, which precedes the provisions distributed in 66 articles (in 6 different chapters) and 8 important annexes.

Chapter I (Articles 1–6) relates to the General Provisions, including the scope, definitions, the application of stricter measures, the 3Rs principles, the permitted purpose of experimental procedures, and the methods of euthanasia (called "killing"). Article 6 on "methods of killing" refers to Annex IV, which lists the accepted methods by species.

Chapter II (Articles 7–11) establishes the provisions on the use of certain animals in procedures, such as endangered species, nonhuman primates, animals taken from the wild, animals required to be purpose-bred, and stray and feral animals of domestic species. Article 10 on purpose-bred animals refers to Annex I, which lists the 12 different species that must be purpose-bred to be used in procedures, and to Annex II, which lists the species of nonhuman primates that are subjected to the second generation requirement. The provisions on nonhuman primates are particularly important because they are especially restrictive.

Requirements on the conduct of procedures are detailed in Chapter III (Articles 12–19). Articles refer to issues such as the choice of methods (by application of the 3Rs principle), provision of anesthesia and analgesia, the mandatory classification of the severity of procedures, the reuse of animals, the sharing of organs and tissues, and the potential setting free of animals and rehoming. The requirement for the classification of severity of procedures of Article 15 is particularly important. Examples of procedure classification can be found in Annex VIII.

Chapter IV (Articles 20–45) is the most extensive. Its first section (Articles 20–33) focuses on the requirements for breeders, suppliers, and users. In addition to the required administrative authorization, they refer to installations and equipment, personnel (competence and several required positions including the designated veterinarian), Animal Welfare Body (AWB, the institutional oversight body), records, special considerations for nonhuman primates, cats and dogs, and care and accommodation. Articles 22 and 33 refer to Annex III for installations and equipment, and care and accommodation, respectively. Annex III evidences the importance of the Appendix A of ETS 123, because it contains part of its provisions, especially the minimum cage sizes (in an engineering-based approach). However, it does not contain the extensive performance-based standards by species of the Appendix A. In contrast with the old Directive, where the provisions of Appendix A were in the form of recommendations, Directive 2010/63/EU requires the obligatory compliance with these minimum (housing) standards.

Minimum requirements for training and education of personnel are set out in Annex V. However, the EC has no competence to define standards in education, as it is at Member State level. Therefore Annex V only includes general topics, which are not even assigned to the specific categories of responsible personnel.

Section II of Chapter IV (Articles 34–45) defines the inspections and controls by the competent authorities, and Section III focuses on the requirements for research projects. Section III is particularly important, as it describes how the projects have to be evaluated and authorized before being initiated. Annex VI lists the elements that project applications must describe to be evaluated as required in Article 37.

The alternative approaches are addressed in Chapter V (Articles 46–49), which also includes the requirement for Member States to establish national committees for the protection of animals used for scientific purposes (Article 49). Article 48 refers to Annex VII, which defines the duties and tasks of the Union Reference Laboratory (the Commission's Joint Research Centre) to coordinate and promote the development of alternative methods.

The final Chapter VI (Articles 50–66) contains the Final Provisions, such as the requirement for reporting the use of animals in procedures, the safeguard clauses, and the definition of competent authority (CA).

The provisions of Directive 2010/63/EU have been transposed into the legislation of the EU Member States. Although there is a common legislative general framework within the EU, there may be some differences in the interpretation and implementation of some of the Directive provisions. In terms of collaborative research, it is recommended that specific regulations be reviewed at national level. A complete list of all Member States legislation concerning implementation of the Directive is available at http://eur-lex.europa.eu/legal-content/EN/NIM/?uri=CELEX:32010L0063&qid=1439308631281.

To help harmonizing the implementation of the Directive, the EC produced a Question and Answer (Q&A) document (http://ec.europa.eu/environment/chemicals/lab_animals/pdf/qa.pdf) providing guidance on specific questions and established several Expert Working Groups (EWGs) with experts from professional associations in the field, NGOs, animal protection groups, and representatives of Member States, which have produced Consensus documents on several areas and articles of the Directive. These Consensus documents are not legally binding, but can be endorsed by the National Contact Points (responsible persons for the implementation of the Directive in each Member State), and are having impact on how the Directive is being implemented in practice. The list of topics covered is summarized in Table 5.1. All guidance documents are available at http://ec.europa.eu/environment/chemicals/lab_animals/interpretation_en.htm.

TABLE 5.1 Consensus Documents on Implementation of Directive 2010/63/EU

Subject Matter	Related Articles	Contents of the Document
Animal welfare bodies and national committees	26, 27, 49	Guidance and principles of good practice on the design and functioning of animal welfare bodies and national committees for the protection of animals used for scientific purposes
Education and training	23, 24, 25, (34), (38)	Education and training framework with a modular learning outcome–based training structure, principles and criteria for supervision, competence assessment, continued professional development and for a mutual approval/accreditation of courses
Genetically altered animals	1 (2), 3(1), 4(3), 17(1)	The principles of creation, establishment, and maintenance of genetically altered animal lines and how these are considered within project authorization and statistical reporting
Inspections and enforcement	34, 60	Guidance and principles of good practice on the planning and execution of an effective inspection and enforcement program
Nontechnical project summaries	43	Guidance on the drafting and publication of nontechnical project summaries including a template and an illustrative example
Project evaluation/ retrospective assessment	37, 38, 39	Guidance for performance of project evaluation including harm–benefit assessment and retrospective assessment of projects

TABLE 5.1 Consensus Documents on Implementation of Directive 2010/63/EU—cont'd

Subject Matter	Related Articles	Contents of the Document
Severity assessment	4(3), 15(1), 16(1,d), 54(2)	The severity assessment framework from project planning, monitoring, and assessing the severity through to final assignment of actual severity
Severity assessment illustrative examples	4(3), 15(1), 16(1,d), 54(2)	Illustrative examples of practical application of the severity assessment process
Various articles in the Directive	1(5), 3, 16, 40, 41	Practices that are exempted from the scope; understanding of definitions for a procedure and project; use, reuse, and continued use; multiple generic projects and complex or multidisciplinary projects

Available at http://ec.europa.eu/environment/chemicals/lab_animals/interpretation_en.htm.

Directive 2010/63/EU establishes a deadline for its own review, which is November 10, 2017 (Article 58). Depending on the outcome of this review, amendments could be proposed, as appropriate.

Although the legal framework at the EU level covers a majority of the animal research conducted in Europe, there are many countries outside the EU that have other specific legislation in place or no legislation on laboratory animal protection at all. For example, countries such as Switzerland and Norway have modern regulations (in the case of Norway following the EU Directive), while others lack them or have outdated ones.

The last revision of the Swiss Animal protection legislation concluded in 2008 with amendments in 2010 and 2012.[6,7] Main innovations relate to the use and care of genetically altered (GA) animals (mainly rodents and fish), the licensing of animal facilities and accreditation of laboratories producing GA organisms, the obligation of a detailed phenotyping of GA animals with constraints, and finally the reporting not only of the number of animals used in experiments but also produced in all Swiss animal facilities, with emphasis on the GA animals, with and without a constrained phenotypes. New articles on transparency of research using animals have been added to the law. Finally a third layer of legislation, the administrative ordinance,[8] has been added to the law (principles driven) and the ordinance for animal protection (more practically driven), which resembles a guideline but with legal enforcement power (see for some examples, The Principles section in this chapter). This last piece of legislation is issued by the CAs, the Federal Veterinary Office, and does not need to go through the whole political validation process, thus allowing faster legal changes.

In addition to the political organizations that frame the regulatory care and use of laboratory animals in Europe, other professional organizations have issued important guidelines that have significant influence on the ground. The most representative is the Federation of European Laboratory Animal Science Associations (FELASA, www.felasa.eu), an organization that comprises many of the national/regional laboratory animal science associations in Europe. FELASA has published a number of guidelines and recommendations resulting from the collaborative work of highly respected laboratory animal science professionals.[9] Especially noteworthy are the recommendations on health monitoring and on training and education. The health monitoring recommendations are widely followed by breeders and users across Europe, and the training and education recommendations are served in the past as the basic scheme for some regulations at national level. FELASA has also established joint working groups with the American Association for Laboratory Animal Science (AALAS) trying to harmonize practices on both continents. This collaboration has already produced recommendations on health monitoring for rodent transportation[10] and on harm–benefit analysis.[11,12]

All taken together, European regulations and guidelines compose a detailed framework for the protection, care, and use of laboratory animals. The umbrella organizations (CoE and EU) have allowed the development of a framework based on the same principles but that still has some practical differences across countries. The areas addressed by the main European regulations and guidelines are described in the following sections.

THE PRINCIPLES

The European framework is evidently based on the principles of 3Rs. Although historically these principles were not mentioned explicitly in the main regulations, they were included implicitly in the various articles. ETS 123[1] is a very good example. Replacement is addressed in Article 6.1, which states that "A procedure shall not be performed for any of the purposes referred to in Article 2, if another scientifically satisfactory method, not entailing the use of an animal, is reasonably and practicably available." Also related to replacement is Article 6.2 in which it is stated that "Each Party should encourage scientific research into the development of methods which could provide the same information as that obtained in procedures."

The principles of reduction and refinement can be identified in several articles, such as Article 8 (anesthesia and analgesia) and Article 7, which is the most obvious example: "When a procedure has to be performed, the choice of species shall be carefully considered and, where required, be explained to the responsible authority; in a choice between procedures, those should be selected which use the minimum number of animals, cause the least pain, suffering, distress or lasting harm and which are most likely to provide satisfactory results." Also, Article 5 and the related entire Appendix A[3] with all the guidelines for accommodation and care can be considered a call to refinement.

However, the significance of the 3Rs principles is more evident in the recent legislation at the EU level. In fact, Directive 2010/63/EU[2] explicitly requires the implementation of 3Rs practices, with particular emphasis on replacement. This emphasis is evident from the preamble of the Directive, where it is stated that the Directive is a significant step toward achieving the final goal of full replacement of procedures as soon as it is scientifically possible (Recital 10). Recital 11 acknowledges that the care and use of laboratory animals is governed by the internationally established principles of the 3Rs and that these principles should be considered systematically when implementing the Directive.

After the recitals, Article 4 of the Directive is explicit on the 3Rs principles (the title of Article 4 is "Principle of Replacement, Reduction and Refinement") and defines specific requirements. Replacement is addressed in Article 4.1: "Member States shall ensure that, wherever possible, a scientifically satisfactory method or testing strategy, not entailing the use of live animals, shall be used instead of a procedure." Reduction is addressed in Article 4.2: "Member States shall ensure that the number of animals used in projects is reduced to a minimum without compromising the objectives of the project." In addition, Refinement is addressed in Article 4.3: "Member States shall ensure refinement of breeding, accommodation and care, and of methods used in procedures, eliminating or reducing to the minimum any possible pain, suffering, distress or lasting harm to the animals."

In addition to Article 4, other clear references to the 3Rs can be found along this Directive. Article 13 ("Choice of methods") states that "Member States shall ensure that a procedure is not carried out if another method or testing strategy for obtaining the result sought, not entailing the use of a live animal, is recognised under the legislation of the Union." It also states that the procedures to be used are those that "(a) use the minimum number of animals; (b) involve animals with the lowest capacity to experience pain, suffering, distress or lasting harm; (c) cause the least pain, suffering, distress or lasting harm; and are most likely to provide satisfactory results."

The same principles are linked to the practices relating to the ethical review process (ERP), which, although not mentioned as such (the word "ethical" is not used), include oversight, project (ethical) evaluation, and more technical evaluation at procedure level. One of the tasks assigned to the AWB (Article 27) is to "advise the staff on the application of the requirement of replacement, reduction and refinement, and keep it informed of technical and scientific developments concerning the application of that requirement"; another task is to "follow the development and outcome of projects, taking into account the effect on the animals used, and identify and advise as regards elements that further contribute to replacement, reduction and refinement." In addition to this, the project (ethical) evaluation (Article 38) shall include "an assessment of the compliance of the project with the requirement of replacement, reduction and refinement"; and one of the areas of expertise required for the evaluation by the CA is "the areas of scientific use for which animals will be used including replacement, reduction and refinement in the respective areas." The "elements that may contribute to the further implementation of the requirement of replacement, reduction and refinement" have to be part of the retrospective assessment of projects (Article 39) and "a demonstration of compliance with the requirement of replacement, reduction and refinement" is to be part of the nontechnical summaries required for the project authorization process (Article 43). Also, the requirement of 3Rs is one of the elements listed in Annex V as minimum requirements for training and education of personnel.

Other articles develop requirements on specific areas related to the 3Rs, such as accommodation and care (Article 33), anesthesia (Article 14), and reuse (Article 16). Chapter V is entirely dedicated to avoidance of duplication of procedures and alternative approaches to animal use.

The same 3Rs principles govern the legislation at the national level, both within and outside of the EU, and they represent the common area of all European regulations on animal protection, care, and use. The differences between European countries are not in the principles, but may be in some of the engineering standards used for their implementation. The new regulatory framework in the EU has helped harmonizing the legislation to this respect within the EU, and other countries out of the EU may follow a similar approach.

For example, the Swiss legislation shares these principles and the performance approach contained in the Directive; however, we may find a few cases where the engineering standard stance is still at work, besides the classical example of housing of laboratory animals. The administrative ordinance, *Ordinance of the Federal Veterinary Office on laboratory animal husbandry, the production of genetically modified animals and the methods of animal experimentation* (RS 455.163),[8] is mostly oriented toward engineering standards. The ordinance gives, for example, the required weekly frequency of checking the animals in the facility; the list of allowed and forbidden identification methods for rodents, i.e., ear tags for mice and rats are forbidden in Switzerland; the list of procedures that can be performed in a housing room; what kind of biopsy can be taken from GA rodents; how phenotyping of GA rodents can be performed, etc. The legislator has created this new piece of legislation because the guidelines that were recommended to the researchers were not well followed. The only way to introduce these refinements has been through legal enforcement.

SCOPE AND APPLICABILITY: ACTIVITIES, ANIMALS, AND INSTITUTIONS COVERED

To describe just the scope of the European legal framework is one thing. To fully appreciate the scope, one should include some of the motivation behind it. The motivation and rationale behind European Directive 2010/63/EU can be found in its 56 recitals.[2] As for ETS 123, its reasons and aims are described in a preamble.[1] Although they cover similar issues, the Convention's preamble uses more generic terms, while the Directive's recitals are more specific. The preamble of the Convention recalls that the aim of the CoE is to achieve greater unity between its members and that it wishes to cooperate with Member States in the protection of the live animals used for experimental and other scientific purposes. It recognizes that a human has a moral obligation to respect all animals and to have due consideration for their capacity for suffering and memory. It accepts nevertheless that a human in his/her quest for knowledge, health, and safety has a need to use animals where there is a reasonable expectation that the results will extend knowledge or be to the overall benefit of man or animal, just as he uses them for food, clothing, and as beasts of burden. It resolves to limit the use of animals for experimental and other scientific purposes with the aim of replacing such uses wherever practical, in particular by seeking alternative measures, and encouraging the use of these alternative measures. Its intention is to adopt common provisions to protect animals used in those procedures from the unnecessary infliction of pain, suffering, distress, or lasting harm.

Recital 1 of Directive 2010/63/EU[2] states that it should provide more detailed rules to reduce disparities among Member States, i.e., establish harmonization across Member States regarding the national levels of protection of animals used for scientific procedures to ensure the proper functioning of the internal market. Since the previous Directive 86/609/EEC,[4] it is noted that "new scientific knowledge is available in respect of factors influencing animal welfare as well as the capacity of animals to sense and express pain, suffering, distress and lasting harm. It is therefore necessary to improve the welfare of animals used in scientific procedures by raising the minimum standards for their protection in line with the latest scientific developments" (Recital 6).

Recital 12 specifies that animals are sentient and have intrinsic value that must be respected. This recital is a reminder of Chapter II, Article 13 of the Treaty on the Functioning of the European Union (TFEU) introduced in 2009, after the Lisbon Treaty came into force.[13] This article specifies that "In formulating and implementing the Union's agriculture, fisheries, transport, internal market, research and technological development and space policies, the Union and the Member States shall, since **animals are sentient beings**, pay full regard to the welfare requirements of animals, while respecting the legislative or administrative provisions and customs of the Member States relating in particular to religious rites, cultural traditions and regional heritage." Recital 12 has not only consequences for working with laboratory animals but will also affect other activities involving companion, farm, and wild animals. Introduction of concepts such as sentience and intrinsic value in the EU framework may in the near future bring new elements in the harm/benefit assessment of projects, going beyond the pathocentric view of the costs to the animal. An example of such conceptual shift is to be found in Switzerland. This country, since 1999, bears in its National Constitution, Article 120, that we have to respect the "dignity of the living organism." Consequence is that in the list of potential costs that procedures inflict to the animal are not only pain, suffering, distress, and lasting harm but also include procedures that can "...profoundly alter their appearance or capacities, or unduly instrumentalize them" (Animal Welfare Act, Article 3).[6] Consequently, the Swiss Animal Welfare Act protects both the dignity and the welfare of the animal.

The Directive should be reviewed regularly in light of evolving science and animal protection measures (Recital 10). The care and use of live animals for scientific procedures is governed by the 3Rs of laboratory animal science: replacement, reduction, and refinement. To ensure that the way in which animals are bred, cared for, and used in procedures within the EU is in line with that of the other international and national standards applicable outside the Union; the principles of the 3Rs should be considered systematically when implementing 2010/63/EU (Recital 11). The implementation of the 3Rs in projects is promoted by comprehensive project evaluation, taking into account ethical considerations in the use of animals, and project authorization (Recital 38). Furthermore, it is thought that the welfare of the animals used in procedures is highly dependent on the quality and professional competence of the personnel involved. Hence, it is essential and should be ensured that staff is adequately educated, trained, and competent (Recital 28).

Chapter 1 of 2010/63/EU deals with the general provisions. The subject matter and scope are laid out in Article 1. First it is explicated that the Directive establishes measures for the protection of animals used for scientific or educational purposes. To that end, it lays down rules on (Article 1.1):

a. the replacement and reduction of the use of animals in procedures and the refinement of the breeding, accommodation, care, and use of animals in procedures;
b. the origin, breeding, making, care of accommodation, and killing of animals;
c. the operations of breeders, suppliers, and users;
d. the evaluation and authorization of projects involving the use of animals in procedures.

The Directive applies where animals are used or intended to be used in procedures or bred specifically so that their organs or tissues may be used for scientific purposes. A procedure means any use, invasive or noninvasive, of an animal for experimental or other scientific purposes, with known or unknown outcome, or educational purposes, which may cause the animal a level of pain, suffering, distress, or lasting harm equivalent to, or higher than, that caused by the introduction of a needle in accordance with good veterinary practice (Article 3.1). The definition of a procedure according to Convention ETS 123 Article 1.2 is different from that of the definition in the Directive and reads as follows: "procedure means any experimental or other scientific use of an animal which may cause it pain, suffering, distress or lasting harm, including any course of action intended to, or liable to, result in the birth of an animal in such conditions, but excluding the least painful methods accepted in modern practice (that is 'humane' methods) of killing or marking animals."

Directive 2010/63/EU applies until the animals have been killed, rehomed, or returned to a suitable habitat or husbandry system. The Directive applies also when anesthetic, analgesic, or other methods have been applied successfully for the elimination of pain, suffering, distress, or lasting harm (Article 1.2).

The following animals are included in Directive 2010/63/EU (Article 1.3):

1. live nonhuman vertebrate animals, including:
   a. independently feeding larval forms; and
   b. fetal forms of mammals as from the last third of their normal development;
2. live cephalopods.

The Directive also applies to animals in procedures, which are at an earlier stage of development than the last third of their normal development, if the animal is to be allowed to live beyond that stage of development and, as a result of the procedure performed, is likely to experience pain, suffering, distress, or lasting harm after it has reached that stage of development (Article 1.4).

Directive 2010/63/EU shall apply without prejudice to the Directive of the European Council 76/78/EEC on the approximation of the laws of the Member States relating to cosmetic products (Article 1.6).[14]

The following purposes are covered and accepted by 2010/63/EU (Article 5):

1. basic research;
2. translational or applied research with any of the following aims:
   a. the avoidance, prevention, diagnosis, or treatment of disease, ill-health, or other abnormality or their effects in human beings, animals, or plants;
   b. the assessment, detection, regulation, or modification of physiological conditions in human beings, animals, or plants; or
   c. the welfare of animals and the improvement of the production conditions for animals reared for agricultural purposes;
3. for any of the aims in point (2) in the development, manufacture, or testing of the quality, effectiveness and safety of drugs, foodstuffs and feedstuffs, and other substances or products;
4. protection of the natural environment in the interests of the health or welfare of human beings or animals;
5. research aimed at preservation of the species;
6. higher education or training for the acquisition, maintenance, or improvement of vocational skills; and
7. forensic inquiries.

Although the Directive is specifically meant for the protection of animals used for scientific purposes, Article 1.5 has been included to explicate those areas where the Directive does not apply. Those areas are covered by other pieces of (inter)national legislation. These areas are as follows:

a. nonexperimental agricultural practices;
b. nonexperimental clinical veterinary practices;
c. veterinary clinical trials required for the marketing authorization of a veterinary medicinal product;
d. practices undertaken for the purposes of recognized animal husbandry;
e. practices undertaken for the primary purpose of identification of an animal;
f. practices not likely to cause pain, suffering, distress, or lasting harm equivalent to, or higher than, that caused by the introduction of a needle in accordance with good veterinary practice.

One of the purposes of 2010/63/EU is harmonization of legislation across the Member States of the EU. However, Article 2 allows for stricter national measures to be maintained, while observing the general rules laid down in the TFEU.[13] These measures had to be in force already on November 9, 2010, and aimed at ensuring more extensive protection of animals falling within the scope of 2010/63/EU than those contained in that Directive (Article 2.1). However, Member States shall not prohibit or impede the supply or use of animals bred or kept in another Member State in accordance with 2010/63/EU, nor shall it prohibit or impede the placing on the market of products developed with the use of such animals in accordance with 2010/63/EU.

The scope and applicability of the ETS 123 are very similar to that of the Directive albeit in more condensed, and perhaps again in more generic terms, and are laid down in Article 1.1 and 2 of ETS 123. The Convention applies to any animal used or intended for use in any experimental or other scientific procedure where that procedure may cause pain, suffering, distress, or lasting harm. It does not apply to any nonexperimental agricultural or clinical veterinary practice (ETS 123, Article 1.1). According to ETS 123, a procedure may be performed for any one or more of the following purposes only and subject to the restrictions laid down in the Convention:

a. avoidance or prevention of disease, ill-health, or other abnormality, or their effects, in man, vertebrate or invertebrate animals or plants, including the production and the quality, efficacy, and safety testing of drugs, substances, or products; diagnosis of treatment of disease, ill-health, or other abnormality, or their effects, in man, vertebrate or invertebrate animals or plants;
b. detection, assessment, regulation, or modification of physiological conditions in humans and vertebrate and invertebrate animals or plants;
c. protection of the environment;
d. scientific research;
e. education and training; and
f. forensic inquiries.

The similarities between the ETS 123 and the Directive are evident. However, the different wording is a potential source of differences in interpretation. Although the English versions are leading, translations are another source of interpretational differences. The EC is, therefore, maintaining an active Q&A document on the legal understanding of Directive 2010/63/EU.[15] Furthermore the EC together with EWGs drafts and publishes Consensus guidance documents on particular areas and articles of 2010/63/EU. These documents impact on the elaboration in practice of 2010/63/EU. The Consensus guidance documents are available at http://ec.europa.eu/environment/chemicals/lab_animals/interpretation_en.htm.

REQUIREMENTS FOR BREEDERS, SUPPLIERS, AND USERS

In this section, the requirements for breeders, suppliers, and users to comply with the Directive 2010/63/EU[2] are reviewed. The stated purpose of the Directive is to establish measures for the protection of animals used for scientific or educational purposes and, in particular, the importance of replacement and reduction of the use of animals in procedures and the refinement of the breeding, accommodation, care, and use of animals in procedures. There is a general requirement that Member States shall permit animals belonging to the common laboratory species (listed in Annex 1) to be used only if they have been specifically bred for use in procedures; they may be obtained directly from the breeder or from a supply establishment. In addition, scientific procedures may only be carried out at authorized establishments; however, an exemption may be given on the basis of scientific justification.

Authorization of Breeders, Suppliers, and Users

The requirements for all establishments at which animals are used, bred, or supplied for use for scientific procedures are harmonized, although requirements relating to the provision of space for animals have been adapted from the European Convention for the Protection of Vertebrate Animals Used for Experimental and Other Scientific Purposes (ETS 123),[1] which was drafted to take into account of the fact that breeding animals and their litters do not always require the same provision as groups of older animals. The Directive also recognizes that many Member States have already introduced substantial safeguards for animal welfare in breeding, supplying, and user establishments and provides for Member States to maintain existing requirements exceeding those laid down in this Directive, provided this does not inhibit or impede the marketing of products developed involving the use of such enhanced conditions.

The Directive defines different categories of establishment as follows:

- *Establishment*—any installation, building, group of buildings, or other premises and may include a place that is not wholly enclosed or covered and mobile facilities.

- *Breeder*—any natural or legal person breeding animals referred to in Annex I (a list of purpose-bred animals) with a view to their use in procedures or for the use of their tissue or organs for scientific purposes, or breeding other animals primarily for those purposes, whether for profit or not.
- *Supplier*—any natural or legal person, other than a breeder, supplying animals with a view to their use in procedures or for their tissue or organs for scientific purposes, whether for profit or not.
- *User*—any natural or legal person using animals in procedures, whether for profit or not.

It will be evident that these definitions seek to ensure that an individual is identified—sometimes called the Institutional Official (or in the UK the Establishment Licence Holder or Certificate Holder) who can be held accountable in law in the event of noncompliance, thereby instituting a firmer line of accountability.

Establishments belonging to each of these categories are required to obtain prior authorization and to register with the relevant CA (Article 20). Authorization is only to be granted if the breeder, supplier, or user establishment is in compliance with the requirements of this Directive and may be granted for a limited period. Authorization shall specify the person responsible for ensuring compliance with the requirements of this Directive; the designated veterinarian (or other suitably qualified expert); and the persons responsible for overseeing the welfare and care of the animals, education, competence, and supervision of the staff and ensuring availability of information specific to the species housed (Articles 24(1) and 25), which is discussed further in "Institutional and Designated Personnel Responsibilities" section. Sufficient staff must be maintained on-site and must be adequately educated and trained before performing any of the following functions:

- carrying out procedures on animals;
- designing procedures and projects;
- taking care of animals; or
- killing animals.

These needs are discussed in greater detail in "Education, Training, and Competence of Personnel" section.

There is an additional requirement for Member States to ensure that persons with responsibility for overall implementation of the projects and those who undertake them shall ensure that the projects are carried out in accordance with the project authorization or, in the cases referred to in Article 42, in accordance with the application sent to the CA or any decision taken by the CA and that any unnecessary pain, suffering, distress, or lasting harm inflicted on an animal in the course of a procedure is alleviated.

The CA must be notified of any changes of the person or persons referred to above, and authorization must be renewed if there are any significant changes to the structure or the function of an establishment of a breeder, supplier, or user that could have an adverse effect on animal welfare. In particular, Article 22 of the Directive requires that the animals' environment is maintained so as to comply with the criteria set out in Annex III (which draws very heavily on the revised Appendix A of the European Convention for the Protection of Vertebrate Animals Used for Experimental and Other Scientific Purposes).[3]

Member States are required to ensure that all establishments of breeders, suppliers, and users are equipped with installations and equipment suited to the species of animals housed and, where procedures are carried out, to the performance of those procedures. The design, construction, and method of functioning of these installations and equipment must be such as to ensure that the welfare of animals is not compromised and that procedures are carried out as effectively as possible, with the aim at obtaining reliable results using the minimum number of animals and causing the minimum degree of pain, suffering, distress, or lasting harm.

Authorization may be suspended or withdrawn if the breeder, supplier, or user no longer complies with the requirements set out in the Directive. In such cases the CA is required to take appropriate remedial action or to require such action to be taken or to suspend or withdraw the authorization. In the latter case the CA must ensure that the welfare of the animals is not adversely affected (Article 21).

Detailed requirements are laid down for annually reporting the number of animals used in procedures and there is a new requirement for publication of nontechnical summaries (considered elsewhere in this text).

Penalties

Member States are required to lay down rules on penalties applicable to infringements of the national provisions adopted (Article 60). These must be effective, proportionate, and dissuasive and notified to the EC. Although it is likely that different sanctions might be applied by each Member State, the EC intends to review these using its regular National Contact Point meetings as a vehicle.

Inspections

Articles 34 (Inspections by the Member States) and 35 (Controls of Member State Inspections) describe the requirements for inspections by CAs. Prior to introduction of Directive 2010/63/EU, there had been substantial differences in the nature and frequency of inspections carried out by Member States and those that were performing more frequent inspections than are stipulated by the Directive can continue their practices in accordance with Article 2 (Stricter National Measures).

The Directive stipulates that inspection frequency must be based on species and number of animals, history of compliance, and the number and types of project undertaken. At least one-third of user establishments shall be inspected each year, which in practice could mean only one inspection every 3 years. In the case of breeder, supply establishments, and users of nonhuman primates, inspections are to be carried out at least once a year. An appropriate proportion of the inspections must be carried out without prior warning. Records of all inspections shall be kept for at least 5 years.

Article 35 states that if the Commission has reason for concern, it may, without prior warning, undertake controls of the infrastructure and operation of national inspections in Member States, which are required to give all necessary assistance to the experts of the Commission in carrying out their duties.

The EC established an EWG whose objectives were to develop guidance and principles of good practice with respect to the requirements of the Directive for inspection and enforcement to facilitate the implementation of the Directive. The consensus document produced offers guidance to the CAs on the design of an inspection program, planning and conducting an inspection visit, and reporting on inspections.[16]

Reporting—Annual Statistics

Because one of the objectives of Directive 2010/63/EU is to increase transparency concerning the use of animals within Member States, the EC was anxious to ensure the collection and publication of meaningful statistics that could inform public interest and debate. An EWG was established to consider how this might best be achieved, and the discussions held were served as the basis for the processes described below. It was agreed that each 5 years, information from Member States concerning implementation of the Directive will be submitted to the EC.

Article 54.2 sets out the requirements for annually reporting the number of animals used in procedures. Member States are required to publish statistical information on the use of animals in procedures (including the actual severity incurred) and the origin and species of nonhuman primates used in procedures. Information will be reported also on each reuse of an animal. Further details of the statistical requirements have been published on the EC website (http://ec.europa.eu/environment/chemicals/lab_animals/statistics_en.htm). There has been a shift from prospective to retrospective reporting, and reports include the number of animals and the number of procedures or animal uses (http://ec.europa.eu/environment/chemicals/lab_animals/pdf/table-2003_with_bookmarks.pdf). The severity of each procedure needs to be recorded at its termination and reported annually. In the case of studies continuing beyond one calendar year, all of the animals may be accounted for together in the year in which the last procedure ends. This system of retrospective reporting of severity has been in use in the Netherlands and Switzerland for several years, but the new requirements laid down in 2010/63/EEC differ somewhat. The Netherlands had used a six-scale severity classification system rather than the Directive's four (nonrecovery, mild, moderate, and severe). The Swiss system provided information on the "animal," whereas the Directive requires both numbers of "procedures" and "animals" to be reported.

The statistics include animals used for the generation of a GA strain, but subsequent breeding in itself would not be regarded as a scientific experiment. To take into account of political and public concern, a 1-year snapshot of all GA animals not reported in annual statistics will be taken every 5 years as part of the implementation report (see below). No annual statistical data collection will be made for genotyping carried out under project authorization but this will be included in the 1-year snapshot. Reporting on all genotyping methods will also be included in the 5-yearly implementation report.

Animals killed for tissues and sentinels used for health monitoring are excluded from the provision of statistical data reporting unless the killing is performed under a project authorization using a method not included in Annex IV (Methods of Killing Animals) of the Directive. Surplus animals that are killed are not included apart from GA animals bred under project authorization and exhibiting a harmful phenotype.

Larval forms are counted only when they become capable of independent feeding. Fetal and embryonic forms of mammalian species are excluded.

Reporting—Implementation of the Directive

Article 54.1 introduces a new requirement for Member States for periodic (5 yearly) reporting of information relating to implementation of the Directive. In addition to the GA cases mentioned above, there will be a need to report on progress with a number of articles, for example, the breeding strategy for nonhuman primates (Article 28), nontechnical project summaries (Article 43), and avoidance of duplication of procedures (Article 46).

Reporting—Periodic Reports

Additional requirements for reporting are contained in Article 58 (Review). There is now a requirement for the EC to review the Directive before November 10, 2017, taking into account advancements in the development of alternative methods not entailing the use of animals, in particular, the use of nonhuman primates. Amendments will be proposed where appropriate. The EC, in consultation with Member States and stakeholders, will also conduct periodic thematic reviews of the implementation of the 3Rs in procedures, paying specific attention to nonhuman primates, technological developments, and new scientific and animal welfare knowledge.

Reporting—Nontechnical Project Summaries

Directive 2010/63/EU (Article 43) requires the publication of nontechnical project summaries providing information on the objectives of the project, the predicted harm and benefits, the number and types of animals to be used, and a statement of compliance with the requirement of the 3Rs. Lay summaries have been published in the UK by the Home Office for a number of years and have proved valuable during the ERP within establishments. Further information on nontechnical project summaries is included in "Institutional and Designated Personnel Responsibilities" section. An EC EWG dedicated to nontechnical project summaries produced a consensus document with guidance on the drafting and publication of nontechnical project summaries, including a template and an illustrative example.[17]

NONHUMAN PRIMATES: SPECIAL CONSIDERATIONS AND RESTRICTIONS

As stated above, the revision of Directive 86/609/EEC[2] was the result of a 10-year-long process during which viewpoints of various stakeholders were examined and discussed. One major matter of discussion was the continued need of use of nonhuman primates in biomedical research. Due to their genetic proximity to human beings and their highly developed social skills, their use raises specific ethical and practical challenges in terms of meeting their behavioral, environmental, and social needs in a laboratory environment. Moreover, as stated in the Directive (Recital 17), the use of nonhuman primates is of the greatest concern to the public.

The 2011 statistics of the EU reveal that the total number of animals used amounts to 11.5 million animals; nonhuman primates represent 0.05% of that number.[18] In Europe, various countries have banned the use of Great Apes for more than 10 years, and this nonuse, already noted in the previous statistical reports,[19] is confirmed in 2011. Compared to the United States, very few nonhuman primates are used in biomedical research in Europe.[20]

Of those, the European 2011 statistical report shows that 56% of the nonhuman primates were used for the purposes of toxicological and other safety evaluations, 10% in the area of basic research, and 23% in applied research, development, and quality control of products and devices for human medicine and dentistry and for veterinary medicine.

In preparation of the revision of Directive 86/609/EEC,[4] after lengthy discussions and debates,[21] the allowed use of nonhuman primates was clearly defined in Article 8 of Directive. Specimens of nonhuman primates shall not be used in procedures with the exception of those procedures meeting the following conditions: Only when there is scientific justification that the purpose of the procedure cannot be achieved by the use of species other than nonhuman primates, and on condition that procedures are undertaken with a view to the avoidance, prevention diagnosis or treatment of debilitating or potentially life-threatening clinical conditions in human beings, or for fundamental research or for research aimed at the preservation of the species. These conditions may raise discussions on the extent to which basic research could be included.

The use of great apes is prohibited, although under exceptional circumstances, a safeguard clause, as defined in Article 55 (2), could provisionally allow a Member State to use great apes when there are justifiable grounds for believing that action is essential for the preservation of the species, or in relation to an unexpected outbreak or a life-threatening or debilitating clinical condition in human beings.

Directive 2010/63/EU requires that any work involving the use of nonhuman primates is subject to a retrospective assessment at the end of the project. The use of wild-caught nonhuman primates for scientific purposes is prohibited, and specific measures are required to move toward using only second (or higher)-generation purpose-bred nonhuman primates. Recital 49 of the Directive states that "Technical and scientific advancements in biomedical research can be rapid, as can the increase in knowledge of factors influencing animal welfare. It is therefore necessary to provide for a review of this Directive. Such review should examine the possible replacement of the use of animals, and in particular non-human primates, as a matter of priority where it is possible, taking into account the advancement of science…." The legal obligation for the review is embedded in the first paragraph of Article 58: "The Commission shall review this Directive by November 10, 2017, taking into account advancements in the development of alternative methods not entailing the use of animals, in particular of non-human primates, and shall propose amendments, where appropriate."

In view of the above, the EC has asked again the Scientific Committee on Health and Environmental Risks (SCHER) to issue a scientific opinion (http://ec.europa.eu/health/scientific_committees/consultations/calls/scheer_call_info_01_en.htm), updating the SCHER opinion of January 13, 2009, on the need for nonhuman primates in biomedical research, production, and testing of products and devices.

Capture of Wild-Caught Animals

Article 4.2 in Appendix A of ETS 123,[1] which is thus applicable in Directive 2010/63/EU,[2] explicitly details precautions to be taken when capturing animals from the wild. Humane methods, used by competent personnel and ensuring minimal impact on the remaining wildlife and habitat, should be applied. Appropriate transport containers and means of transport should be available at capture sites, in case animals need to be moved for examination or treatment. When injured or in poor health, animals should be examined by a competent person as soon as possible and, if needed, treated by a veterinarian or euthanized. Special considerations are to be given to the acclimatization, quarantine, housing, and husbandry of these animals.

ETS 123 also requires the fate of wild-caught animals to be given due consideration, taking into account possible welfare issues associated with any subsequent release into the wild. Under the species-specific section (Section F), the requirement for group housing of nonhuman primates in an enriched environment is justified. This section also states that captive-bred nonhuman primates should be used and, where practicable, reared on-site to avoid transport stress. When imported, nonhuman primates should be obtained as offspring from established colonies with high welfare and care standards.

Directive 2010/63/EU is far more restrictive, as Article 9 states that *only when* competent authorities grant exemptions, can animals taken from the wild be used in procedures. Article 10 requires certain species to be purpose-bred when used in procedures, as listed in Annex I. The same article refers to Annex II for applicable dates for the use of captive-bred nonhuman primates or originating from self-sustaining colonies only. As of January 1, 2013, only captive-bred marmosets (*Callithrix jacchus*) can be used; the deadline for similar requirements for cynomolgus (*Macaca fascicularis*), rhesus (*Macaca mulatta*), and other species of nonhuman primates is set at the date of 5 years after the publication of a feasibility study (to be published no later than November 10, 2017), provided the study does not recommend an extended period. This feasibility study is to be conducted by the Commission in consultation with Member States and stakeholders.

Use and Reuse of Nonhuman Primates

Conditions of authorized reuse of animals in procedures, as defined in Article 16, are applicable to nonhuman primates: Only when the first procedure was classified as mild or moderate, after advice of the designated veterinarian and when it is demonstrated that the animal's general state of health and well-being has been fully restored, may an animal be reused in a further procedure classified as mild, moderate, or nonrecovery.

In exceptional circumstances, after a veterinary examination and after derogation by the CA, reuse of an animal, provided it has not been used more than once in a procedure classified as severe, may be allowed. Whether this is a restriction compared to ETS 123 is debatable. Article 10 of ETS 123 allowed that reuse, on condition that the animal had returned to good health and well-being, and either, was subjected to a nonsurvival procedure or to a further minor procedure only.

Specific Conditions for Establishments Using Nonhuman Primates

Article 34 requires Member States to perform annual inspections of all breeders, suppliers, and users of nonhuman primates. Records of these inspections are to be kept for 5 years.

Article 39 requires Member States to perform retrospective assessment of all projects using nonhuman primates.

GENETICALLY ALTERED ANIMALS: SPECIAL CONSIDERATIONS

Since its introduction, technology enabling the alteration of the genome of plants and animals including vertebrates has attracted much interest from society at large and has led to widespread debate. At the basis of the discussions are safety and ethical issues concerning interference with the organism's integrity, also addressed as manipulation of what makes an individual, and its genetic makeup.

Although the technology of creating GA animals is well established, specialized laboratories are needed to apply the technology successfully. Once the GA animals are generated, the founder animals and their offspring are kept in specialized facilities. Technical advances and availability of detailed genomic information of an increasing number of species are contributing factors to the increasing demand for GA animals in biomedical research.

In 1998, Directive 98/44/EC on the legal protection of biotechnological interventions was published.[22] It was recognized then that biotechnology and genetic engineering are playing an increasingly important role in a broad range of industries and that the protection of biotechnological interventions would certainly be of fundamental importance for the Community's industrial development (Recital 1, 98/44/EC). Patentability is a major subject of this Directive.

In this context of biotechnology and its importance for industrial development, the *ordre public* and morality in particular to ethical and moral principles are considered relevant (Recital 39, 98/44/EC). Special consideration is given to the genetic alteration of animals, which are likely to cause them suffering without any substantial medical benefit in terms of research, prevention, diagnosis, or therapy to humans or animals and to the resulting animals. Those are excluded from patentability (Recital 45, 98/44/EC).

While the biotechnological technology and processes are subjects of 98/44/EC, the protection of the welfare of GA animals resulting from such processes is a subject of both Directive 2010/63/EU of the EU[2] and, in similar but more general terms, the European Convention for the Protection of Vertebrate Animals Used for Experimental and Other Scientific Purposes of the CoE, ETS 123.[1] GA animals include transgenic, knockout, and naturally occurring or induced mutant animals, and other forms of genetic alteration. For the purposes of Directive 2010/63/EU, the creation and maintenance of GA animal lines are considered procedures, hence fall within the scope of this Directive, when there is a chance of causing a level of pain, suffering, distress, or lasting harm equivalent to, or higher than, that caused by the introduction of a needle in accordance with good veterinary practice (Article 3.1). With regard to new GA lines, a procedure shall be deemed to end when progeny are no longer observed or expected to experience pain, suffering, or lasting harm equivalent to, or higher than, that caused by the introduction of a needle (Article 17.1).

The severity classification of procedures has been exemplified in Annex VIII of the Directive. The power of annexes to directives is that they may be updated without having to amend the core of a Directive. In Annex VIII, four severity categories have been defined: nonrecovery, mild, moderate, and severe. Breeding GA animals that are expected to have no clinically detectable adverse phenotype may be classified as "mild." Breeding of GA animals that are expected to result in a phenotype with moderate effects and the creation of GA animals through surgical procedures are classified as "moderate."

Through the Directive, Member States are obliged to collect and make publicly available, on an annual basis, statistical information on the use of animals in procedures including information on the actual severity of the procedures; Member States had to submit that statistical information to the EC by November 10, 2015, and have to do it every year thereafter (Article 54.2). An EWG for the statistical reporting established by the EC concluded that some further understanding was needed as to how GA animals are to be considered.[23] In this context, the relevant articles of Directive 2010/63/EU are the abovementioned Articles 3.1 and 17.1 and Articles 1.2 and 4.3. In Article 1.2, the elimination of pain, suffering, distress, or lasting harm by the successful use of anesthesia, analgesia, or other methods shall not exclude the use of an animal in procedures from the scope of the Directive. Article 4.3 stipulates that Member States shall ensure refinement of breeding, accommodation and care, and methods used in procedures, eliminating or reducing to the minimum any possible pain, suffering, distress, or lasting harm to the animals. As for GA animals, an animal with a harmful phenotype is to be understood as an animal that is likely to experience, as a consequence of the genetic alteration, pain, distress, suffering, or lasting harm equivalent to, or higher than, that caused by the introduction of a needle in accordance with good veterinary practice. Therefore, an EWG on severity assessment focused on GA animals. It was concluded that the creation of a new GA line is regarded as a procedure, hence requiring a project authorization until such time when the line is "established." A new strain or line of GA animals is considered to be "established" when transmission of the genetic alteration is stable, which will be a minimum of two generations, and an initial welfare assessment is completed (Table 5.2).

Welfare assessment should include an assessment of general health, welfare, and behavior together with a review of production parameters such as breeding and growth performance, which will ideally be compared with an appropriate non-GA background strain. This should be done on animals of representative age groups:

- soon after birth, around weaning and again following sexual maturity, and at additional time points as considered by a prospective review of the potential impact of the gene alteration, e.g., where there is an age-dependent onset of disease;
- a minimum of seven males and seven females sampled from more than one litter;
- data from a minimum of two breeding cycles (from F2 onwards); and
- comparisons made wherever possible with similar non-GA animals.

TABLE 5.2 Key Elements of a Genetically Altered Rodent Welfare Assessment Scheme

Criteria	What to Look for
Overall appearance	Is the animal morphologically "normal"? Are there any malformations or any other indicators that the phenotype has been affected? For example, skeletal deformity or hydrocephalus.
Size, conformation, and growth	Are there any deviations from expected size or growth curve?
Coat condition	Is there any piloerection, areas of fur loss, loss of whiskers, or barbering? Is the skin/fur in good condition?
Behavior—posture, gait, activity, and interactions with the environment	Do they exhibit the full repertoire of behaviors appropriate for the strain/species, including social interactions, grooming, walking, running, digging, and climbing? Are these normal? Is the animal hunched or reluctant to move? Is movement impaired or is there any difficulty with orientation? Are there any signs of rigidity or tremors? Are there any abnormal activity levels? Prolonged inactivity could indicate chronic stress or depression (anhedonia) and/or sickness/pain, particularly if linked with a hunched posture and/or rough or unkempt coat. Unusual activity, such as hyperactivity, could indicate stereotypy or other behavioral abnormality.
Clinical signs	For example, nasal or ocular discharge, swollen or closed eyes, increased respiratory rate, dyspnea, seizures/twitches/tremors, increased vocalization with handling, overgrown teeth, presence of tumors, and neurological or musculoskeletal abnormalities. Is metabolism impaired, for example, increased or decreased food or water intake, excessive urination? Consistency of feces.
Relative size	Any unusual changes in size of the animals should be noted, and comparisons should be made within the litter. It may be helpful to generate a growth curve for the line.
Numbers	Where death occurs, it is important to maintain accurate records such that any pre- or postweaning losses can be investigated. Where appropriate (e.g., higher than anticipated mortality rate), postmortem examinations should be carried out to help determine the cause of death. A review of fertility can also be helpful in assessment of whether or not the modification is having an effect, e.g., conception rates, abortions, stillbirths.
Additional Considerations for Assessment in Neonatal Animals	
Color of pups (for neonate only)	Do any pups show evidence of abnormal skin color (e.g., anemia, poor circulation)?
Activity of pups (for neonate only)	Is there any abnormal activity, e.g., reduced wriggling? Does righting reflex intact?
Milk spot (for neonate only)	Do any pups fail to show presence of a milk spot? Is there any evidence of mismothering?
Litter	Litter sizes, litter homogeneity, development and growth of pups

The use of animals for the maintenance of colonies of GA established lines with a likely harmful phenotype is considered a procedure, hence it requires project authorization. GA animals requiring a specific, intentional (nonaccidental) intervention to induce gene expression can be considered as having a nonharmful phenotype until deliberate induction of transgene expression. Therefore, their breeding is not regarded as a procedure and does not require project authorization. The breeding and maintenance of GA lines, which retain a risk of developing a harmful phenotype (e.g., age onset of disease or tumors; risk of infection due to compromised immune system) regardless of the applied refinement (e.g., barrier conditions, culling at early age), require project authorization as the application of refinement does not eliminate the risk. If welfare issues are later identified, these should be reviewed to consider whether the welfare problems may be attributed to the genetic alteration. If so, these should be reclassified as "harmful phenotypes" and brought under project authorization.

In Switzerland with the 2008 revision of the Animal Welfare Act,[7] the creation of GA is no longer considered as a procedure per se, i.e., requesting a license to perform such procedure. Currently, the Swiss legislator delivers a license to a specific technological platform applying a set of specific procedures to create GA lines whatever the kind and number of GA lines they will produce; it is similar to the accreditation of a laboratory. This applies only to the mouse, the rat, the guinea pig, the rabbit, and the zebra fish. It is then the responsibility of the researcher and the animal house to report any harmful phenotype of a GA line to the CAs. Following the announcement and the burden that the line endures, an authorization to pursue the breeding and use of the line will be granted. The main reason to this change of legislation has been the huge amount of paperwork for licensing every GA line creation and breeding that needed to be done compared to the relatively low number of animal protection issues that these lines have generated. Creation and production of GA animals are reported annually under the laboratory animal housing form, which counts all animals weaned in the animal house or imported from abroad, distributed according to their species, the wild-type form, the GA lines without harmful phenotype, and the ones with constraints. A proportion of these animals, part of an experiment, will be included in the annual report of animals having been under a procedure.

With regard to statistical reporting in the EU, GA animals are reported either when used for the creation of a new line; when used for the maintenance of an established line bred under project authorization and that exhibits harmful phenotype; or when used in procedures. During the creation of a new line, all animals carrying the genetic alteration should be reported. In addition, those used for superovulation, vasectomy, and embryo implantation should equally be reported (these may or may not be genetically altered themselves). Genetically normal animals (wild-type offspring) produced as a result of creation of a new GA line should not be reported.

Genetic characterization is an essential part of GA colony management. GA animals need to be genetically characterized for scientific purposes, i.e., to settle scientific uncertainty or to confirm scientific suitability. The genetic characterization of GA animals requires sampling of the animals for genotyping. The severity classification of tissue sampling methods such as fecal pellets and hair follicles could be regarded as below threshold. In contrast, more invasive methods such as blood sampling and tail, ear, or phalange clipping should be regarded as mild or above. By definition of a procedure as per Article 3.1, tissue sampling with methods that are below minimum threshold does not fall within the scope of the Directive. On the other hand, above-threshold methods of tissue sampling carried out for the sole purposes of genetic characterization are to be considered as a procedure. Some practices for the primary purpose of identification, which allow also genotyping to be performed (for example ear-punching in mice), would not fall within the scope as per Article 1.5e.

Although GA animals are only specifically mentioned in Article 17.1 (the end of a procedure), the consequences of Directive 2010/63/EU for the breeding and maintenance of GA animals are considerable and important for protecting the welfare of these animals.

INSTITUTIONAL AND DESIGNATED PERSONNEL RESPONSIBILITIES

Directive 2010/63/EU[2] considerably strengthens the provisions of the European Convention for the Protection of Vertebrate Animals used for Experimental and Other Scientific Purposes (ETS 123),[2] as elaborated by Directive 86/609/EEC.[4] Directive 2010/63/EU recognizes that effective harmonization of the breeding and use of animals for scientific purposes requires a sound regulatory framework at both national and institutional levels. Details of the agreed mechanism for creating these structures are laid down; however, it is anticipated that each Member State will have the freedom to permit establishments to apply these requirements in ways that are responsive to their varying size and complexity. In this context, the word "establishment" means any installation, building, group of buildings, or other premises and includes places that are not wholly enclosed or covered as well as mobile facilities.

It was also recognized that there are differing national frameworks that need to be respected. Each Member State is required to ensure that there is a national CA that has the legislative powers and responsibilities to ensure compliance of all establishments in which animals are bred or used for scientific purposes with the appropriate institutional management organization. In brief, this requires a person to be responsible for overall compliance and one or several persons to be responsible for the welfare and care of the animals and provision of prompt veterinary attention when needed. In establishments where scientific procedures are carried out on animals, authorities have to be issued to the persons who design and oversee the conduct of procedures and those who carry them out. In addition, specific requirements are laid down for ensuring that there is an information culture to encourage introduction of the 3Rs at all levels and that all staff working with animals are competent and appropriately trained. It is also required that measures should be taken to avoid any potential conflicts of interest that may interfere with (or have the potential to hinder) the conduct of these activities.

To facilitate the integration of these various activities, establishments are required to set up a committee known as the Animal Welfare Body, which is required to oversee and monitor all activities relating to the welfare of animals housed or used in the establishment including their acquisition and eventual disposal and to provide advice on promoting implementation of the 3Rs. Member States are required to ensure that the number of animals used in projects is reduced to a minimum, without compromising the scientific objectives of the work.

Interesting to note, Switzerland has taken another approach to ensure that the animal welfare is optimally managed from birth to death. Instead of setting up at institutional level an overseeing body, the legislator has attributed the animal welfare responsibility during the breeding and the maintenance of animals to the head of the animal facility, a legally defined function not found in the European Directive, and during the experiences to the person responsible for the scientific project (like the European Directive defines it). This division of responsibility at the level of animal welfare management will be legally reproduced at the level of the day-to-day care of animals. Animal attendants or animal caretakers have the responsibility of the welfare of the animals during their breeding and their maintenance. This responsibility is then passed to the persons carrying out procedures when the animals are recruited for scientific investigations.

The Person Responsible for Overall Compliance

Although this individual occupies a central position in setting the general ethos and culture of the organization and ensuring that the welfare of animals kept or used is safeguarded within the framework of the 3Rs, Directive 2010/63/EU does not detail the qualifications required or responsibilities of this role. The requirement is laid out in Article 20.2 (Authorization of Breeders, Suppliers and Users) and in Article 24.1.a (Specific Requirements for Personnel), but the role would appear to be largely administrative, which is possibly why it is also a requirement that an AWB be established, which can provide valuable support and advice. In the UK the Home Office (the relevant CA) expects the person responsible for overall compliance to have undertaken relevant ethical and legislative training or to already have equivalent knowledge. Persons newly appointed to this role are recommended to take additional relevant training.

Breeders, suppliers, and user establishments should operate only if they are authorized by the CAs and shall specify the person responsible for ensuring compliance with the provisions of this Directive. This person should ensure that there are adequate facilities in place to meet the accommodation requirements of the animal species concerned and to allow procedures conducted there to be performed efficiently and with the least distress to the animals.

Persons responsible for ensuring compliance may have additional responsibilities placed on them under national legislation—for example, in the UK they must ensure that animals are sourced and transported in conformity with legal requirements including EC Regulation 1/2005 on the protection of animals during transport.[24]

The Person(s) Responsible for Welfare and Care

The requirement to identify a person or persons with administrative responsibility for the care of the animals and the proper functioning of the equipment was originally set out in ETS 123 (Article 20a) and Directive 86/609/EEC (Article 19.2a). Directive 2010/63/EU (Article 24.1a) retains the obligation that one or several persons on-site be identified as having responsibility for overseeing the welfare and care of animals in the establishment. There is, however, a requirement that the AWB shall include at least the person or persons responsible for the welfare and care of the animals. Details of the role of these persons and the training considered appropriate are not included, but the consensus document on education and training produced by the EC proposes a suggested profile for this role, initial training, and continued professional development,[25] and the Member States may specify these issues in national legislation. The role is carried out by Animal Welfare Officers in Germany and the Netherlands, the person(s) responsible for animal welfare in Spain, and Named Animal Care and Welfare Officers in the UK; these positions are given wider-reaching responsibilities, and formal educational requirements are laid down.

The Designated Veterinarian

In contrast to the position in the United States, the role of the veterinarian in Europe has always been seen as advisory and as providing support rather than direction to the routine care and use of animals. Directive 2010/63/EU (Article 25) requires Member States to ensure that each breeder, supplier, and user establishment ensures the availability of a designated veterinarian with expertise in laboratory animal medicine (or occasionally, if more appropriate a suitably qualified expert) charged with advisory duties in relation to the well-being and treatment of the animals. The possibility of having in some cases a suitable qualified expert who might not be a veterinarian was included for the benefit of institutions using only nontraditional species (e.g., fish) where the expertise can be found in persons having different academic backgrounds. However, it is a matter for discussion whether the lack of a veterinarian would be acceptable. In addition to their clinical role in providing appropriate veterinary treatment for animals that may require it, this person is required to provide input to the AWB in support of the 3Rs, for example, by predicting the likely outcomes of procedures on animals and recommending ways of alleviating these. In many establishments the veterinarian undertakes additional responsibilities, for example, in relation to the transport of animals, rehoming, or reuse.

The Person Responsible for the Scientific Project

The principal focus of the controls outlined in Directive 2010/63/EU is the personnel carrying out experimental procedures—both those who design experimental programs and those who conduct them. Member states are required to ensure, through authorization or by other means, that individuals have achieved a satisfactory level of competence before authorization is granted (Article 23.2b). The pathway by which this is achieved is laid out in greater detail in "Education, Training, and Competence of Personnel" section. To ensure that investigations are carried out in a rigorous and appropriate way, Member States are required to ensure that the person responsible for the scientific project shall have received instruction in a scientific discipline relevant to the work being undertaken and have specific knowledge about the biology of the species being used.

The person responsible for the scientific project is required to seek prior authorization from the relevant CA and to ensure that the project is carried out in accordance with that authorization; in addition, he or she must ensure that in the event of noncompliance all appropriate measures to rectify it are taken, recorded, and reported as necessary (Article 24.2). If the process of project approval had involved a simplified administrative procedure (Article 42), the investigation must be conducted in accordance with the application submitted, taking into account of any decision taken by the CA, and in the event of noncompliance, appropriate measures must be taken to rectify the matter and recorded.

In some Member States, including the UK, the person responsible for the scientific project is required to undertake additional responsibilities including preparation of the nontechnical summary, the supervision of persons carrying out procedures involved, and maintaining detailed records of the sources and numbers of animals used, procedures carried out, and the eventual fate of the animals. These are then used in compiling reports to the CA to enable publication of the statistics of animal use in that Member State.

The Person(s) Carrying Out Procedures

Persons carrying out procedures (users) have a major role in ensuring that during the conduct of investigations, the welfare of animals will be safeguarded as far as possible. They are required to work within the confines of the authorization granted to the person responsible for the project. Directive 2010/63/EU stipulates that they should be adequately educated and trained before they perform any task (Article 23.2b) and in addition that they be supervised in the performance of their tasks until they have demonstrated the requisite competence.

No specific educational requirements are laid down as a prerequisite for persons carrying out procedures; however, the consensus document on education and training[25] proposes the academic qualifications, training, and learning outcomes for this function, and Member States are required to develop formal educational requirements.

The Person(s) Responsible for Education and Competence

Directive 2010/63/EU emphasizes the importance of education, training, supervision, and competence in ensuring a high standard of welfare of the animals (Recital 28). To this end, each breeder, supplier, and user is required to appoint one or several persons on-site to take responsibility for ensuring that staff engaged in the animal care and use program are adequately educated, competent, and continuously trained. In addition, they should be supervised until they have demonstrated the requisite competence (Article 24.1c). Member States are free to determine where this responsibility lies and it could form part of an existing role, e.g., designated veterinarian or the person(s) responsible for welfare and care. Ultimately, the accountability lies with the person responsible for overall compliance in the establishment. In the UK, for example, this role has been formalized and the person is called the "Named Training and Competence Officer." In large establishments, such as some universities and pharmaceutical companies, this role is likely to be identified as a new, stand-alone role. Tasks could include establishing and maintaining a database of education and training attainment of individuals, recommending suitable training (including continuing education, continuing professional development (CPD)) for personnel involved in animal-related work and liaising with those designing projects on the suitability of trained persons. It is not foreseen that this individual should necessarily conduct the training.

The Person Responsible for Access to Species-Specific Knowledge

A new role has been established in Directive 2010/63/EU for a person to ensure that personnel involved in the care of animals have access to information specific to the species housed in the establishment (Article 24.1b). Specifically this person is expected to advise staff on the application of the requirement of 3Rs and to keep them informed of technical and scientific developments concerning the application of those requirements; this might include, for example, advising the establishment about information resources required relevant to the species and techniques used at the establishment, alerting relevant stuff to the availability of training and workshops and liaising with the person responsible for education and competence in the development of appropriate educational materials. The role could be combined with that of a different responsibility, such as the person responsible for education and competence or the person responsible for welfare and care. In the UK, this new role has been given the title of "Named Information Officer."

The Animal Welfare Body

The Directive aspires to ensure that throughout the Member States, animal welfare considerations are given the highest priority in the context of animal care, breeding, and use. In recognizing the importance of a broad institutional approach in developing a pervasive culture of care at establishments, the Directive requires breeders, suppliers, and users to establish an AWB, which will bring together expertise from a number of knowledgeable individuals to provide oversight and direction of institutional activities and provide tools for the practical application and timely implementation of technical and scientific developments to enhance the lifetime experience of the animals. All AWBs are required to include at least the person or persons responsible for the welfare and care of the animals and, in the case of user establishments, a scientific member; it must also receive input from the designated veterinarian (or the designated nonveterinary expert). In practice, most establishments include a broader range of members depending, among other things, on their size and complexity, area of research, number and type of projects and procedures, and species and number of animals used. Most include one or more lay members (without a specific function within the animal care and use program), and it is recommended that membership should be sufficiently flexible to ensure all aspects are covered. Roles of the AWB are described in Oversight and Ethical Review Process section.

Parallels are often drawn between the AWB and the Institutional Animal Care and Use Committee (IACUC) that is part of the regulatory framework in the United States. However, there are substantial differences between the two, because the latter is based on a much wider committee membership, holistically oversees the entire animal care and use program, including arrangements for the protection of the health of staff, has an executive role in authorizing the conduct of procedures, and has a legal obligation to report annually to the NIH Office of Laboratory Animal Welfare (OLAW).

Each AWB is expected to develop its own effective terms of reference, clarifying its level of authority within the establishment and its roles and responsibilities; these terms should be endorsed and visibly supported by management within the establishment. To ensure that deliberations and decisions taken by the AWB are widely disseminated and incorporated into the institutional culture of care, all staff should be made aware of the existence and role of the AWB, its stance on standards and policies in the establishment, and should be encouraged to contribute suggestions and raise any matters of concern.

OVERSIGHT AND ETHICAL REVIEW PROCESS

The oversight and ERP have been traditionally developed at the national level in Europe until the publication of Directive 2010/63/EU.[2] ETS 123[1] does not require explicitly this kind of process, and the old Directive 86/609/EEC[4] followed the scheme of ETS 123 closely, only requiring the Member States to designate the authorities responsible for verifying that the provisions of the Convention or Directive be met, which did not include requirement for ethical review. Therefore, in many EU countries there has not been a legal mandate for ERP until the provisions of the new Directive entered into force on January 1, 2013. However, many countries or even regions within countries had established their own systems.[26] Some of these countries were not affected by the Directive because they do not belong to the EU (i.e., Switzerland), and in others the development of ERP was a voluntary process, as the old Directive did not impede the implementation of national stricter measures. The development of stricter measures happened in a number of EU Member States, such as the UK, the Netherlands, Germany, Spain, and others. Because each Member State established a particular process, there have been traditionally many different systems in place across Europe, some based more on personnel and processes at the institutional level, and others giving more predominance to the oversight by public CAs. In all cases there was some kind of combination of institutional and government oversight activities.

Then Directive 2010/63/EU was published, and although after its transposition into national legislation was effected, systems have tended to be more harmonized within the EU, there are still important differences in the way the oversight and ethical review systems are implemented because of different interpretations of the Directive by Member States.

The Directive establishes many requirements to ensure oversight and ERP are conducted effectively. First, it is important to know the concept of *"Competent Authority" (CA)* because it plays an essential role in the oversight and ERP activities. The CA is defined in Article 3 as "authority or authorities or bodies designated by a Member State to carry out the obligations arising from this Directive." Although one may easily understand that the CA is the responsible person(s) at regional or national government level, this may not be entirely true in all cases, because Article 59 allows Member States to "designate bodies other than public authorities for the implementation of specific tasks laid down in this Directive, only if there is proof that the body: (a) has the expertise and infrastructure required to carry out the tasks; and (b) is free of any conflict of interests as regards the performance of the tasks. Bodies thus designated shall be considered competent authorities for the purposes of this Directive." This article means that if the requirements are met, public CA can designate institutional or external bodies or committees (that can be private) to be also CA with the legal competence to perform some of the activities related to the ERP, and this may be the main reason for the existence of different systems especially with regard to the project evaluation.

The first position required by the Directive who is related to oversight at each breeder, supplier, and user is the *person(s) responsible for overseeing the welfare and care of the animals in the establishment* (Article 24.1a), who in addition to this general function is required to be a member of the *AWB* (Article 26).

The AWB plays a crucial role in oversight, is required for all establishments, and, in addition to the person(s) responsible for overseeing the welfare and care of the animals in the establishment, has to include at least (in case of a user establishment) a scientific member and in all cases receive input from the designated veterinarian. The AWB could be interpreted up to some extent as the equivalent to the IACUC in the United States; however, there may be significant differences in composition and function. Especially noteworthy is that the designated veterinarian is given advisory duties only in relation to the well-being of animals (Article 25), and his/her presence as an AWB member is not mandatory (only input is needed). However, some countries consider that the position of the designated veterinarian should be stronger and require this figure to be part of the AWB.

The AWB tasks have to do with oversight and ERP, especially if we understand ERP as a set of several activities and not only the project or protocol evaluation. The tasks that as a minimum have to be performed by the AWB are "(a) advise the staff dealing with animals on matters related to the welfare of animals, in relation to their acquisition, accommodation, care and use; (b) advise the staff on the application of the requirement of replacement, reduction and refinement, and keep it informed of technical and scientific developments concerning the application of that requirement; (c) establish and review internal operational processes as regards monitoring, reporting and follow-up in relation to the welfare of animals housed or used in the establishment; (d) follow the development and outcome of projects, taking into account the effect on the animals used, and identify and advise as regards elements that further contribute to replacement, reduction and refinement; and (e) advise on rehoming schemes, including the appropriate socialisation of the animals to be rehomed" (Article 27). Clearly the intention of the Directive is to make the AWB become the tool to create a culture of care at the institution; however, it may be a matter for discussion if this can be done when institutions only comply with the minimum required composition of only two individuals at a user establishment. One of the EC EWGs produced a document with guidance and principles of good practice on the design and functioning of AWBs (and national committees), which was endorsed by the national CAs.[27] In this document it is recommended to take into account the complexity of the institution for the structure of the AWB and highlights the benefits of a wider structure. The document emphasizes the essential role an AWB should have to implement an institutional culture of care. It is important to note that the endorsement of these documents by the national CAs does not mean they become legally binding.

The project evaluation and/or authorization functions are not assigned to the AWB, but to the CA (Articles 36 and 38). There are Member States that have kept these functions at the level of the public CA exclusively, and others have opened the possibility to assign the project evaluation (and in the case of Belgium, also the authorization) to other designated CAs including institutional AWBs or ad hoc ethics committees, and/or external bodies/committees. In all cases, the project evaluation must be transparent and performed in an impartial manner (Article 38). The possibility of institutional committees performing the project evaluation is an object of controversy, because in addition to the transparency and impartiality, the designated CA must be free of any conflict of interest. A number of Member States consider that these requirements cannot be achieved at the institutional level and they retain the evaluation at the public CA level. On the other side, several other countries allow the delegation, which in many cases is a pragmatic decision based on the lack of resources to perform this function effectively at the government level. The different processes in place at national level have been discussed elsewhere.[28,29]

At this point it is important to clarify the concept of project, which "means a program of work having a defined scientific objective and involving one or more procedures" (Article 3.2). The Directive uses the term "project" instead of "protocol," which is more often used in other areas of the world such as the United States. Regardless of which type of CA performs the project evaluation, the project application shall include (Article 37) at least the project proposal, a nontechnical project summary (see below), and information on a number of items that are listed in Annex VI of the Directive: (1) relevance and justification of the following: (a) use of animals including their origin, estimated numbers, species, and life stages; (b) procedures; (2) application of methods to replace, reduce and refine the use of animals in procedures; (3) the planned use of anesthesia, analgesia, and other pain-relieving methods; (4) reduction, avoidance, and alleviation of any form of animal suffering, from birth to death where appropriate; (5) use of humane endpoints; (6) experimental or observational strategy and statistical design to minimize animal numbers, pain, suffering, distress, and environmental impact where appropriate; (7) reuse of animals and the accumulative effect thereof on the animals; (8) the proposed severity classification of procedures; (9) avoidance of unjustified duplication of procedures where appropriate; (10) housing, husbandry, and care conditions for the animals; (11) methods of killing; and (12) competence of persons involved in the project.

Concerning the severity classification, Annex VIII in the Directive describes the four categories (mild, moderate, severe, and nonrecovery) and offers examples for each of them. However, the potential inconsistency in applying these criteria, which may have significant impact in the project evaluation processing, led the EC to establish an EWG that produced guidance on the severity assessment framework from project planning, monitoring, and assessing the severity through to final assignment of actual severity,[30] and offered illustrative examples of practical application of the severity assessment process.[31] Both documents have been endorsed by the national CAs.

The nontechnical summaries (Article 43), which must be part of the application, have to be published by the Member States because one of the intentions of the EU is to be transparent in the use of animals for research purposes. To avoid confidentiality issues, the nontechnical summaries shall be anonymous and shall not contain the names and addresses of the user and its personnel.

They have to include (safeguarding intellectual property and confidential information) (1) information on the objectives of the project, including the predicted harm and benefits and the number and types of animals to be used and (2) a demonstration of compliance with the requirement of 3Rs. The publication of these summaries is done for authorized projects only. Another of the EC EWGs produced a document, also endorsed by the national CAs with guidance on the drafting and publication of nontechnical project summaries, including a template and an illustrative example.[17] The document considers that a word limit of 500 (one A4 page) should be sufficient to include all necessary information for all but the most complex projects, and that publication should be accessible for a period of 5 years.

The project evaluation and authorization process has to be performed within a defined timeframe, which is 40 working days. This period can be extended with 15 extra days in the case of complex multidisciplinary projects. The decision regarding authorization must be communicated to the applicant within that period (Article 41), and very importantly, the project cannot be started until the authorization is granted and communicated, even if the CA does not respond within the established timeframe. Only projects containing procedures are classified as "nonrecovery," "mild," or "moderate" and not using nonhuman primates, which are necessary to satisfy regulatory requirements, or which use animals for production or diagnostic purposes with established methods (basically regulatory testing) can be subjected to a simplified authorization administrative procedure (Article 42) if Member States decide to do so.

The project authorization has to include (1) the user who undertakes the project; (2) the persons responsible for the overall implementation of the project and its compliance with the project authorization; (3) the establishments in which the project will be undertaken, where applicable; and (4) any specific conditions following the project evaluation, including whether and when the project shall be assessed retrospectively (Article 40).

Once granted, the project authorization is valid for a period not exceeding 5 years (Article 40.3). The amendment or renewal of the project authorization is required "for any change of the project that may have a negative impact on animal welfare" and "any amendment or renewal of a project authorisation shall be subject to a further favourable outcome of the project evaluation" (Article 44).

The project evaluation may not be only performed prospectively. All projects using nonhuman primates and projects involving procedures classified as "severe" shall undergo a retrospective assessment. This retrospective assessment is the responsibility of the designated CA, who shall evaluate (1) whether the objectives of the project were achieved; (2) the harm inflicted on animals, including the numbers and species of animals used, and the severity of the procedures; and (3) any elements that may contribute to the further implementation of the requirement of 3Rs (Article 39).

Another of the EC EWGs produced a document, also endorsed by the national CAs, with guidance for performance of project evaluation including harm–benefit assessment and retrospective assessment of projects.[32] This document proposes the principles for an effective project evaluation process (availability of suitable expertise, impartiality, proportionality, consistency, efficiency, transparency, access to appeal process, understanding of the context and criteria, sufficient resources, and knowledge of local culture and practices), with special emphasis on the harm–benefit analysis, and offers guidance on more practical details (use of template, level of detail, etc.). It focuses also on the benefits and when and how to perform the retrospective assessment.

Although apparently the retrospective assessment of projects is to be performed only for certain projects, in practice the actual severity of all procedures inflicted to all animals have to be assessed, because this must be part of the statistical information that Member States have to submit to the EC every year (Article 54). The requirement to report the actual severity of procedures is forcing institutions to ensure a close oversight on the animals is implemented. This requirement may be linked to one of the tasks assigned to the AWB (following the development and outcome of projects, taking into account the effect on the animals used), but the cooperation of the researchers may be essential to perform this task effectively.

The ERP is not limited to the project evaluation. The follow-up of the development and outcome of projects, the retrospective assessment, and the need to categorize the actual severity of procedures, and not only the one assigned prospectively, may be linked to the US concept of postapproval monitoring. Also, in addition to the functions (described above) assigned to the AWB, there are also other responsible persons related to oversight functions. These include the already cited person(s) responsible for overseeing the welfare and care of the animals that all institutions must have and the persons responsible for the overall implementation of the project and its compliance with the project authorization. This last category of person has to be defined in the project authorization process.

On top of this entire process, Directive 2010/63/EU requires Member States to establish national committees to advise the CAs and AWBs on matters dealing with the acquisition, breeding, accommodation, care, and use of animals in procedures and ensure sharing of best practice (Article 49). One of the EC EWGs offers guidance on how these national committees should perform their advisory functions in an effective and consistent manner.[27]

It has been described how the EU Directive frames a complex scenario for the oversight and ERP in the EU, in which a combination of parties play different roles. The transposition of the Directive into Member States' national legislation has led to some differences in the way they are performed across Europe. These differences are based mainly on how the balance between the functions retained by the public CA or delegated to other CA (mainly at the institutional level) is established. Another crucial reason for the difference in how the process is implemented in practice relates to the capacity (either in time, resources, or even in knowledge) of the public CA having these responsibilities. While some countries have well-developed, experienced, and resourced public bodies, others lack them. It is in these last cases where designation of CA other than public bodies is more common.

Outside of the EU, the situation is diverse. There are countries with no specific requirements on ERP (for example, the Russian Federation), and others such as Switzerland have well-developed systems. In Switzerland, there are regional ethical committees in charge of the protocol review, which advise the Cantonal Authority whether or not the experiments should be authorized. In addition to evaluation of projects, the committee gives also recommendations to the Cantonal Authority on the breeding of GA animals, which present a negative phenotype. Moreover in their duties, committee members participate in inspection of animal facilities and audit experiments. On the committee sit representatives of the scientific institutions, members of animal protection association, and usually a veterinarian and an ethicist. Committee members such as scientists have the obligation to attend continuing education.

The Swiss law has also planned a national committee for animal experimentation, which can participate in controversial cases and is instrumental for legislation revisions. The committee is constituted by a maximum of nine members with at least a cantonal representative, experts in animal experiments, laboratory animal facilities, and animal welfare issues. The federal committee works closely with the national ethics committee for biotechnology in the nonhuman field. FELASA reviewed the principles and practices across Europe before the publication of Directive 2010/63/EU. In this report[26] (full report available online at http://www.felasa.eu/recommendations/reports/principles-and-practice-in-ethical-review-of-animal-experiments-across-euro/), in addition to the review of the situation of ethical review at that time across Europe, a set of 30 recommendations is offered for the conduct of effective ethical review in practice. Some of the requirements on ethical review present in the new European Directive are very similar to recommendations in the FELASA document. Because the Directive still allows room for different ways of implementation, the recommendations in the FELASA document may be a very useful guidance for AWB, ethics committees, and responsible personnel at institutional level.

REUSE OF ANIMALS

The ability to reuse animals, often seen by the general public, is a method used to minimize the numbers of animals used in scientific procedures. The situation, however, is much more complex than this because in evaluating proposals for the conduct of experiments in animals, it is the lifetime experience of the animal that is taken into account rather than the consequences of that one procedure. Generally, from the point of view of the 3Rs, it is preferable to use two animals rather than to carry out the same procedure twice on the same one.

Before describing the regulatory requirements concerning reuse, it is important that the two terms, "continued use" and "reuse," are clearly defined.

"Continued use" is an administrative concept that describes the use of an animal in a series of related procedures or experiments that are interlinked such that, if carried out on different animals, the scientific objective could not be achieved. The use of the same animal must be essential to achieve the objectives of the second or subsequent experiments. Perhaps the commonest example is the breeding of transgenic mice (which is regulated as a scientific procedure), which are subsequently used in procedures to investigate a disease state. Another example would be the preparation of an ovariectomized mouse in one laboratory, which is transferred to another for subsequent scientific study. Reuse is a term defined in Article 16 of Directive 2010/63/EU[2] as a procedure carried out on "an animal already used in one or more procedures, when a different animal on which no procedure has previously been carried out could also be used." The criterion for determining whether an animal is being reused is whether the use of a naive animal for the second or subsequent procedure would still achieve the scientific objective. An example of reuse would be the use of a sheep to study the metabolism of a dietary supplement that may cause adverse effects, after which it is further used to provide blood to make diagnostic plates for bacteriology. The two studies are not related. Any naive sheep could have been used for the second study. The second use would be reuse of that animal. The distinction between continued use and reuse has been elaborated by the UK Home Office.[33]

The matter of reuse is considered in Article 16 of Directive 2010/63/EU, paragraph 1 of which requires that "Member States shall ensure that an animal already used in one or more procedures, when a different animal on which no procedure has been carried out could also be used, may only be re-used in a new procedure provided that the following conditions are met:

- The actual severity of the previous procedures was "mild" or "moderate"
- It is demonstrated that the animal's general state of health and well-being has been fully restored
- The further procedure is classified as "mild", "moderate" or "nonrecovery" and
- It is in accordance with veterinary advice, taking into account the lifetime experience of the animal."

In coming to a decision about whether reuse can be justified, the veterinarian must take into account of the degree of suffering experienced by the animal from the time when it was first protected under the Directive to the time of the proposed reuse and the likely impact of that further use in the proposed regulations procedure.

In exceptional circumstances, Article 16 does allow that after a veterinary examination of the animal, the CA may allow reuse of an animal, provided the animal has not been used more than once in a procedure entailing severe pain, distress, or equivalent suffering.

It would be possible to reduce the number of animals used in procedures by performing procedures on them more than once, where this does not detract from the scientific objective or result in poor animal welfare. However, the benefit of reusing animals should be balanced against any adverse effects on their welfare, taking into account the lifetime experience of the individual animal. In view of this potential conflict, the reuse of animals should be considered on a case-by-case basis. Reuse can only be justified if the animals remain fit and healthy, and the accumulation of discomfort from unrelated studies is effectively restricted. However, reuse can sometimes have overall positive welfare benefits, for example, when radio transmitters are surgically implanted into animals to monitor relatively mild effects after administration of potential therapeutic agents—alternatively the surgical procedure would need to be repeated to prepare a naive animal for a second such experiment; animals might be trained to carry out maneuvers in certain procedures or to become acclimatized to particular experimental situations. All reuse should be authorized by the relevant CA and reported in the annual reports that establishments are required to submit. It is also appropriate for the AWB to have oversight of the general policy on reuse and its implementation within the establishment.

A particular case of reuse, which is encouraged by the Directive, is the sharing of organs and tissues within establishments, so naive animals do not need to be killed unnecessarily for the conduct of in vitro investigations. In this instance there is no adverse welfare impact because the animals have been humanely killed at the end of the primary investigation. In circumstances where tissues or organs are not required immediately on termination of the original procedure, animals may be maintained for a short period of time providing this does not cause suffering or require measures that would otherwise be regarded as regulated.

SETTING FREE/REHOMING

ETS 123[1] does not specifically mention conditions required to be fulfilled for setting free or rehoming of animals after use in procedure, although Article 12 allows for such conditions: "Notwithstanding the other provisions of this Convention, where it is necessary for the legitimate purposes of the procedure, the responsible authority may allow the animal concerned to be set free provided that it is satisfied that the maximum practicable care has been taken to safeguard the animal's well-being. Procedures that involve setting the animal free shall not be permitted solely for educational or training purposes."

Recital 26 of Directive 2010/63/EU[2] introduces the concept of rehoming cats and dogs, taking into account the high level of public concern as to the fate of such animals. It states that, in case Member States allow rehoming, a scheme must be in place providing adequate socialization to those animals to ensure success, avoid unnecessary distress to the animals, and guarantee public safety.

Article 19 of Directive 2010/63/EU explicitly allows for a possibility for setting free or returning animals (used or intended to be used in procedures) to a suitable habitat or husbandry system, appropriate to the species, provided that the following conditions are met: (1) the state of health of the animal allows it; (2) there is no danger to public health, animal health or the environment; and (3) appropriate measures have been taken to safeguard the well-being of the animal.

The Directive thus is not specific on which species, or on which procedures, are concerned. Moreover, rehoming is offered as a possibility, not a requirement. Article 29 states that "Where Member States allow rehoming, the breeders, suppliers and users from which animals are intended to be rehomed shall have a rehoming scheme in place that ensures socialisation of the animals that are rehomed. In the case of wild animals, where appropriate, a programme of rehabilitation shall be in place before they are returned to their habitat." Thus, on condition that Articles 19 and 29 are met, it is left up to Member States to decide whether they want to implement these, and if yes, for which species. Some Member States can, and will allow rehoming of, e.g., animals that have undergone surgical procedures (e.g., explants of an implant), whereas in other Member States this will not be allowed.

Article 27 states that one of the tasks of the AWB is to advise on rehoming schemes, including the appropriate socialization of the animals to be rehomed. In case a Member State requires/allows implementation, this becomes another major responsibility of the AWB. This task will definitely call for financial and other resources, as it would not only require a careful selection of suitable animals, but also availability of appropriate means to allow for adapted socialization before rehoming, careful selection of retained possibilities of rehoming and, maybe more importantly, clearly defined alternative options in case of failure of the retained choice.

Conditions for setting free animals after use in procedures, as applicable to wild animals, require that a program of rehabilitation be in place before they are returned to their habitat (Article 29). For nonhuman primates, this might at the time of writing imply being placed in a suitable husbandry system such as a zoo, or returned to a suitable habitat, such as release into the wild. However, once Annex II of the Directive 2010/63/EU becomes applicable, only allowing the use of captive-bred nonhuman primates or originating from self-sustaining colonies, the option of releasing nonhuman primates into the wild might no longer be an option as these animals might not be able to cope with those unknown conditions. It thus remains to be seen whether Member States would still allow these animals to be placed in zoos.

OCCUPATIONAL HEALTH AND SAFETY

In any working environment, one should be safe and feel safe. The employer's responsibility is to provide a safe working place. It is the responsibility of the employee to act responsibly within this safe working place.

In February 1988, the European Parliament adopted four resolutions following the debate in the internal market on worker protection. These resolutions specifically invited the EC to draw up a framework Directive to serve as a basis for more specific directives on minimum health and safety requirements at the workplace (https://osha.europa.eu/en/safety-and-health-legislation/european-Directives). The important driver was the notion that the incidence of accidents at work and occupational diseases was considered to be too high and that preventive measures had to be introduced or improved without delay to safeguard the safety and health of workers and ensure a higher level of protection. These issues are not specifically addressed in Convention ETS 123[1] or in Directive 2010/63/EU.[2]

In 1989, the framework Directive on the Introduction of Measures to Encourage Improvements in the Safety and Health of Workers at Work, 89/391/EEC, was published by the Council of the European Communities.[34] According to Article 16.1 of this framework Directive 89/391/EEC, the Council, acting on a proposal from the EC, should adopt individual directives, inter alia, in the areas listed in the annex of 89/391/EEC, such as the following:

- workplaces,
- work equipment,
- personal protective equipment,
- work with visual display units,
- handling of heavy loads involving risk of back injury,
- temporary or mobile work sites, and
- fisheries and agriculture.

To date, 20 individual directives within the meaning of Article 16.1 of 89/391/EEC have been published.[34–53]

In 1994, the Council of the European Union published Directive 94/33/EC on the Protection of Young People at Work.[54] Children and adolescents were considered specific risk groups. Following Article 15 of 89/391/EEC, particularly sensitive risk groups have to be protected against the dangers, which specifically affect them. Directive 94/33/EC applies to any person under 18 years of age having an employment contract or an employment relationship defined by the law in force in a Member State and/or governed by the law in force in a Member State. It contains articles on obligations on employers, vulnerability of young people and the consequential prohibition of work, working time, night work, rest period, breaks, and annual rest.

89/391/EEC: Framework Directive

Important considerations of the framework Directive 89/391/EEC were that Member States have a responsibility to encourage improvements in the safety and health of workers on their territory. At the same time, the Member States' legislative systems covering safety and health at the workplace were considered to differ widely and needed to be improved. It was anticipated that national provisions on the subject, which often include technical specifications and/or self-regulatory standards, were likely to result in different levels of safety and health protection and allowed for competition at the expense of safety and health.

Directive 89/391/EEC contains general principles concerning the prevention of occupational risks; the protection of safety and health; the elimination of risk and accident factors; the informing, consultation, balanced participation in accordance with national laws and/or practices; and training of workers and their representatives, as well as general guidelines for the implementation of said principles. The Directive applies to all sectors of activity, both public and private (industrial, agricultural, commercial, administrative, service, educational, cultural, leisure, etc.). In Section II of the Directive, the employer's obligations are listed; Section III deals with the workers' obligations.

Section II: Employers' Obligations

Among the general provisions of Section II, it is stated that the employer shall have a duty to ensure the safety and health of workers in every aspect related to the work. Furthermore, the workers' obligations in the field of safety and health at work shall not affect the principle of the responsibility of the employer (Articles 5.1 and 5.3). The general obligations on employers are described in Article 6: Within the context of his responsibilities, the employer shall take the measures necessary for the safety and health protection of workers, including prevention of occupational risks and provision of information and training, as well as provision of the necessary organization and means. The employer shall be alert to the need to adjust these measures to take into account of changing circumstances and aim to improve existing situations (Article 6.1). The employer shall implement these measures on the basis of the following general principles of prevention (Article 6.2):

- avoiding risks;
- evaluating risks that cannot be avoided;
- combating the risks at source;
- adapting the work to the individual;
- adapting to technical progress;
- replacing the dangerous by the nondangerous or the less dangerous;
- developing a coherent overall prevention policy;
- giving collective protective measures priority over individual protective measures; and
- giving appropriate instructions to the workers.

Without prejudice to other provisions of Directive 89/391/EEC, the employer shall, taking into account the nature of the activities of the enterprise and/or establishment, evaluate the risks to safety and health of workers. Subsequent to this evaluation and as necessary, the preventive measures and the working and production methods implemented by the employer must assure an improvement in the level of protection afforded to workers and be integrated into all activities of the undertaking and/or establishment and at all hierarchical levels (Article 6.3a); where the employer entrusts tasks to a worker, he should take into considerations the worker's capabilities as regards health and safety

(Article 6.3b). The employer should ensure that the planning and introduction of new technologies are the subjects of consultation with the workers and/or their representatives, as regards the consequences of the choice of equipment, the working conditions, and the working environment for the safety and health of workers (Article 6.3c). The employer should take appropriate steps to ensure that only workers who have received adequate instructions may have access to areas where there is serious and specific danger (Article 6.3d). Where several undertakings share a workplace, the employers shall cooperate in implementing the safety, health, and occupational hygiene provisions; taking into account the nature of the activities, shall coordinate their actions in matters of the protection and prevention of occupational risks; and shall inform one another and their respective workers and/or their representatives of these risks (Article 6.4). In no circumstances may measures related to safety, hygiene, and health at work involve the workers in financial cost (Article 6.5). Furthermore, the employer shall be in possession of an assessment of the risks to safety and health at work, including those facing groups of workers exposed to particular risks (Article 9.1a); decide on the protective measures to be taken and, if necessary, the protective equipment to be used (Article 9.1b); keep a list of occupational accidents resulting in a worker being unfit for work for more than three working days (Article 9.1c); and draw up, for the responsible authorities and in accordance with national laws and/or practices, reports on occupational accidents suffered by his workers (Article 9.1d). In Article 10 the obligation of informing workers is described. The consultation and participation of workers is described in Article 11 and the training of workers in Article 12.

Section III: Workers' Obligations

It is each worker's responsibility to take care as far as possible of his/her own safety and health and that of other persons affected by his/her acts or commissions at work in accordance with his training and the instructions given by his/her employer (Article 13.1). To this end, workers must in particular "make correct use of machinery, apparatus, tools, dangerous substances, transport equipment and other means of production; make correct use of the personal protective equipment supplied to them and, after use, return it to its proper place; refrain from disconnecting, changing or removing arbitrarily safety devices fitted, e.g., to machinery, apparatus, tools, plant, and buildings and use such safety devices correctly; immediately inform the employer and/or the workers with specific responsibility for the safety and health of workers of any work situation they have reasonable grounds for considering represents a serious and immediate danger to safety and health and of any shortcomings in the protection arrangements; cooperate, in accordance with national practice, with the employer and/or workers, for as long as may be necessary to enable any tasks or requirements imposed by the competent authority to protect the safety and health of workers to be carried out, and to enable the employer to ensure that the working environment and working conditions are safe and pose no risk to safety and health within their field of activity" (Article 13.2).

Individual Directives

The more specific nature of the individual directives makes them more or less relevant in the context of animals used for scientific purposes. The relevance is dependent on the type of enterprise and facilities among others. However, an attempt is made to rank them into three categories: those of general relevance, relevant for certain enterprises, irrelevant in this context.

Generally Relevant

Those of general relevance are those dealing with the workplace (Directive 89/654/EEC)[35]; the use of work equipment by workers at work (Directive 2009/104/EC)[36]; the use by workers of personal protective equipment at the workplace (Directive 89/656/EEC)[37]; the manual handling of loads where there is a risk particularly of back injury to workers (Directive 90/269/EEC)[38]; the work with display screen equipment (Directive 90/270/EEC)[39]; the provision of safety and/or health signs at work (Directive 92/58/EEC)[43]; pregnant workers who have recently given birth or are breastfeeding (Directive 92/85/EEC)[44]; risks arising from vibration (Directive 2002/44/EC)[50]; risks arising from noise (Directive 2003/10/EC)[51]; and young people at work (Directive 94/33/EEC)[54].

Relevant for Certain Enterprises

Relevant for certain enterprises are those dealing with risks related to exposure to carcinogens or mutagens at work (Directive 2004/37/EC)[40]; exposure to biological agents at work (Directive 2000/54/EC)[41]; work on board fishing vessels (Directive 93/103/EEC)[47]; risks related to chemical agents at work (Directive 98/24/EEC)[48]; risks from explosive atmospheres (Directive 1999/92/EC)[49]; risks arising from electromagnetic fields (Directive 2004/40/EC has been repealed on June 13, 2013, by 2013/35/EU)[52,55]; and risks arising from artificial optical radiation (Directive 2006/25/EC).[53]

Irrelevant in This Context

Irrelevant are those dealing with safety and health requirements at temporary or mobile construction sites (Directive 92/57/EEC)[42]; improving the safety and health protection of workers in the mineral-extracting industries through drilling (Directive 92/91/EEC)[45]; and improving the safety and health protection of workers in surface and underground mineral-extracting industries (Directive 92/104/EEC).[46]

Concluding Remarks

Neither the Convention ETS 123[1] nor the Directive 2010/63/EU[2] deals with occupational health and safety. The framework Directive 89/391/EEC, together with the individual directives, forms a European legislative core on these topics. The framework Directive encourages in more general terms improvements in the safety and health of workers at work. The individual directives provide minimum health and safety requirements for specified subjects including Directive 94/33/EC on the protection of young people at work. Control of laboratory animal allergy is not among these specified subjects. At the national level, there is to varying degree, legislation in place and guidance documents being published. Links to the national laws implementing the different directives are listed per Directive on the website of the European Agency for Safety and Health at Work (https://osha.europa.eu/en).

EDUCATION, TRAINING, AND COMPETENCE OF PERSONNEL

Background

One key principle underlying introduction of Directive 2010/63/EU[2] is that public concerns over animal experimentation must be recognized and that appropriate measures taken to ensure that high welfare standards are set in place. This in turn requires a high level of professionalism among all those concerned with the care and use of laboratory animals and to this, the concept of competence is central. The emphasis on competence is intended both to satisfy ethical and societal concerns about animal well-being and to facilitate the movement of personnel between Member States. In addition, scientific investigations designed and carried out by competent personnel are less likely to be flawed or to lack rigor. Consequently, the Directive introduces a number of important concepts into the care and use of laboratory animals. Specifically, it requires appropriate prior education and training for those who care for or kill animals, conduct scientific procedures or design such procedures and projects and that such personnel should be appropriately supervised until they have demonstrated competence.

Although the term competence is not defined within this Directive, there is no shortage of definitions from elsewhere; however, many of these are conflicting. In the recommendation adopted by the European Parliament and of the Council for the Establishment of the European Qualifications Framework for Lifelong Learning (2008),[56] it is defined as "the proven ability to use knowledge, skills and personal, social and/or methodological abilities, in work or study situations and in professional and personal development," and this document recommends a tiered structure representing eight different levels of competence.

The Council of Europe Treaty Series 123—the Convention for the Protection of Vertebrate Animals used for Experimental and other Scientific Purposes (1986)[1]—stipulated that persons who carry out procedures, or take part in procedures, or take care of animals used in procedures (including supervision) shall have had "appropriate education and training" (Article 26). The Convention also made provision for the use of animals in the course of education and training (Article 25) providing this is "absolutely necessary for the purpose of the education or training concerned" and only if the objective cannot be achieved in any other way. The use of animals in the course of teaching and instruction was required to be reported to the responsible authorities and to comply with national legislation. In the explanatory report that accompanied the Convention,[57] the phrase "appropriate education and training" is interpreted as meaning "education and training appropriate for the task at hand." This very loose requirement was considerably expanded by the publication in 1993 by the CoE of specific guidelines for the Education and Training of Persons Working with Laboratory Animals.[58] This guidance identified four categories of persons: those taking care of animals (Category A), those carrying out procedures (Category B), persons responsible for directing animal experiments (Category C), and specialists in laboratory animal science (Category D). The guidelines were intended to provide a basis for topics to be included in educational and training programs required for persons working with laboratory animals, to the extent necessary for the task at hand, so as to develop the appropriate level of training by the Convention. A list of topics was presented appropriate to the training of personnel in each of the four categories. To facilitate movement of personnel between countries, those providing training were to be encouraged to issue a document proving the attendance at the training programs and giving information about them.

The guidance was developed and expanded over successive years, for example, the publication of recommendations by FELASA for the education and training of persons involved in each of the four categories mentioned above.[59–63]

Directive 86/609/EEC

Directive 86/609/EEC[4] had made relatively little reference to the education and training or competence of personnel involved in the care and use of animals. Article 7 required that experiments should be performed only by a competent authorized person, or under the direct responsibility of such a person, and Article 14 required persons conducting experiments on animals or taking care of them to have appropriate education and training; persons carrying out or supervising the conduct of experiments were required to have received instruction in a relevant scientific discipline, be capable of handling and taking care of laboratory animals and have attained a level of training sufficient for carrying out their tasks. The focus here is clearly on the person who conducts and supervises the experiments.

As a consequence, several Member States (including The Netherlands, Germany, France, and the UK) had introduced recommendations or mandatory requirements for those involved in the conduct of experimental procedures and in some cases also those caring for animals (animal technicians and caretakers).

However, the content, structure, and delivery of education and training varied between Member States. In 2002, FELASA proposed a scheme for the accreditation of training based on its earlier recommendations[64] and this had an important influence in harmonizing the provision of education and training over the following years. This quality assurance system, introduced in 2003, was intended to assist in the development of uniform, high-quality educational programs. The system was widely recognized and operated in a number of European countries; however, uptake in Member States that recently joined the EU was patchy.

Directive 2010/63/EU

Directive 2010/63/EU[2] places much greater emphasis on competence, rather than the education and training of personnel involved at all levels in the care and use of animals for scientific purposes (the EC has no legal power concerning education, which is for the Member States to decide), but no guidance is given as to what this competence entails. Specifically, it argues (Recital 15) that the welfare of animals used in procedures is highly dependent on the quality and professional competence of the personnel supervising procedures, as well as of those performing procedures or supervising those taking care of the animals on a daily basis. In view of this, Member States are required to ensure that all such personnel are adequately educated, trained, and competent and that they are supervised until they have obtained and demonstrated the requisite competence. This offers a unique opportunity to establish a harmonized framework across Europe that assures the competence of persons in all Member States who are involved in the care, breeding, and use of animals for scientific procedures. Although the legal authority of the EC did not extend to the specification of educational requirements, each Member State was required to publish its own minimum requirements. To assist in this, Recitals 50 and 51 provided for the Commission to adopt nonbinding guidelines on the requirements with regard to education, training, and competence of breeders', suppliers', and users' staff and to adopt relevant delegated acts in accordance with Article 290 of the TFEU.[13]

The Directive emphasizes the importance of competence in certain key areas, including the killing of animals (Article 6), working with animals from the wild (Article 9), and undertaking of decisions to keep an animal alive at the end of a procedure (Article 17). The principal requirements with regard to the competence of personnel are laid out in Articles 23 and 24. Member States are required to ensure that each breeder, supplier, and user has sufficient members of staff on-site and that they are adequately educated and trained before they perform any of the following functions:

- carrying out procedures on animals (function A);
- designing procedures and projects (function B);
- taking care of animals (function C); or
- killing animals (function D).

In addition, staff carrying out function (A), (C), or (D) shall be supervised in the performance of their tasks until they have demonstrated the requisite competence. Persons carrying out function (B) are required to have received instruction in a scientific discipline relevant to the work being undertaken and to have relevant species-specific knowledge.

Member States are obliged to develop and publish appropriate education and training programs, based on a list of topics laid out in Annex V so as to ensure the achievement of at least those minimum requirements and to introduce requirements for obtaining, maintaining, and demonstrating requisite competence for the functions set out in paragraph 2.

Annex V lists the relevant elements referred to above:

- National legislation in force relevant to the acquisition, husbandry, care, and use of animals for scientific purposes
- Ethics in relation to human–animal relationship, intrinsic value of life, and arguments for and against the use of animals for scientific purposes
- Basic and appropriate species-specific biology in relation to anatomy, physiological features, breeding, genetics, and genetic alteration
- Animal behavior, husbandry, and enrichment
- Species-specific methods of handling and procedures, where appropriate
- Animal health management and hygiene
- Recognition of species-specific distress, pain, and suffering of most common laboratory species
- Anesthesia, pain-relieving methods and killing
- Use of humane endpoints
- Requirement of 3Rs
- Design of procedures and projects, where appropriate

To support the development of appropriate competencies, each breeder, supplier, and user is required to appoint one or several persons on-site with responsibility for ensuring that the staff are adequately educated, competent, and continuously trained and that they are supervised until they have demonstrated the requisite competence.

Modular Training—The European Commission Expert Working Group

During 2012, the EC established an EWG tasked with developing a nonbinding training framework within the EU that would assure the competence of staff caring for or using animals in procedures and facilitate the free movement of personnel within the EU. This framework was to take into account of the training, supervision, assessment of competence, and lifelong learning requirements of persons fulfilling tasks within the four functions identified above. This would affirm the EC's commitment to uniform, high standards of animal welfare and assisting the movement of trained persons within Europe. The text of the

Directive makes specific reference to education and training requirements over and above those of the previous Directive. A major challenge is to ensure consistency across Europe because many Member States have already introduced programs of education and training at a national level and the duration, content, and form of these vary considerably.

The EWG published a working document in 2014,[25] recommending a training program consisting of 22 modules based on the minimum knowledge and skills required and capable of delivering the training considered necessary. The content (syllabus) of each module was described in terms of learning outcomes rather than educational input, and the various elements listed in Annex V of Directive 2010/63/EU were incorporated into the relevant modules. Examples were given of assessment criteria for learning outcomes using this approach.

Several different types of module were proposed:

- core modules relative to all four functions, including legislation, ethics, basic biology, and animal care management recognition of pain and euthanasia;
- function-specific, prerequisite modules relevant to function A, such as basic biology and the conduct of minimally invasive procedures;
- function-specific, prerequisite modules relevant to function B, including ethics, the conduct of minimally invasive procedures the design of procedures and projects;
- function-specific, prerequisite modules relevant to function C, including basic biology and methods of humane killing;
- function-specific, prerequisite modules relevant to function D, including basic biology;
- additional task-specific modules, such as anesthesia, principles of surgery, advanced husbandry practices, and project evaluation; and
- other additional modules (e.g., information provision and retrieval, environmental issues and lifetime learning strategies).

These modules were seen as constituting a framework, which would assure the competence of staff caring for or using animals in procedures and facilitate the free movement of personnel within the EU. The description of the contents of each module focuses on learning outcomes, thereby providing reassurance that the trainee has achieved an appropriate level of understanding to meet the learning criteria. It is also envisaged that a period of supervision will follow to reenforce understanding and to verify that the tasks/duties/procedures are conducted to an appropriate standard; advice is provided on the implementation of such supervision.

The working document proposes principles for the assessment of learning outcomes. It stipulates that criteria for assessment should

- be objective and transparent;
- be comprehensible and clear unambiguous;
- provide reliable results;
- be economically applicable;

- have clear pass–fail criteria; and
- identify critical elements that must be understood.

After initial training, individuals are expected to maintain competence through a process of lifelong learning (CPD). Allowance is made to waive the need for initial training requirements (apart from training in national legislation in the case of persons who have previously worked in a different Member State) for individuals conducting any of the four functions A to D and who are already trained and experienced in their areas of work. The way that this training fits together as part of an educational learning process for persons working in different functions is shown in a diagram produced by the Commission.[65]

The working document addresses the education and training requirements for all persons with special responsibilities, including overall compliance, welfare and care, and education and competence, and for the designated veterinarian and persons responsible for the inspection of establishments. It should be noted that other bodies have released recommendations for the education and training of some of these persons. One of these described the roles, responsibilities, and training required for veterinarians, by a joint working group of the European Society of Laboratory Animal Veterinarians (ESLAV), the European College of Laboratory Animal Medicine (ECLAM), the UK Laboratory Animal Veterinary Association (LAVA), and the Association for European Veterinarians in Education, Research and Industry (EVERI).[66] Another by a joint committee of the Laboratory Animal Science Association (LASA), the Institute of Animal Technology (IAT), and the LAVA outlined guiding principles for Named Training and Competency Officers (NTCOs), Named Information Officers (NIO), and Home Office Liaison Contacts (HOLC—a UK-specific role) working under the UK Animals (Scientific Procedures) Act 1986.[67]

Following the recommendations developed by the EC EWG, FELASA has proposed modifications to its accreditation scheme to accommodate the changed modular structure, and some national organizations (including LASA) have developed their own guidance on the development of competence that builds on the EWG guidance by summarizing principles of good practice for supervision and assessment of competence.[68]

Lifelong Learning

An important aspect of ensuring competence of those working with animals to be used for scientific purposes is to make provision for competence to be maintained by a process of lifelong learning or CPD; this is a prerequisite for many professional, medical, and scientific disciplines but is less common for those involved in laboratory animal work. Article 23.3 of the Directive requires Member States to publish minimum standards with regard to education and training and the requirements for obtaining, *maintaining*, and demonstrating requisite competence for the various functions. Switzerland (out of the EU) has such a mandatory requirement for CPD. In 2010, FELASA proposed Guidelines for Continuing Education for all Persons Involved in the Care and Use of Animals for Scientific Purposes,[69] which has been endorsed by the EC. It proposed that compliance with the scheme would be by self-regulation ideally including oversight by the CA during inspection. The principal recommendations are as follows:

- Personnel working with animals should maintain state-of-the-art knowledge and skills.
- CPD should be available and organized in a flexible way.
- CPD should commence when a person starts working with animals and continue through the working career.
- The system should be based on the award of credits over a period of time.
- There should be a process for review and endorsement of CPD activities but up to 50% of credits can be achieved from activities recognized by the institution.
- The operation of the scheme should be reviewed.
- Joint courses should be encouraged between countries.

Training Records

As part of the initiative to ensure that training provision is recognized throughout the Member States, the EWG proposed a training record template to provide a level of reassurance regarding the skills and competence of individuals and thereby facilitate staff movement and safeguard animal welfare. It is recommended that the person assessing competency should not be the same as the one delivering training, and it is seen as important that not only the training but subsequent supervision and lifelong learning are included in the record. In practice, it is likely that most establishments will choose to maintain their records electronically, but in such cases care needs to be taken to ensure confidentiality and appropriate backup should be available in the case of IT failure. The person responsible for education and compliance at the establishment would be expected to have a thorough understanding of local measures for managing training and assuring personnel competence, including arrangements for training and supervision. This individual would also be in a position to offer advice about the maintenance of records and to advise on the training needs of persons who have recently joined a particular establishment.

Quality Assurance

To achieve transferability between Member States, there needs to be a uniform quality of training, which is accurately documented. It is very difficult to apply a "quality standard" to training, including assurances relating to content, delivery, and outcome. Within the UK, three accrediting bodies have been approved by the Home Office (the CA) to oversee mandatory training courses, and the activities of these have been adapted to take into account of the changed requirements of the Directive. The mechanism of assessment used by course providers in each of these accreditation schemes differs, although not substantially. The CA in Switzerland similarly manages education and training across establishments. FELASA has recently updated its accreditation program to meet the requirements of the Directive,[70] and it is anticipated that other accreditation schemes will be introduced in the future, so there is a pressing need to make an arrangement to minimize divergence of processes in different Member States.

Platform and Information Portal on Education and Training

The EWG recommended that an EU Platform should be established to address the modular training framework and to enable the sharing of information and communication between bodies undertaking accreditation or course approval, course providers, and Member State authorities.

It was envisaged that this Platform (www.etplas.eu) would conduct business electronically and be made up from balanced representation of the various members. Its purpose would be the establishment of conditions for approval or accreditation of courses, dissemination of information about the availability and status of courses, maintaining criteria for modules and evolving these as necessary, sharing information on standards for supervision and assessment, sharing information on mechanisms for recording and assessment, and the provision of contact details to facilitate liaison. The establishment of such a body would provide a flexible framework within which Member States could develop education and training activities, which would be broadly harmonized across the EU. It was not intended that the Platform would add bureaucracy or exercise any control over approval or accreditation processes and mechanisms; its function would be to facilitate and inform. The future of this proposal is not yet clear.

TRANSPORT

Appendix A of ETS 123[3] states that "for animals, transportation is a stressful experience which should be mitigated as far as possible. Animals should be transported in accordance with the principles of the ETS on the Protection of Animals during International Transport (ETS No 65 and ETS No 193),[71,72] having regard to the Resolution on the Acquisition and Transport of Laboratory Animals, adopted by the May 1997 Multilateral Consultation of the Partiers to Convention ETS No 123. Both sender and recipient should agree on the conditions of transport, departure and arrival times to ensure that full preparation can be made for the animals' arrival." Appendix A of ETS 123 further lists conditions of animals to be taken into account to allow for transport and requires that, in case of transport of sick or injured animals, a competent person shall confirm these animals are fit for transport. The route should be planned to ensure that the transport is carried out efficiently to minimize journey time, from loading to unloading, and to avoid delays to limit any stress and suffering of the animals.

Care is needed to ensure that animals are maintained under suitable conditions for the species and that measures are taken to minimize sudden movements, excessive noise, or vibrations during transport. ETS 123 specifies conditions to be fulfilled for transport of nonhuman primates, amphibians, fish, and reptiles.

Directive 2010/63/EU,[2] however, only consecrates one article (Article 23) to transport of animals: "Member States shall ensure that animals are transported under appropriate conditions." Annex III does mention transport, but only refers to transport container requirements.

Whereas the vast majority of animals used in research in Europe are purpose bred in facilities within Europe, very few institutions only use animals bred on-site; some animal models (some transgenic lines, nonhuman primates) are only available in non-European countries. Transport thus is a very important issue, and all steps should be taken to ensure that the time taken to transport animals between breeders and users is kept to a minimum, thus avoiding unnecessary stress.

Transportation of animals is governed by a number of international bodies and regulation. ETS 193 clearly states that animals shall be transported without delay to their place of destination. However, that condition can only be fulfilled if the various aspects of the process are clearly understood. Animal transportation not only requires careful planning but also assures that all the required conditions are fulfilled. It includes checking authorizations for transporters, knowledge of design, and construction of means of transporters, fulfilling the various conditions of the preparation for transport (planning, presence of attendants, certificates of fitness for transport), inspections needed and health certificates required, arrangements for loading and unloading, and transport practices including special provisions for transport by rail, by road, by water, in road vehicles or rail wagons on roll-on/roll-off vessels, or by air.

In the EU, all these requirements have been specifically defined in the Council Regulation (EC) 1/2005 on the protection of animals during transport and related operations.[24] At the EU level, *regulations* do not need transposition into the national legislation (directives are different and need to be transposed), and therefore, they enter into force automatically in all Member States. This Regulation 1/2005 applies to the transport of live vertebrate animals in general; thus it also applies to laboratory animals. It includes specific requirements on documentation, transporters (must be authorized), inspections, emergency measures, penalties, and training of personnel. It also includes a number of technical rules (Annex I) related to fitness of animals, means of transport, transport practices, transport vessels, watering, long journeys, space allowances, logs (Annex II), forms (Annex III), and training (Annex IV).

There are two important requirements in the Regulation 1/1005 that may specifically impact the transport of laboratory animals. One refers to the obligatory successfully completed training courses (Annex IV) that road drivers and attendants have to follow. The courses have to include technical and administrative aspects of Community legislation concerning protection of animals during transport and more particularly items such as animal physiology, animal behavior, practical aspects of handling of animals, impact of driving behavior on welfare, emergency care, and safety consideration for personnel handling animals. Many animals are routinely transported by "common" couriers, which may have problems implementing these requirements. Another important requirement is in Annex I, Chapter V, where it is stated that for several species (equidae, bovine, ovine, caprine, and porcine), when transport time exceeds the maximum for each species, animals must be unloaded, fed, and watered and be rested for at least 24 h. In the case of pigs, the maximum transport period is 24 h, which may be exceeded in the transportation of specific pathogen free (SPF) animals (i.e., minipigs). Indeed, to follow the regulation in cases of long road transportation, animals would have to be unloaded and the retention of the SPF condition of the animals would be at stake.

The International Air Transport Association (IATA) updates the Live Animals Regulations annually, and IATA member airlines and many countries agree to comply with these regulations to ensure safe and humane transport of animals. In September 2008, an interesting initiative was taken by the World Organization for Animal Health (OIE), which signed a joint IATA-OIE agreement[74] to cooperate, through formal and informal consultations, on issues of common interest including the transport of live animals. IATA and OIE have also started a dialogue on problems affecting the international transport of research animals by air and possible solutions.

Air transportation of animals not only promotes good research but also promotes good animal welfare. The IATA recently stated, "In today's modern world, carriage of live animals by air is considered the most humane and expedient method of transportation over long distances." IATA ensures that both safety and animal welfare are addressed in all regulatory issues pertaining to transportation of live animals by air.

However, over the last years, ongoing campaigns by animal rights activists have targeted the transport of live animals for biomedical research, with the objective to stop research. As a result, a significant number of European and International airline companies, together with ferry operators, no longer transport live animals for research purposes. Undoubtedly this will result in undermining animals' welfare, either because of prolonged transport or because the studies will be conducted outside of European authorities jurisdiction and scrutiny.

Several initiatives are ongoing in the biomedical community to raise awareness at the level of legislators and airlines CEOs, as thousands of research programs at institutions worldwide are dependent on the shipment of laboratory animals. In June 2015, FELASA issued a statement on Animal Transportation stating "It is essential that any transport is conducted according to law, best practice and applying humane principles, thus assuring the safety and welfare of animals."[73]

The outcome of a joint AALAS-FELASA Working Group, set up in 2014 and charged to formulate recommendations related to good transport practices of laboratory animals, will be a very welcomed reference document.

HOUSING AND ENRICHMENT

The natural histories of the species that are used for experimental purposes are fascinating. What is the historical background of our laboratory mouse (*Mus musculus*) before it ended up in the cages in our animal facilities? What are the characteristics of the species? How is that for species more recently introduced as experimental animals, like the zebra fish (*Danio rerio*)? One thing is certain: Their captive environment in the laboratory does not compare to anything in their natural habitat. Some suggest that the animals born and bred in our facilities for generations do not know any better, but the generally accepted obligation is to ensure the optimal well-being of the animals in our care. Proper housing conditions and the application of enrichment are among the basic requirements enforced through (inter)national legislation. In 1986, the CoE published ETS No. 123, the European Convention for the Protection of Vertebrate Animals used for Experimental and Other Scientific Purposes,[1] and its Appendix that contains guidelines for the accommodation and care of these animals was revised in 2005.[3] These guidelines served as the basis for the requirements on accommodation and care as stipulated in the European Directive 2010/63/EU of the European Parliament and of the Council on the Protection of Animals Used for Scientific Purposes.[2]

European Treaty Series 123

Article 5.1 deals with the animal's housing and environment: "Any animal used or intended for use in a procedure shall be provided with accommodation, an environment, at least a minimum degree of freedom of movement, food, water and care, appropriate to its health and well-being. Any restriction on the extent to which an animal can satisfy its physiological and ethological needs shall be limited as far as practicable. In the implementation of this provision, regard should be paid to the guidelines for accommodation and care of animals set out in appendix A to this convention."

European Treaty Series 123, Appendix A

In the introduction to Appendix A, it is stated that it provides guidelines for the accommodation and care of animals, based on present knowledge and good practice. The object is thus to help authorities, institutions, and individuals in their pursuit of the aims of the CoE in this matter. A general section and species-specific sections follow the introduction.

The general section provides guidelines on accommodation, housing, and care relevant to all animals used for experimental and other scientific purposes. Supplementary guidance concerning commonly used species is presented in specific sections. Where no information is included in these specific sections, the provisions of the general section apply. The species-specific sections are based on proposals made by expert groups on rodents, rabbits, dogs, cats, ferrets, nonhuman primates, farm species, minipigs, birds, amphibians, reptiles, and fish. In addition to these proposals, the expert groups also submitted background information to support their proposals, based on scientific evidence and practical experiences. This background information is the sole responsibility of the respective expert groups and is separately available. For some groups of species, namely amphibians, reptiles, and fish, these explanatory documents also provide additional information on less commonly used species not referred to in the species-specific provisions.[75]

In case behavioral or breeding problems occur or further information on specific requirements for other species is required, there is the obligation to seek advice from experts and care staff specialized in the species concerned to ensure that any particular species' needs are adequately addressed.

ETS 123 Appendix A includes advice about the design of appropriate animal facilities and provides recommendations and guidance about how the welfare provisions contained within the Convention can be met. It is important to understand that the standards of space and dimensions given represent minimum allowances. These may have to be increased in some circumstances, as environmental requirements for individual animals might vary according to, for example, species, age, physiological conditions, stocking density, and whether the animals are kept as stock, for breeding or experimental purposes, whether for long or short term.

Environmental enrichment is also an important aspect contributing to the well-being of the animals. It has been suggested to change the term "environmental enrichment" to "environmental refinement" when applied to laboratory animals since "environmental enrichment" is used in neuroscience in the context of novelty-induced stimulation and regularly changing items, primarily to measure the effects of brain neuronal plasticity. In contrast, "environmental enrichment" in laboratory animal science is coined as a term for the enhancement of welfare by means of appropriate enrichment focused on animal needs.[76] However, "environmental enrichment" will still be used here since that is the term used in legislative documents as well as in most guidelines.

The General Section of European Treaty Series 123, Appendix A

The general section starts with the functions and general design of the physical facilities. It covers not just construction and maintenance but also security (item 1.1). It deals with these issues down to the level of the holding rooms (item 1.2), the general- and special-purpose procedure rooms (item 1.3), and the service rooms (item 1.4).

The general section continues with "the environment and its control" (item 2). Topics are ventilation (item 2.1), temperature (item 2.2), humidity (item 2.3), lighting (item 2.4), noise (item 2.5), and alarm systems (item 2.6).

Next are "education and training" (item 3) and "care" (item 4). Under care, item 4.5 is on "housing and enrichment." The introduction to this item describes the preconditions set for housing, or rather "lodging," animals: "all animals should be allowed adequate space to express a wide behavioral repertoire. Animals should be socially housed wherever possible and provided with an adequately complex environment within the animal enclosure to enable them to carry out a range of normal behaviors. Restricted environments can lead to behavioral and physiological abnormalities and affect the validity of scientific data. Consideration should be given to the potential impact of the type of accommodation, and that of the environmental and social enrichment programs, on the outcome of scientific studies, in order to avoid the generation of invalid scientific data and consequential animal wastage. The housing and enrichment strategies used in breeding, supplying and user establishments should be designed to fulfil the needs of the species housed and to ensure that the animals can make the best use of the space available. Their design should also take into account the need to observe the animals with minimum disruption and to facilitate handling. Suggested minimum animal enclosure sizes and space allowances are included in the subsequent individual species sections. Unless otherwise specified, additional surface areas provided by enclosure additions, such as shelves, should be provided in addition to the recommended minimum floor areas" (item 4.5.1).

Housing is addressed in item 4.5.2: "Animals, except those, which are naturally solitary, should be socially housed in stable groups of compatible individuals. Single housing should only occur if there is justification on veterinary or welfare grounds. Single housing on experimental grounds should be determined in consultation with the animal technician and with the competent person charged with advisory duties in relation to the well-being of the animals. In such circumstances, additional resources should be targeted to the welfare and care of these animals. In such cases, the duration should be limited to the minimum period necessary and, where possible, visual, auditory, olfactory, and tactile contact should be maintained. The introduction or re-introduction of animals to established groups should be carefully monitored by adequately trained staff, to avoid problems of incompatibility and disrupted social relationships. The possibility of social housing should be promoted by purchasing compatible individuals when procuring animals of gregarious species."

Enrichment is addressed in item 4.5.3: "All animals should be provided with sufficient space of adequate complexity to allow expression of a wide range of normal behaviour. They should be given a degree of control and choice over their environment to reduce stress-induced behaviour. This may be achieved by using appropriate enrichment techniques, which extend the range of activities available to the animals and increase their coping activities. In addition to social activities, enrichment can be achieved by allowing and promoting physical exercise, foraging, manipulative, and cognitive activities, as appropriate to the species. It is advisable to allow the animals to exercise at every possible opportunity. Environmental enrichment in animal enclosures should be appropriate to the species-specific and individual needs of the animals concerned. Forms of enrichment should be adaptable so that innovation based on new understanding may be incorporated. The enrichment programme should be regularly reviewed and updated. The staff responsible for animal care should understand the natural behaviour and biology of the species, so that they can make sensible and informed choices on enrichment. They should be aware that all enrichment initiatives are not necessarily to the advantage of the animal and therefore should monitor their effects and adjust the programme as required."

And *animal enclosures* in item 4.5.4: "Animal enclosures should not be made out of materials detrimental to the health of the animals. Their design and construction should be such that no injury to the animals is caused. Unless they are disposable, they should be made from materials that will withstand cleaning and decontamination techniques. In particular, attention should be given to the design of animal enclosure floors, which should be appropriate to the species and age of the animals and be designed to facilitate the removal of excreta."

The Species-Specific Section of European Treaty Series 113, Appendix A

The species-specific section consists of 11 subsections each addressing a specific order of animals. The subsections contain considerations valid for the order, and where appropriate, additional order or species-specific provisions are included:

A. species-specific provisions for rodents;
B. species-specific provisions for rabbits;
C. species-specific provisions for cats;
D. species-specific provisions for dogs;
E. species-specific provisions for ferrets;
F. (a) species-specific provisions for nonhuman primates with additional provisions for housing and care of (b) marmosets and tamarins; (c) squirrel monkeys; (d) macaques and vervets; and (e) baboons;
G. (a) species-specific provisions for farm animals and minipigs with additional provisions for housing and care of (b) cattle; (c) sheep and goats; (d) pigs and minipigs; and (e) equines, including horses, ponies, donkeys, and mules;
H. (a) species-specific provisions for birds with additional provisions for housing and care of the (b) domestic fowl; (c) domestic turkey; (d) quail; (e) ducks and geese; (f) pigeons; and (g) zebra finch;
I. species-specific provisions for amphibians;
J. species-specific provisions for reptiles; and
K. species-specific provisions for fish.

The items addressed in the general section are filled in with more specific information in the species-specific subsections. Appendix A includes performance standards for each species. For example, acceptable temperature and humidity ranges are specified; in more general terms, requirements for light intensities and color, noise levels, and type of noise are given.

In each subsection, housing, enrichment, and care are dealt with one subheading. Emphasis is given to the enclosures, their dimensions, and flooring.

Recommended minimal enclosure sizes are listed in table format. As an example the table presenting the minimum enclosure dimensions and space allowances for mice (ETS 123, Appendix A, Table A.1) is copied here (Table 5.3).

In 2007, the FELASA published the Euroguide.[77] It is an abbreviated version of the revised Appendix A of ETS 123 published in 2005. The primary consideration was to produce a concise, "user-friendly" reference for anyone working with laboratory animals and also for those who provide services related to the animals' housing and care. It should be understood, however, that the approved Appendix A is the definitive document that should be consulted for points of certitude on the accommodation and care of animals.

TABLE 5.3 Mice: Minimum Enclosure Dimensions and Space Allowances

	Body Weight (g)	Minimum Enclosure Size (cm^2)	Floor Area per Animal (cm^2)	Minimum Enclosure Height (cm)
In stock and during procedures	Up to 20	330	60	12
	Over 20 to 25	330	70	12
	Over 25 to 30	330	80	12
	Over 30	330	100	12
Breeding		330 (for a monogamous pair (outbred/inbred) or a trio (inbred)). For each additional female plus litter, 180 cm^2 should be added		12
Stock at Breeders[a]				
Enclosure size 950 cm^2	less than 20	950	40	12
Enclosure size 1500 cm^2	less than 20	1500	30	12

[a] *Postweaned mice may be kept at these higher stocking densities, for the short period after weaning until issue, provided that the animals are housed in larger enclosures with adequate enrichment. These housing conditions should not cause any welfare deficit such as increased levels of aggression, morbidity or mortality, stereotypes and other behavioral deficits, weight loss, or other physiological or behavioral stress responses.*

Directive 2010/63/EU

Part of the guidelines of Appendix A of ETS 123 (mainly the tables with the enclosures' dimensions) has been included as requirements in Annex III—"requirements for establishments and for the care and accommodation of animals"—of Directive 2010/63/EU of the European Parliament and of the Council on the Protection of Animals Used for Scientific Purposes. Like Appendix A of ETS 123, Annex III has a general section and a species-specific section. Tables with minimum enclosure dimensions and space allowances have been copied in from Appendix A of ETS 123. However, the tables have been complemented with an additional column in which the date of January 1, 2017 is listed. This is the date by which the Member States of the EU had to ensure that the care and accommodation standards as set out in Annex III were applied (Article 33.2). The extensive performance standards as described in Appendix A of ETS 123 have not been adopted in Directive 2010/63/EU.

The final provisions of 2010/63/EU ensure adaptation of annexes to technical or scientific progress (Article 50). The EC may adopt, by means of delegated acts, modifications to that effect. This could mean that the information presented in Annex III may be updated according to technical or scientific advances while Appendix A of ETS 123 remains unchanged. The dates referred to in the species-specific section of Annex III may be subject to change but shall not be brought forward. In line with the intentions of the Convention, the European Directive 2010/63/EU in Article 33.1 states that the Member States shall, as far as the care and accommodation of animal are concerned, ensure that

1. all animals are provided with accommodation, an environment, food, water, and care which are appropriate to their health and well-being;
2. any restrictions on the extent to which an animal can satisfy its physiological and ethological needs are kept to a minimum;
3. the environmental conditions in which animals are bred, kept or used are checked daily;
4. arrangements are made to ensure that any defect or avoidable pain, suffering, distress or lasting harm discovered is eliminated as quickly as possible; and
5. animals are transported under appropriate conditions.

Member States may allow exemptions from the requirements of paragraph 33.1a or paragraph 33.2 for scientific, animal welfare, or animal health reasons.

Zebra Fish

Both Appendix A of ETS 123 and Directive 2010/63/EU include a more general section on fish without addressing the more species-specific requirements. Recommendations and guidance notes are available for further reference. Among those is the Guidance on the Housing and Care of Zebra fish.[78] Also the Guide for the Care and Use of Laboratory Animals provides species-specific information.[79] The report of a joint working group of FELASA and the European Network on Fish Biomedical Models (EuFishBioMed) on zebra fish housing, husbandry, and health monitoring recommendations will be made available through the FELASA website (www.felasa.eu).

Outside the EU, housing and enrichment requirements in Switzerland depend on whether you deal with laboratory animals or pet and wild animals. The Swiss law on animal protection[7] is intended not only for laboratory animals but also for farm, pet, and wild animals. This means that with the exception of the species classified as laboratory animals in Switzerland (mouse, rat, hamster, gerbil and guinea pig, nonhuman primates (marmoset, tamarins/Goeldi's monkey, saimiri, spider monkey, *Cercopithecus* such as vervet monkeys and macaques), and the African clawed frog) all other species that can be used for experiments have to adhere to the minimum requirements of either farm, pet, or wild animals.

For example, researchers working with tree shrew (*Tupaia belangeri*), a species of a similar size as marmosets, will need to follow the wild species housing requirements instead of the laboratory requirements. Concretely it means doubling the cage volume, from 3 to 6 m^3 to house five animals. An interesting case relates to the housing of rodents when considered as wild or pet animals instead of laboratory animals. The minimal requirements are significantly higher, a fact that animal protection organizations use to argue against the laboratory animal requirements. It can be argued, however, that animals in the laboratory context are usually of better health conditions, under stricter hygienic rules, and are taken care of in a much more professional manner thanks to professional animal keepers, technicians, and veterinarians.

Conclusion

Minimum cage sizes or space allocations to be provided per animal are generally not only based on professional judgment but are also based on other considerations as Switzerland shows. This does not necessarily mean that they are backed up by scientific evidence nor that animal well-being is guaranteed when legal requirements are met. Further research is needed to elucidate the effect of captivity and space confinement under varying experimental conditions on the welfare of the different animal species, strains, including genetically modified animals, in our care. Careful observation and evaluation of the animals' well-being on a day-to-day basis remains the most important.

HUSBANDRY AND ENVIRONMENT

There is a close relationship between ETS 123[1] and Directive 2010/63/EU[2] with regard to guidelines or requirements on husbandry and environment. Both documents have an appendix or annex dealing with these topics, among other important issues such as housing and enrichment. In the ETS, it is the Appendix A, entitled "Guidelines for Accommodation and Care Of Animals,"[3] and in the

Directive it is the Annex III, entitled "Requirements for Establishments and for the Care and Accommodation of Animals." The article in ETS 123 referring to the Appendix A is Article 5 of Part II (General Care and Accommodation), and the article in the Directive referring to Annex III is Article 33, which mandates, among other things, that Member States ensure that "all animals are provided with accommodation, an environment, food, water and care which are appropriate to their health and well-being" (Article 33.1a); and "the environmental conditions in which animals are bred, kept or used are checked daily" (Article 33.1.c). The close relationship between the husbandry and environment guidelines of ETS 123 Appendix A and the requirements of the Directive Annex III is that these last ones have been extracted from those in the Appendix A. In most cases, the contents in the Directive Annex III are only a short summary (mostly, engineering standards concerning cage dimensions, for a total of 28 pages) of those in ETS 123 Appendix A, which generally contains more performance standards (a total of 109 pages). The main difference is that ETS 123 Appendix A is more intended as a guideline (the term "should" is used), while in the Directive the more mandatory term "shall" is used. All EU Member States must comply with the minimum cage standards in the Directive since January 1, 2017.

Husbandry (sanitation, feeding, watering, and bedding) is addressed very briefly. There is little more emphasis on the environmental conditions, such as temperature, relative humidity, ventilation, lighting, and noise. As the requirements in the Directive are similar, although shorter, to the guidelines of ETS 123 Appendix A, the description below focuses first on the Directive requirements, which have been transposed into the legislation of the EU Member States, and secondly includes some of the performance standards given in ETS 123 Appendix A.

Sanitation is addressed very shortly and generically in the General Section of Annex III of Directive 2010/63/EU. First, when dealing with "Holding Rooms" (1.2), establishments are explicitly required to have a regular and efficient cleaning schedule for the rooms and to maintain satisfactory hygienic standards. The need to have walls and floors (the ETS 123 Appendix A also refers to ceilings) surfaced with materials resistant to damage by sanitation procedures is also included. Other minor references to sanitation address the existence of cleaning and washing areas large enough and the separation of the flow of clean and dirty equipment to prevent cross-contamination; the mandate that all feed hoppers, troughs, or other utensils used for feeding be regularly cleaned and, if necessary, sterilized; and the need to clean enrichment devices and sleeping areas. There are no more detailed or specific requirements on sanitation. ETS 123 Appendix A contains some additional guidelines in a performance-based approach (General Section 4.9): "A very high standard of cleanliness and order should also be maintained in holding, washing and storage rooms. Adequate routines for the cleaning, washing, decontamination and, when necessary, sterilisation of enclosures and accessories, bottles and other equipment should be established and carried out." It is also recommended that cleaning and disinfection regimens should not be detrimental to animal health or welfare, be recorded, and follow well-established procedures. The importance to address behavioral needs such as odor-marking when establishing cleaning frequency is recognized, and a performance approach is given with regard to frequency of sanitation: "decisions on frequency of cleaning should be based on the type of animal enclosure, the type of animal, the stocking density, and the ability of the ventilation system to maintain suitable air quality."

With regard to *feeding*, the General Section of Directive Annex III (subsection 3.4) requires that the form, content, and presentation of the diet meet the nutritional and behavioral needs of the animal, as well as be palatable and noncontaminated. Practices to minimize or avoid contamination or deterioration during production, transport, and storage are required. In addition to the sanitation of the feed utensils, the access of all animals to food must be ensured. ETS 123 Appendix A contains a few more guidelines addressing opportunities for foraging, information on the feedbags (production and expiry dates), and conditions for storage (cool, dark, dry, vermin-proof, and also refrigeration for perishable feed). However, there are no specifications on nutritional quality, sterilization treatments, or other practices.

Watering is briefly addressed too in subsection 3.5 of Directive Annex III. The only requirements for drinking water are that "(a) Uncontaminated drinking water shall always be available to all animals; and (b) When automatic watering systems are used, they shall be regularly checked, serviced and flushed to avoid accidents. If solid-bottomed cages are used, care shall be taken to minimise the risk of flooding." There are almost no extra guidelines in ETS 123 Appendix A with regard to drinking water either. There are no indications on specific quality parameters, frequency of changing bottles, or treatment regimens of water.

There is only one reference to *bedding*, in subsection 3.6 of Directive Annex III, where bedding materials adapted to the species are required. ETS 123 Appendix A extends more on the functions of the bedding: "to absorb urine and faeces, and thus facilitate cleaning; to allow the animal to perform certain species-specific behaviour, such as foraging, digging or burrowing; to provide a comfortable, yielding surface or secure area for sleeping; and to allow the animal to build a nest for breeding purposes." Also on the recommended characteristics for the bedding materials: "dry, absorbent, dust-free, non-toxic and free from infectious agents or vermin and other forms of contamination. Materials derived from wood that has been chemically treated or containing toxic natural substances as well as products which cannot be clearly defined and standardised should be avoided."

The European documents are not very thorough with regard to *environmental conditions* either. However, there is a positive aspect in the way this is addressed in the Directive Annex III, as it does not refer to engineering standards and offers a performance approach, although the requirements are extremely succinct. For *ventilation*, subsection 2.1 of the General Section states that "Insulation, heating and ventilation of the holding room shall ensure that the air circulation, dust levels, and gas concentrations are kept within limits that are not harmful to the animals housed." There is nothing more on the subject. On the other hand, the ETS 123 Appendix A does contain on one side some good performance standards: "The purpose of the ventilation system is to provide sufficient fresh air of an appropriate quality and to keep down the levels and spread of odours, noxious gases, dust and infectious agents of any kind. It also provides for the removal of excess heat and humidity." But this performance approach is merged with a partially engineering-based recommendation that 15–20 air changes would be normally adequate, but that a lesser number (i.e., 8–10) may be enough depending on the density of animals.

Again in a more performance rather than engineering approach, the only requirement in the Directive for *temperature and relative humidity* in the holding rooms is that they have to be adapted to the species and age groups housed, and be measured and logged on a daily basis. There are no specific ranges required. For its part, ETS 123 Appendix A recommends specific temperature ranges for adult, normal animals, and considers that "new-born, young, hairless, newly-operated, sick or injured animals will often require a much higher temperature level." It also acknowledges the importance of potential changes in animal's thermal regulation and the effect of temperature on behavior and metabolism of animals and therefore in experimental results. In practice, European institutions generally follow the recommended ranges in the species-specific section of ETS 123 Appendix A (Table 5.4). No recommended ranges for farm animals (except for pigs and minipigs) are offered, as it is explained they may be dependent on a number of factors including, for example, breed, age, caloric intake, weight, stage of lactation, and type of environment. Ranges for some avian species are also given depending on the age of the animals, and for some amphibians and reptiles. With regard to relative humidity, only a specific comment in the General Section of ETS 123 Appendix A states that "For some species, such as rats and gerbils, the relative humidity may need to be controlled within a fairly narrow range to minimise the possibility of health or welfare problems, whereas other species, such as dogs, tolerate well wide fluctuations in humidity levels." However, recommended ranges are also given for some species in the species-specific section (Table 5.4).

Lighting is limited in the Directive to a few requirements on the need of an appropriate dark/light cycle (that may be based on natural light), regular photoperiods and intensity of light (no ranges given), and the need to consider sensitivity of albino animals. These requirements are clearly extracted from the same recommendations in ETS 123 Appendix A.

The reference to *noise* is even shorter in both the Directive and ETS 123. The few requirements/recommendations focus on avoiding effects on animal welfare, and isolating as much as possible animal areas from noise, and also from that noise above the hearing range of humans, which Appendix A states is above 20 kHz.

TABLE 5.4 ETS 123 Appendix A Recommended Temperature and Relative Humidity Levels for Several Species

Species	Temperature (°C)	Relative Humidity (%)
Rodents	20–24	45–65 (35–55 for gerbils)
Rabbits	15–21	≥45
Cats/dogs	15–21	No need to control
Ferrets	15–24	No need to control
Marmosets and tamarins	23–28	40–70
Squirrel monkeys	22–26	40–70
Vervets	16–25	40–70
Macaques	21–28	40–70
Baboons	16–28	40–70

The last page of the species-specific section of the Directive Annex III (subsection 11) includes the requirements for the *fish environment*. Without detailing ranges on water parameters, it is required to control the water quality and keep parameters within acceptable levels that sustain normal activity and physiology for a given species and stage of development. The water parameters mentioned are oxygen, nitrogen compounds, pH, and salinity. Other environmental factors mentioned very briefly are temperature, lighting, and noise. Some more information on the importance and effects of these parameters (including CO_2) is given in the species-specific section of ETS 123 Appendix A.

VETERINARY CARE

Veterinarians play an essential role in ensuring that high animal welfare standards are met. Often decisions concerning animal welfare are based on expert judgment especially in areas where clear clinical signs are not always available, such as assessment of pain, distress, and suffering. Veterinarians are specifically trained in diagnosing and treating diseases in animals. As such, they are recognized by the CA and have not only legal rights but also duties and responsibilities in relation to public health and the animals they care for.

Whereas ETS 123[1] clearly identifies requirement for the care and use of animals in procedures, respecting the Five Freedoms (freedom of movement, hunger, thirst, disease and ill-health, and behavioral needs), the education and training requirements of the person responsible for ensuring/supervising the implementation of this requirement are not spelled out, and arrangements are left to each Party.

Although wording used in ETS 123 to describe the role of the veterinarian might at first glance seem limited, specific and major responsibilities in decision taking and care of animals are attributed to the veterinarian in Articles 11 and 20. Article 11 specifies that the decision whether an animal shall be kept alive or killed by a humane method at the end of a procedure and the supervision of the care that animal (when kept alive) is to receive is the responsibility of a veterinarian or other competent person. Article 20 requires user establishments to ensure that a veterinarian *or other competent person* be charged with advisory duties in relation to the well-being of animals. In 2009, more detailed European guidelines for the veterinary care were produced by FELASA, ECLAM, and ESLAV.[80]

Directive 2010/63/EU[2] confirms the essential role of the veterinarian in the protection of animals used for scientific procedures. Indeed, Article 25 states "Member States shall ensure that *each* breeder, supplier and user has a *designated veterinarian with expertise in laboratory animal medicine, or a suitably qualified expert where more appropriate*, charged with advisory duties in relation to the well-being and treatment of the animals."

This provision "where more appropriate" lays out two conditions that are explained in the Q&A website of the EC (http://ec.europa.eu/environment/chemicals/lab_animals/pdf/qa.pdf): The expert has to be suitably qualified and more appropriate to the institution than a designated veterinarian. Such could be the case when the user/breeder/supplier is dealing with more rare species, and veterinarians specialized in these specific species are not available.

The primary role of the designated veterinarian is thus to oversee the well-being and clinical care of animals used for scientific procedures. That responsibility extends to promoting and monitoring animal well-being at all times and during all phases of animal use and the animal's life, as also described in the FELASA/ECLAM/ESLAV Guidelines for the Veterinary Care of Laboratory Animals[79]:

> *Traditional veterinary care such as acute medical treatment, while still important, takes up a relatively small proportion of the laboratory animal veterinarian's time. Adequate veterinary care in laboratory animal science encompasses several aspects which all reflect the complexity of the veterinarian's role in this specialty. These include the following:*

- *All activities directly related to the animals to promote their welfare, such as during transportation, health monitoring and health management, husbandry, selection of environmental enrichment, surgery, anesthesia, analgesia and euthanasia.*
- *Scientific activities, often as a scientific collaborator and advisor in laboratory animal science.*
- *Activities related to regulatory and administrative compliance. The veterinarian must be knowledgeable about relevant legislation, including any appropriate ethical review process.*

- *Education and training of personnel and guidance of administrative staff, animal care staff and scientists to the benefit of the animals, the science and the institution.*

Article 24.1a of Directive 2010/63/EU requires establishments to identify a person(s) responsible for overseeing the welfare and care of the animals. In addition to this general function, and according to Article 26, this person is required to be a member of the AWB the tasks of which are outlined in this chapter in Oversight and Ethical Review Process section. This function is separated from the designated veterinarian; however, apparently there is no prohibition for the same person to hold both roles.

Depending on the complexity of the establishment, the species housed, and the techniques used, the designated veterinarian can be a major asset and an added value to the AWB. Indeed, in his/her role as designated veterinarian, he/she will have the best inside knowledge on the adequacy of veterinary care provided, the pain/distress/welfare of animals involved in experiments, the competence of personnel, and the type and severity of procedures used at the institution. Moreover, as the project evaluation requires expertise in areas of his/her competence and as the process should allow for complete transparency and be performed in an impartial manner (Article 38), the designated veterinarian could be the crucial link.

Some countries have therefore opted to require the designated veterinarian to be an integral part of the AWB. In other countries, institutions are free to choose that option. In any case the AWB is to seek input from the designated veterinarian or the expert referred to in Article 25, record the advice and decisions taken regarding that advice, and keep those records available for competent authorities for at least 3 years.

The designated veterinarian/AWB in their day-to-day responsibilities will evaluate/advise compliance with the Directive and implementation of the 3Rs. They could therefore also act as the contact person for the institution and have a privileged liaison with the inspectors. Procedures may be very specific to an institution; the designated veterinarian/AWB may give the necessary insight into questions raised during inspections; the same may be true for deviations with, e.g., housing requirements. Exchange of such specific information could allow for the building of a relationship of mutual trust and respect.

Harmonized inspections and establishing guidance for these is one of the aims of the EC. Veterinarians working in the field of Laboratory Animal Science and Medicine are often subject to different certification/accreditation of the animal care and use programs they supervise. Their willingness to share their experience and participate in the preparation of guidelines and training of inspectors has recently been communicated to the EC.

Compared to the US applicable policies and guidelines,[78,81] Directive 2010/63/EU does not impose explicit authority of the designated veterinarian. Although recital 30 states that "...appropriate veterinary care should be available at all times...," Article 25 gives the designated veterinarian: "advisory duties in relation to the well-being and treatment of the animals." With regard to the AWB, Article 26 indicates that "The animal welfare body shall also receive input from the designated veterinarian..." but does not require explicitly that he/she be a member. While some Member States have maintained the specific role and duties of the attending veterinarian as already existing in their legislation (e.g., the UK), other Member States have used the occasion of the transposition of the Directive into national legislation to reinforce the role of the designated veterinarian. References to the need of veterinary intervention can be found in Article 9 related to animals taken for the wild: "Any animal found, at or after capture, to be injured or in poor health shall be examined by a veterinarian..."; Article 16 on reuse, where the veterinary advice and examination is required to allow animal reuse; and Article 17: "At the end of a procedure, a decision to keep an animal alive shall be taken by a veterinarian."

The recommendations and requirements of ETS 123 and Directive 2010/63/EU do not include a detailed or organized program of veterinary care as described in other international documents. Areas related to veterinary care, such as anesthesia and analgesia, and euthanasia, are referred to in different sections of the documents. For example, Article 14 of the Directive focuses on anesthesia and analgesia to ensure that pain, suffering, and distress are kept to a minimum. Euthanasia is addressed in Article 6 and Annex IV and is discussed in other sections of this chapter.

In 2014, a joint effort was produced by an ESLAV/ECLAM/LAVA/EVERI working group, which formulated the recommendations for the roles, responsibilities, and training of the laboratory animal veterinarian and the designated veterinarian under Directive 2010/63/EU.[65]

This document provides professional guidance on the role and postgraduate training of laboratory animal veterinarians, which may be working as designated veterinarian under Directive 2010/63/EU. It also aims at advising employers, regulators, and other persons working under the Directive on the role of the designated veterinarian. According to it, the role and responsibilities of the designated veterinarian include the development, implementation, and continuing review of an adequate program for veterinary care at establishments breeding and/or using animals for scientific purposes.

CONDUCT OF EXPERIMENTAL PROCEDURES

The European Convention for the Protection of Vertebrate Animals used for Experimental and Other Scientific Purposes (ETS123)[1] set down basic standards for the conduct of scientific procedures in Part III, Articles 6 to 12. These provisions are considerably expanded in Directive 2010/63/EU[2] to include additional measures to promote refinement, and it also introduces a system of classification of the severity of procedures.

The Directive defines a procedure as any use of an animal for experimental or other scientific purposes, or educational purposes, which may cause the animal a level of pain, suffering, distress, or lasting harm equivalent to, or higher than, that caused by the introduction of a needle in accordance with good veterinary practice (Article 3(1)). The operative word here is "may" and if the outcome is uncertain, the work must be assumed to require authorization. The definition includes "any course of action intended, or liable, to result in the birth or hatching of an animal or the creation and maintenance of a genetically modified animal line in any such condition, but excludes the killing of animals solely for the use of their organs or tissues." It therefore includes techniques used for the genetic manipulation of protected animals. The breeding of colonies of GA, established lines likely to exhibit a harmful phenotype requires project authorization (which may be part of a multiple generic authorization (Article 40.4)), whereas the breeding and maintenance of colonies of GA, established lines without a likely harmful phenotype is not considered a procedure. In the case of new genetically modified animal lines, when the progeny are no longer observed or expected to experience pain, suffering, distress, or lasting harm (equivalent to, or higher than, that caused by the introduction of a needle), their further breeding, care, and husbandry do not require authorization. Further guidance on scientific work with GA animals is available in a working document on GA animals approved by the national CAs in 2013.[23]

Choice of Methods

The Directive recognizes that animals have an intrinsic value, which must be respected, and that they should always be treated as sentient beings and used only to benefit human health, animal health, or the environment (Recital 12). Emphasis is placed on meticulous application of the 3Rs, and Article 4 prohibits the use of animals if there is already available a scientifically satisfactory method or testing strategy, not entailing the use of live animals; it also requires that the number of animals used should be reduced to a minimum without compromising the objectives of the project. The use of animals for educational purposes is prohibited except for their use specifically in higher education or training for the acquisition, maintenance, or improvement of vocational skills (Article 5). Moreover, Member States are required to ensure that measures for breeding, accommodation, and care, and the methods used in procedures, eliminate or minimize any possible pain, suffering, distress, or lasting harm to the animals.

Death as the endpoint of a procedure must be avoided as far as possible and replaced by early and humane endpoints. Where it is impossible to avoid this, the procedure shall be designed so as to result in the deaths of as few animals as possible, to reduce the duration and intensity of suffering by the animal to the minimum possible and as far as possible to ensure a painless death.

Member States have raised several questions in relation to the interpretation of wording of the Directive, and in particular concern has been expressed about the implications of the 3Rs approach in relation to the development and testing of pharmaceutical substances within the EU. There is a helpful review of many of these questions and the responses to them are available on the EC's website (http://ec.europa.eu/environment/chemicals/lab_animals/pdf/qa.pdf).

For example, it is recognized that many States outside the EU require in vivo testing as part of the pharmaceutical registration process, and this may conflict with the requirement of article 13 of Directive 2010/63/EU that no procedure is to be carried out if another method or testing strategy for obtaining the result sought, but not entailing the use of a live animal, is recognized under the legislation of the EU. The EC has ruled that if such a request is received, the only situation where the animal method could be authorized would be a case where the nonanimal alternative fails to provide "the result sought"; in any case, authorization should not be granted if national legislation of the Member State concerned prohibits it. The concept of "result sought" should here be interpreted as meaning "endpoint"; obtaining information from the animal test could not be argued to be a result in itself. For example, if there is a regulatory requirement to identify only whether a particular compound is toxic and there is a replacement alternative recognized by the EU legislation, the procedure should not be authorized. However, if additional information is required (for example, the level or nature of toxicity) and there is no suitable replacement alternative, the animal method could be authorized.

At the end of a procedure, i.e., when all relevant observations have been made, animals likely to remain in moderate or severe pain, suffering, distress, or lasting harm shall be killed as soon as possible (17(1)). Any decision to keep an animal alive shall be taken by a veterinarian or other competent person and it must receive care and accommodation appropriate to its state of health.

Article 14 of Directive 2010/63/EU requires that unless it is inappropriate, procedures must be carried out under general or local anesthesia and that analgesia or other appropriate methods are used to ensure that pain, suffering, and distress are kept to a minimum. Procedures that involve serious injuries that may cause severe pain shall not be carried out without anesthesia.

In determining the appropriateness of using anesthesia, account should be taken of whether the anesthesia is considered to be more traumatic to the animal than the procedure itself and whether it is incompatible with the purpose of the procedure. Drugs such as neuromuscular blocking agents that inhibit or restrict the ability of animals to show pain without an adequate level of anesthesia or analgesia are not permitted unless a scientific justification is provided, accompanied by details of the anesthetic or analgesic regimen. If an animal is likely to suffer pain once anesthesia has worn off, it shall be treated preemptively with analgesics or other appropriate pain-relieving methods provided such treatment does not compromise the purpose of the procedure. On completion of the procedure, appropriate action shall be taken to minimize the suffering of the animal.

Classification of the Severity of Procedures

Article 15 of Directive 2010/63/EU sets out the requirement to classify the severity of procedures to ensure that Member States develop a harmonized approach to monitoring and reporting adverse effects. Before the transposition of the Directive, several Member States (including the Netherlands and the UK) together with Switzerland (in this case the Directive makes no change) had a national requirement to prospectively estimate the anticipated severity of procedures and to include this in the project proposal submitted for authorization. However, there was no consistency in the process across those countries that have such a requirement. The Netherlands had developed a six-scale classification system, while Switzerland and other countries adopted a four-scale system, and even in these there were variations in the definitions of the lowest severity category.

In the EU and Switzerland there is the obligation to assess retrospectively the severity procedures imposed to the animals for reporting in the statistics to be shared with the public. In the EU, all projects using nonhuman primates and projects involving procedures classified as "severe" shall undergo retrospective assessment.

The Directive introduced four severity categories that are applied to procedures and to take into account of the manipulations of an animal and any interventions. All procedures are classified as "nonrecovery," "mild," "moderate," or "severe" on a case-by-case basis, using assignment criteria that are set out in Annex VIII.

Member States are required to ensure that procedures are not performed if they involve severe pain, suffering, or distress that is likely to be long-lasting and cannot be ameliorated. However, a safeguard clause in Article 55.3 permits a Member State to adopt a provisional measure allowing such a procedure where there are exceptional and scientifically justifiable reasons. In such cases, Member States may nevertheless refuse to authorize the use of nonhuman primates.

Annex VIII of the Directive comprises three sections, a description of the severity categories, the criteria to be considered during the assignment process, and a number of examples illustrative of each category of severity.

Section I: Severity Categories

Nonrecovery

Procedures performed entirely under general anesthesia from which the animal does not recover consciousness shall be classified as "nonrecovery."

Mild

Procedures on animals as a result of which the animals are likely to experience short-term mild pain, suffering, or distress, as well as procedures with no significant impairment of their well-being or general condition shall be classified as "mild."

Moderate

Procedures on animals as a result of which the animals are likely to experience short-term moderate pain, suffering, or distress, or long-lasting mild pain, suffering, or distress as well as procedures that are likely to cause moderate impairment of their well-being or general condition shall be classified as "moderate."

Severe

Procedures on animals as a result of which the animals are likely to experience severe pain, suffering, or distress, or long-lasting moderate pain, suffering, or distress or procedures that are likely to cause severe impairment of the well-being or general condition of the animals shall be classified as "severe."

To comply with the requirements that death should not routinely be used as an endpoint, it is important to develop expertise for particular programs of work in predicting the clinical signs, behavioral, physiological, and biochemical changes that indicate that death is inevitable; these can be used to refine the humane endpoints as the work progresses. Staff education and training and the frequent monitoring of the condition of animals at times of expected high incidence of morbidity and death will contribute to ensuring timely application of humane endpoints. Practical information on the setting and use of humane endpoints has been developed by the 3Rs-Centre, Utrecht Life Sciences.[82]

Section II: Criteria for Assignment of Severity

Assignment of the severity category within a defined procedure must take into account any intervention or manipulation of an animal. It shall be based on the most severe effects likely to be experienced by an individual animal after applying all appropriate refinement techniques. When assigning a procedure to a particular category, the type of procedure and a number of other factors shall be taken into account. All these factors shall be considered on a case-by-case basis. Each category is based on the most severe effects likely to be experienced by an individual animal after application of all appropriate refinement techniques.

Categories are assigned on the basis of the types of procedures and other factors, which an animal will experience and shall be considered on a case-by-case basis. Factors to be considered include the type of manipulation (including handling); the nature of pain; and suffering, distress, and lasting harm resulting from the procedure, its intensity, duration, frequency, and the range of techniques employed. Consideration must also be given to the cumulative suffering within the procedure, including restricting or depriving the animal of its ability to express natural behavior resulting from shortcomings in the housing, husbandry and care standards and the applicable, the need for frequent capture, handing and restraint etc. In several countries (including the UK), consideration also needs to be given to the life experience of the animal, which may include factors relating to its acquisition and transport.

In addition to the factors above, which result from husbandry practices and conduct of the procedure, the following factors should also be taken into account on a case-by-case basis for the purposes of the final severity classification:

- type of species and genotype;
- maturity, age, and gender of the animal;
- training experience of the animal with respect to the procedure;
- if the animal is to be reused, the actual severity of the previous procedures;
- the methods used to reduce or eliminate pain, suffering, and distress, including refinement of housing, husbandry, and care conditions; and
- humane endpoints.

Section III: Examples of Different Types of Procedure

This section provides at least 10 examples of procedures in each of the mild, moderate, and severe categories on the basis of factors related to the type of the procedure. The examples contain no detail and are intended to be broadly indicative and to provide only the first indication about what classification would be the most appropriate for a certain type of procedure; all are subject to modification by details of the procedures being carried out; etc.

Only representative examples are reported here, but the full list can be seen at http://eur-lex.europa.eu/LexUriServ/LexUriServ.do?uri=OJ:L:2010:276:0033:0079:EN:PDF.

1. Mild:
   - Superficial procedures, e.g., ear and tail biopsies, nonsurgical subcutaneous implantation of minipumps and transponders
   - Breeding of GA animals, which is expected to result in a phenotype with mild effects
   - Administration of substances by subcutaneous, intramuscular, intraperitoneal routes, gavage, and intravenously via superficial blood vessels, where the substance has no more than mild impact on the animal, and the volumes are within appropriate limits for the size and species of the animal
2. Moderate:
   - Frequent application of test substances, which produce moderate clinical effects, and withdrawal of blood samples (>10% of circulating volume) in a conscious animal within a few days without volume replacement
   - Surgery under general anesthesia and appropriate analgesia, associated with postsurgical pain, suffering, or impairment of general condition, for example, thoracotomy, laparotomy, or orchidectomy, with effective wound management
   - Breeding of GA animals that are expected to result in a phenotype with moderate effects

3. Severe:
    - Toxicity testing where death is the endpoint, or fatalities are to be expected and severe pathophysiological states are induced, for example, single-dose acute toxicity testing (see OECD testing guidelines)
    - Vaccine potency testing characterized by persistent impairment of the animal's condition, progressive disease leading to death, associated with long-lasting moderate pain, distress, or suffering
    - Surgical and other interventions in animals under general anesthesia, which are expected to result in severe or persistent moderate postoperative pain, suffering, or distress or severe and persistent impairment of the general condition of the animals

An EC EWG has considered the matter of categorizing severity and published a number of recommendations in 2012, including the use of check sheets as a way of attempting to quantify the various categories.[30,31] This document emphasizes that severity assessment should be based on the overall experience of the animal from the start of the procedure to the end and points out that consideration of severity within a procedure should be a continuous process beginning with initial study design, running through the program of work, including day-to-day monitoring of animals during the project and concluding with an "actual" severity assessment on completion of the study. This sequence provides opportunities to identify further refinements for the design and conduct of future studies, which is the fundamental objective of the entire process. The document emphasized the importance of teamwork in ensuring compliance and objective, consistent monitoring.

The UK Home Office has also issued guidance on the assessment and reporting of the actual severity of regulated procedures.[83] This document indicates that the actual severity recorded applies to the animals' experience during the procedure and must reflect the highest severity of the procedure including any accumulation of lesser events and not the severity at the end of the procedure or any estimate of "average" severity. An animal used on a procedure that had been prospectively assessed as moderate might have experienced mild, moderate, or severe suffering, so the prospective assessment should be ignored at this stage. The assessment should be carried out by a competent person who is familiar with the species being assessed, the facilities available, and the procedures to which it has been subjected. In addition, only harms caused by the procedure should be taken into account, and events such as transport or determination of health status including blood sampling performed under veterinary supervision are not included.

In addition, FELASA, ESLAV, and ECLAM have jointly established a working group, which is preparing recommendations on prospective assignment of severity, management of the in-life phase using welfare assessment tools, and assessing severity retrospectively.

EUTHANASIA

Euthanasia is an ethically contentious subject that places moral responsibilities on all those breeding, caring for, and using laboratory animals. There is no ideal way of ending an animal's life, and the method adopted is a compromise between the requirements of the experiment, the facilities and expertise available, and minimizing distress to the animal.

The word "euthanasia" is derived from the Greek "eu" meaning good and "thanatos" meaning death. In this context, "good" should be interpreted as causing minimal pain, distress, and suffering to the animal.

Laboratory animals may be euthanized for various reasons. For example, when a procedure is finished and there is no justification for continuing to house them in laboratory conditions, rehoming is not permitted nor an option, when tissues need to be removed for further analysis or when adverse effects such as pain, distress, and/or suffering can no longer be justified or when animals are no longer suitable for breeding. Euthanasia is difficult to justify ethically when the animals cannot be used in procedures because of inappropriate genotypic and/or phenotypic characteristics including the wrong sex. In those and related cases it is more appropriate to use the terms "humane death" or "humane killing."[84]

In the European Convention for the Protection of Vertebrate Animals used for Experimental and Other Scientific purposes (ETS 123),[1] it is stated that an animal shall not be kept alive if it is likely to remain in lasting pain or distress (Article 11.1). The decision on whether the animal shall be kept alive or not has to be taken by a competent person or the person who is responsible for, or has performed the procedure (Article 11.2). The Appendix A to ETS 123[3] specifies that the killing shall be done by a humane method, in line with the principles set out in the European Commission Recommendations for the Euthanasia of Experimental Animals (Part 1 and Part 2).[85,86]

In a review, some key issues of euthanizing experimental animals were listed[83]:

- All those engaged in animal experimentation must recognize the ethical sensitivity and close regulatory environment in which euthanasia takes place.
- Staff carrying out euthanasia should be trained, competent, experienced, and have access to well-maintained and adequate facilities and equipment.
- When choosing the method, one should consider welfare, research needs, safety, and aesthetical and ethical issues. The method of choice should assure a rapid, stress-free loss of consciousness without subsequent recovery.
- Unnecessary distress should be avoided by using the home cage and established social groupings whenever possible, careful handling, skillful manipulation under circumstances where alarm communication between animals is not possible.
- Establishments should develop a broad strategy to minimize the number of animals euthanized, for example, by encouraging good experimental design, efficient breeding policies, and germline banking.

Some of these considerations have found their way into Directive 2010/63/EU[2] more specifically in Recital 15:

> *The use of inappropriate methods for killing an animal can cause significant pain, distress and suffering to the animal. The level of competence of the person carrying out this operation is equally important. Animals should therefore be killed only by a competent person using a method that is appropriate to the species.*

In Directive 2010/63/EU, killing is listed as an act that needs specific rules as a measure for the protection of animals used in scientific or educational purposes. Article 6 and Annex IV specifically deal with the issue of methods of killing. The animals listed are fish; amphibians; reptiles; birds; rodents; rabbits; dogs, cats, ferrets, and foxes; large mammals; and nonhuman primates. Member States shall ensure that animals are killed with minimum pain, suffering, and distress. They shall ensure that animals are killed in the establishment of a breeder, supplier, or user, by a competent person. However, in the case of a field study, an animal may be killed by a competent person outside of an establishment. The issue of competence is addressed in Article 23.2: the killing should only be done by adequately educated and trained persons. Those persons shall have received relevant instructions and shall have species-specific knowledge. Also, these persons shall be supervised in the performance of their tasks until they have demonstrated the requisite competence.

In relation to the animals covered by Annex IV of Directive 2010/63/EU, the appropriate method of killing as set out in that Annex shall be used. CAs may grant exemptions from these requirements:

- **a.** to allow the use of another method provided that, on the basis of scientific evidence, the method is considered to be at least as humane; or
- **b.** when, on the basis of scientific justification, the purpose of the procedure cannot be achieved by the use of a method of killing set out in Annex IV.

These requirements shall not apply where an animal has to be killed in emergency circumstances for animal welfare, public health, public security, animal health, or environmental reasons.

Annex IV of Directive 2010/63/EU lists the methods accepted for killing animals as anesthetic overdose, captive bolt, carbon dioxide, cervical dislocation, concussion or percussive blow to the head, decapitation, electrical stunning, inert gases, and shooting with a free bullet with appropriate rifles, guns, and ammunition (Annex IV-1). A table with acceptable methods per animal class/species is included (Annex IV-3) (Table 5.5).

The killing of animals shall be completed by one of the following methods:

1. confirmation of permanent cessation of the circulation;
2. destruction of the brain; dislocation of the neck;
3. exsanguination; or
4. confirmation of the onset of rigor mortis.

TABLE 5.5 Euthanasia Methods by Species

Animals-Remarks/ Methods	Fish	Amphibians	Reptiles	Birds	Rodents	Rabbits	Dogs, Cats, Ferrets and Foxes	Large Mammals	Nonhuman Primates
Anesthetic overdose	(1)	(1)	(1)	(1)	(1)	(1)	(1)	(1)	(1)
Captive bolt	No	No	(2)	No	No		No		No
Carbon dioxide	No	No	No		(3)	No	No	No	No
Cervical dislocation	No	No	No	(4)	(5)	(6)	No	No	No
Concussion/percussive blow to the head				(7)	(8)	(9)	(10)	No	No
Decapitation	No	No	No	(11)	(12)	No	No	No	No
Electrical stunning	(13)	(13)	No	(13)	No	(13)	(13)	(13)	No
Inert gases (Ar, N_2)	No	No	No			No	No	(14)	No
Shooting with a free bullet with appropriate rifles, guns, and ammunition	No	No	(15)	No	No	No	(16)	(15)	No

Requirements: (1) shall, where appropriate, be used with prior sedation; (2) only to be used on large reptiles; (3) only to be used in gradual fill, not to be used for fetal and neonate rodents; (4) only to be used for birds under 1 kg, birds over 250 g shall be sedated; (5) only to be used for rodents under 1 kg, rodents over 150 g shall be sedated; (6) only to be used for rabbits under 1 kg, rabbits over 150 g shall be sedated; (7) only to be used for birds under 5 kg; (8) only to be used for rodents under 1 kg; (9) only to be used for rabbits under 5 kg; (10) only to be used on neonates; (11) only to be used for birds under 250 g; (12) only to be used if other methods are not possible; (13) specialized equipment required; (14) only to be used on pigs; (15) only to be used in field conditions by experienced marksmen; (16) only to be used in field conditions by experienced marksmen when other methods are not possible.
From Annex IV of the Directive 2010/63/EU.

The table of Annex IV is the result of literature review and experts' advice. The FELASA (http://www.felasa.eu) together with the ECLAM (http://www.eclam.org), and the ESLAV (http://www.eslav.org) published guidelines on the veterinary care of laboratory animals which includes the Act of euthanasia.[79]

The power of annexes to directives is that they may be updated without having to amend the core of a Directive. Methods of killing animals are subject to active debate and ongoing research.[87] Novel insight and technological advances will lead to refinement. The inclusion of the various methods in an annex to 2010/63/EU allows for relatively easy updating of the methodology of killing animals.

EQUIPMENT AND FACILITIES

The framework for equipment and facilities follows the same pattern as for other described areas: Directive 2010/63/EU[2] incorporates as requirements ("shall") a summary of the recommendations ("should") of ETS 123[1] and more specifically of its Appendix A.[3] In summary, the legal requirements are very brief and general, with very little specifications.

In the Directive, Article 22 is specifically dedicated to "requirements for installations and equipment." This article requires in general terms that all establishments "have installations and equipment suited to the species of animals housed and, where procedures are carried out, to the performance of the procedures." Facilities and equipment are connected in this article to the 3Rs principles, because it also requires their design, construction, and functioning have to ensure procedures be performed effectively, "and aim at obtaining reliable results using the minimum number of animals and causing the minimum degree of pain, suffering, distress, or lasting harm." To achieve this, the brief Article 22 refers to the requirements for establishments and the care and accommodation of animals described in Annex III. The text of Article 22 of the Directive is almost copied literally from Article 19 of ETS 123. Moreover, in ETS 123, the guidelines for equipment and facilities are described in its Appendix A, which serves as the basis for the requirements in the Directive Annex III.

References to equipment and facilities (mainly to facilities) follow the same scheme in both documents, which address functions and general design; holding rooms; general- and special-purpose procedure rooms; and service rooms. The only two pages that include recommendations about these areas in ETS 123 Appendix A are transformed in one page of very general requirements in the General Section (subsection 1) of Directive Annex III.

The only requirements in the Directive Appendix A concerning functions and general design (subsection 1.1) are (a) "all facilities shall be constructed so as to provide an environment which takes into account the physiological and ethological needs of the species kept in them. Facilities shall also be designed and managed to prevent access by unauthorised persons and the ingress or escape of animals"; and (b) "establishments shall have an active maintenance programme to prevent and remedy any defect in buildings or equipment."

With regard to the animal holding rooms (subsection 1.2), requirements are focused on having good and regular sanitation practices, and constructions surfaces and materials that facilitate sanitation and are not detrimental to animal health. Species that are incompatible (e.g., predator and prey) have to be provided with independent environments. ETS 123 Appendix A has some additional recommendations on observation windows, drains (if existing should be covered and filtered), and the possibility of having facilities to perform minor procedures and manipulations inside the holding rooms.

The Directive Annex III (subsection 1.3) indicates that general- and special-purpose procedure rooms shall be available for situations where it is undesirable to carry out the procedures or observations in the holding rooms. Also, there must be areas for separation of sick or injured animals. This is the only potential relation in the Directive between facilities and quarantine; although the concept of quarantine is present, there is no more specific requirement for a quarantine area. This relation is clearer in ETS 123 Appendix A, where it is stated that facilities for newly acquired animals should be provided.

Requirements for service rooms (subsection 1.4) include food and bedding store rooms, which have to be vermin-proof; cleaning and washing areas, which must be large enough for the site needs and allow separation of dirty and clean materials; and surgical areas (when aseptic technique is required) with facilities for postoperative care. These requirements that take only a few sentences mimic the recommendations in ETS 123 Appendix A.

Facilities are expected to have appropriate heating and ventilation systems to ensure appropriate environment (subsection 2.1), as well as illumination (subsection 2.2) with controlled photoperiod, and alarm systems for all environmental control systems including essential services such as heating and ventilation, and illumination.

The few other references to equipment or facilities are scattered throughout the text of the Directive Annex III and ETS 123 Appendix A, such as the indication in Annex III of the need for the floors to be resistant to facilitate removal of excreta; or that protection has to be provided to equipment to avoid being damaged by animals or they do not cause damage to the animals themselves.

In summary, it is evident that the regulations and guidelines on equipment and facilities are brief and general. There are no mention of specialized equipment and facilities such as individually ventilated cages, isolators, imaging facilities, etc. There is not a description of all areas that a facility must/should have or of the different type of facilities depending on its function (breeders, users, etc.), species used, or type of research. On the other side, there is always a connection between the concept of equipment/facilities and the 3Rs principles, and more specifically with behavioral needs of animals. The proof of that is the first sentence of the General Section of both Directive Annex III and ETS 113 Appendix A. In this last one, it reads "All facilities should be so constructed as to provide a suitable environment for the species to be kept, taking into account their physiological and ethological needs." In the Directive Annex III, it reads

"All facilities shall be constructed so as to provide an environment which takes into account the physiological and ethological needs of the species kept in them." This is not only the proof of the main focus of the European framework but also evident of the influence of ETS 123 in the Directive, as has been shown in other sections of this chapter.

REFERENCES

General Framework and Principles

1. Council of Europe. European convention for the protection of vertebrate animals used for experimental and other scientific purposes. *Eur Treaty Ser* 1986:123.
2. The European Parliament and the Council of the European Union. Directive 2010/63/EU of the European Parliament and of the Council of 22 September 2010 on the protection of animals used for scientific purposes. *Off J Eur Union* 2010. L 276/33–79.
3. Council of Europe. Appendix A of the European convention for the protection of vertebrate animals used for experimental and other scientific purposes (ETS No. 123). Guidelines for accommodation and care of animals (Article 5 of the convention). Approved by the multilateral consultation. *Cons* 2006;**123**:3.
4. EEC. Council Directive of 24 November 1986 on the approximation of laws, regulations and administrative provisions of the member states regarding the protection of animals used for experimental and other scientific purposes (86/609/EEC). L358. *Off J Eur Commun* 1986:1–29.
5. Commission. Commission recommendation of 18 June 2007 on guidelines for the accommodation and care of animals used for experimental and other scientific purposes. L 197. *Off J Eur Union* 2007:1–89.
6. Swiss Federal Department of Home Affaseverityirs, Federal Food Safety and Veterinary Office. French version: 455 Loi fédérale sur la protection des animaux (LPA) du 16 décembre 2005 (Etat le 1er mai 2014). German version: Tierschutzgesetz (TschG) vom 16. Dezember 2005 (Stand am 1. Mai 2014).
7. Swiss Federal Department of Home Affairs, Federal Food Safety and Veterinary Office. French version: 455.1 Ordonnance sur la protection des animaux (OPAn) du 23 avril 2008 (Etat le 1er décembre 2015). German version: 455.1 Tierschutzverordnung (TschV) vom 23. April 2008 (Stand am 1. Dezember 2015).
8. Swiss Federal Department of Home Affairs, Federal Food Safety and Veterinary Office. French version: 455.163 Ordonnance de l'OSAV concernant la détention des animaux d'expérience, la production d'animaux génétiquement modifiés et les méthodes utilisées dans l'expérimentation animale (Ordonnance sur l'expérimentation animale) du 12 avril 2010 (Etat le 1er mai 2010).
9. Guillen J. FELASA guidelines and recommendations. *J Am Assoc Lab Anim Sci* 2012;**51**(3):311–21.
10. Pritchett-Corning KR, Prins JB, Feinstein R, Goodwin J, Nicklas W, Riley L. *J Am Assoc Lab Anim Sci* 2014;**53**(6):633–40.
11. Brønstad A, Newcomer CE, Decelle T, Everitt JI, Guillén J, Laber K. Current concepts of harm-benefit analysis of animal experiments – report from the AALAS-FELASA working group on harm-benefit analysis – part I. *Lab Anim* 2016;**50**(1S):1–20.
12. Laber K, Newcomer CE, Decelle T, Everitt JI, Guillén J, Brønstad A. Recommendations for addressing harm-benefit analysis and implementation on ethical evaluation – report from the AALAS-FELASA working group on harm-benefit analysis – part II. *Lab Anim* 2016;**50**(1S):21–42.

13. Consolidated versions of the treaty on european union and the treaty on the functioning of the european union. *Off J Eur Union* 2010;**53**:1–388.
14. OJ L 262, 27.9.1976, p. 169. Directive recast by Regulation (EC) No 1223/2009 of the European Parliament and the Council of 30 November 1009 on cosmetic products (OJ L 342, 22.12.2009, p. 59), which applies from 11 July, 2013.
15. http://ec.europa.eu/environment/chemicals/lab_animals/pdf/qa.pdf.
16. *National Competent Authorities for the implementation of Directive 2010/63/EU on the protection of animals used for scientific purposes. A working document on Inspections and Enforcement to fulfil the requirements under the Directive.* 2014. Available from: http://ec.europa.eu/environment/chemicals/lab_animals/pdf/endorsed_inspection-enforcement.pdf.
17. *National Competent Authorities for the implementation of Directive 2010/63/EU on the protection of animals used for scientific purposes. Working document on non-technical project summaries.* 2013. Available from: http://ec.europa.eu/environment/chemicals/lab_animals/pdf/Recommendations%20for%20NTS.pdf.
18. European Commission SWD (2013) 497 final. Available from: http://eur-lex.europa.eu/resource.html?uri=cellar:e99d2a56-32fc-4f60-ad69-61ead.7e377e8.0001.03/DOC_1&format=PDF.
19. Report from the Commission to the Council and the European Parliament—Fifth Report on the Statistics on the Number of Animals used for Experimental and other Scientific Purposes in the Member States of the European Union (SEC(2007)1455).
20. United States Department of Agriculture. Animal and Plant Health Inspection Service. Annual Report Animal Usage by Fiscal Year, Fiscal year: 2015. Available from: https://www.aphis.usda.gov/animal_welfare/downloads/7023/Annual-Reports-FY2015.pdf.
21. Scientific Committee on Health and Environmental risks SCHER. The need for non-human primates in biomedical research, production and testing of products and devices. Available from: http://ec.europa.eu/environment/chemicals/lab_animals/pdf/scher_o_110.pdf.
22. Directive 98/44/EC of the European Parliament and of the council on the legal protection of biotechnological interventions. *Off J Eur Commun* 1998. L 213/13.
23. *National Competent Authorities for the implementation of Directive 2010/63/EU on the protection of animals used for scientific purposes – CORRIGENDUM of 24 January 2013-working document on genetically altered animals.* 2013. Available from: http://ec.europa.eu/environment/chemicals/lab_animals/pdf/corrigendum.pdf.
24. Council Regulation (EC) No 1/2005 of 22 December, 2004 on the protection of animals during transport and related operations and amending Directives 64/432/EEC and 93/119/EC and Regulation (EC) No 1255/97.
25. National Competent Authorities for the implementation of Directive 2010/63/EU on the protection of animals used for scientific purposes. A working document on the development of a common education and training framework to fulfil the requirements under the Directive-Replacing consensus document of 18–19 September 2013. 2014. Available from: http://ec.europa.eu/environment/chemicals/lab_animals/pdf/Endorsed_E-T.pdf.
26. Smith JA, van den Broek FAR, Cantó Martorell J, Hackbarth H, Ruksenas O, Zeller W. Principles and practice in ethical review of animal experiments across Europe: summary of a report of a FELASA working group on ethical evaluation of animal experiments. *Lab Anim* 2007;**41**:143–60.
27. *National Competent Authorities for the implementation of Directive 2010/63/EU on the protection of animals used for scientific purposes. A working document on Animal Welfare Bodies and National Committees to fulfil the requirements under the Directive.* 2014. Available from: http://ec.europa.eu/environment/chemicals/lab_animals/pdf/endorsed_awb-nc.pdf.

28. Guillén J, Robinson S, Decelle T, Exner C, Fentener Van Vlissingen M. Approaches to animal research project evaluation in Europe after implementation of Directive 2010/63/EU. *Lab Anim Eur* 2015;**44**(1):23–31.
29. Olsson A, Silva S, Townend D, Sandøe P. Protecting animals and enabling research in the European Union: an overview of development and implementation of Directive 2010/63/EU. *ILAR J* 2016;**57**(3):347–57.
30. *National Competent Authorities for the implementation of Directive 2010/63/EU on the protection of animals used for scientific purposes. Working document on a severity assessment framework.* 2012. Available from: http://ec.europa.eu/environment/chemicals/lab_animals/pdf/Endorsed_Severity_Assessment.pdf.
31. *National Competent Authorities for the implementation of Directive 2010/63/EU on the protection of animals used for scientific purposes. Examples to illustrate the process of severity classification, day-to-day assessment and actual severity assessment.* 2013. Available from: http://ec.europa.eu/environment/chemicals/lab_animals/pdf/examples.pdf.
32. *National competent authorities for the implementation of Directive 2010/63/EU on the protection of animals used for scientific purposes. Working document on project evaluation and retrospective assessment.* 2013. Available from: http://ec.europa.eu/environment/chemicals/lab_animals/pdf/Recommendations%20for%20NTS.pdf.
33. https://www.gov.uk/government/uploads/system/uploads/attachment_data/file/470008/Use__Keeping_Alive_and_Re-use_Advice_Note.pdf.
34. The Council of the European Communities. Council Directive 89/391/EEC on the introduction of measures to encourage improvements in the safety and health of workers at work. *Off J Eur Commun* 1989. L 183/1.
35. 89/654/EEC concerning the minimum safety and health requirements for the workplace (first individual Directive within the meaning of article 16(1) of Directive 89/391/EEC).
36. 2009/104/EC concerning the minimum safety and health requirements for the use of work equipment by workers at work (second individual Directive within the meaning of article 16(1) of Directive 89/391/EEC) (codified version).
37. 89/656/EEC on the minimum health and safety requirements for the use by workers of personal protective equipment at the workplace (third individual Directive within the meaning of article 16(1) of Directive 89/391/EEC).
38. 90/269/EEC on the minimum health and safety requirements for the manual handling of loads where there is a risk particularly of back injury to workers (fourth individual Directive within the meaning of article 16(1) of Directive 89/391/EEC).
39. 90/270/EEC on the minimum safety and health requirements to work with display screen equipment (fifth individual Directive within the meaning of article 16(1) of Directive 87/391/EEC).
40. 2004/37/EC on the protection of workers from risks related to exposure to carcinogens or mutagens at work (sixth individual Directive within the meaning of article 16(1) of Directive 89/391/EEC).
41. 2000/54/EC on the protection of workers from risks related to exposure to biological agents at work (seventh individual Directive within the meaning of article 16(1) of Directive 89/391/EEC).
42. 92/57/EEC on the implementation of minimum safety and health requirements at temporary or mobile constructions sites (eighth individual Directive within the meaning of article 16(1) of Directive 89/391/EEC).
43. 92/58/EEC on the minimum requirements for the provision of safety and/or health signs at work (ninth individual Directive within the meaning of article 16(1) of Directive 89/391/EEC).

44. 92/85/EEC on the introduction of measures to encourage improvements in the safety and health at work of pregnant workers who have recently given birth or are breastfeeding (tenth individual Directive within the meaning of article 16(1) of Directive 89/391/EEC).
45. 92/91/EEC concerning the minimum requirements for improving the safety and health protection of workers in the mineral-extracting industries through drilling (eleventh individual Directive within the meaning of article 16(1) of Directive 89/391/EEC).
46. 92/104/EEC on the minimum requirements for improving the safety and health protection of workers in surface and underground mineral-extracting industries (twelfth individual Directive within the meaning of article 16(1) of Directive 89/391/EEC).
47. 93/103/EC concerning the minimum safety and health requirements for work on board fishing vessels (thirteenth individual Directive within the meaning of article 16(1) of Directive 89/391/EEC).
48. 98/24/EC on the protection of the health and safety of workers from the risks related to chemical agents at work (fourteenth individual Directive within the meaning of article 16(1) of Directive 89/391/EEC).
49. 1999/92/EC on minimum requirements for improving the safety and health protection of workers potentially at risk from explosive atmospheres (fifteenth individual Directive within the meaning of article 16(1) of Directive 89/391/EEC).
50. 2002/44/EC on the minimum health and safety requirements regarding the exposure of workers to the risks arising from physical agents (vibration) (sixteenth individual Directive within the meaning of article 16(1) of Directive 89/391/EEC).
51. 2003/10/EC on the minimum health and safety requirements regarding the exposure of workers to the risks arising from physical agents (noise) (seventeenth individual Directive within the meaning of article 16(1) of Directive 89/391/EEC).
52. 2004/40/EC on minimum health and safety requirements regarding the exposure of workers to the risks arising from physical agents (electromagnetic fields) (eighteenth individual Directive within the meaning of article 16(1) of Directive 89/391/EEC) and Directive 2012/11/EU amending Directive 2004/40/EC – 2004/40/EC has been repealed on the 26 June 2013 and replaced by 2013/35/EU.
53. 2006/25/EC on the minimum health and safety requirements regarding the exposure of workers to the risks arising from physical agents (artificial optical radiation) (nineteenth individual Directive within the meaning of article 16(1) of Directive 89/391/EEC).
54. The Council of the European Union. Council Directive 94/33/EC on the protection of young people at work. *Off J Eur Commun* 1994. L216/12.
55. 2013/35/EU on the minimum health and safety requirements regarding exposure of workers to the risk arising from physical agents (electromagnetic fields) (20th individual Directive within the meaning of article 16(1) of Directive 89/391/EEC) and repealing Directive 2004/40/EC.
56. *European Parliament Council recommendation of the European Parliament and of the Council of 23 April 2008 on the establishment of the European qualifications framework for lifelong learning.* 2008. Available from: http://eurlex.europa.eu/LexUriServ/LexUriServ.do?uri=OJ:C:2008:111:0001:0007:EN:PDF.
57. Council of Europe. *Explanatory report to the European convention for the protection of vertebrate animals used for experimental and other scientific purposes.* 1986. Available from: https://rm.coe.int/CoERMPublicCommonSearchServices/DisplayDCTMContent?documentId=09000016800ca438.
58. Council of Europe. *Resolution on education and training of persons working with laboratory animals.* 1993. Available from: http://www.coe.int/t/e/legal_affairs/legal_co-operation/biological_safety_and_use_of_animals/laboratory_animals/Res%20training.asp.

59. Wilson MS, Berge E, Maess J, Mahouy G, Natoff I, Nevalainen T, van Zutphen LFM, Zaninelli P. FELASA recommendations on the education and training of persons working with laboratory animals: categories A and C. Reports of the federation of european laboratory animal science associations working group on education accepted by the FELASA board of management. *Lab Anim* 1995;**29**:121–31.
60. Nevalainen T, Berge E, Gallix P, Jilge B, Melloni E, Thomann P, Waynforth B, van Zutphen LFM. FELASA guidelines for education of specialists in laboratory animal science (category D). Report of the federation of European laboratory animal science associations working group on education of specialists (category D) accepted by the FELASA board of management. *Lab Anim* 1999;**33**:1–15.
61. Nevalainen T, Dontas I, Forslid A, Howard BR, Klusa V, Käsermann HP, Melloni E, Nebendahl K, Stafleu FR, Vergara P, Verstegen J. FELASA recommendations for the education and training of persons carrying out animal experiments. Report of the federation of European laboratory animal science associations working group on education of persons carrying out animal experiments (category B) accepted by the FELASA board of management. *Lab Anim* 2000;**34**:229–35.
62. Nevalainen T, Blom HJM, Guaitani A, Hardy P, Howard BR, Vergara P. Recommendations for accreditation of laboratory animal science education and training. *Lab Anim* 2002;**36**: 373–7.
63. Weiss J, Bukelskiene V, Chambrier P, Ferrari M, van der Meulen M, Moreno M, Sigg H, Yates N. FELASA recommendations for the education and training of laboratory animal technicians: category a. Report of the federation of European laboratory animal science associations working group on education of animal technicians (category A) accepted by the FELASA board of management. *Lab Anim* 2010;**44**:163–9.
64. FELASA Working Group on Accreditation of Laboratory Animal Science Education and Training: Nevalainen (Convenor) T, Blom HJM, Guaitani A, Hardy P, Howard BR, Vergara P. *FELASA recommendations for the accreditation of laboratory animal science education and training*. 2002. Available from: http://www.felasa.eu/recommendations/recommendation/recommendations-for-accreditation-of-laboratory-animal-science-education-an/.
65. http://ec.europa.eu/environment/chemicals/lab_animals/pdf/posters/Poster_Education_Training_vertical.pdf.
66. Poirier GM, Bergmann C, Denais-Lalieve DG, Dontas IA, Dudoignon N, Ehall H, Fentener van Vlissingen JM, Fornasier M, Kalman R, Hansen A, Schueller S, Vergara P, Weilenmann R, Wilson J, Degryse AD. ESLAV/ECLAM/LAVA/EVERI recommendations for the roles, responsibilities and training of the laboratory animal veterinarian and the designated veterinarian under Directive 2010/63/EU. *Lab Anim* 2015;**49**(2):89–99.
67. http://media.wix.com/ugd/a30180_ae0f6d0429594706a60b5326b07594c8.pdf.
68. http://www.lasa.co.uk/wp-content/uploads/2016/09/LASA_supervision_and_competence_2016.pdf.
69. 2010. http://www.felasa.eu/recommendations/guidelines/guidelines-for-continuing-education-for-persons-involved-in-animal-experime/.
70. FELASA Accreditation Board for Education and Training (E&T) Board: Gyger M, Berdoy M, Dontas I, Howard B, Nevalainen T, Santos AI, Vergara P. Available from: http://www.felasa.eu/recommendations/recommendation/recommendations-for-the-accreditation-of-education-and-training-courses-in/; 2015.
71. Council of Europe. European convention for the protection of animals during international transport. *Eur Treaty Ser* 1968;**65**. Available from: http://conventions.coe.int/Treaty/en/Treaties/Html/065.htm.

72. Council of Europe. European convention for the protection of animals during international transport (revised). *Eur Treaty Ser* 2003;**193**. Available from: http://conventions.coe.int/Treaty/en/Treaties/Html/193.htm.
73. http://www.oie.int/en/about-us/key-texts/cooperation-agreements/agreement-with-the-international-air-transport-association-lata/.
74. FELASA. *FELASA statement on animal transportation*. 2015. Available from: http://www.felasa.eu/announcements/felasa-statement-on-animal-transportation.
75. http://wayback.archive-it.org/1365/20090214135802/http:/www.coe.int/t/e/legal_affairs/legal_co-operation/biological_safety%2C_use_of_animals/Laboratory_animals/Revision%20of%20Appendix%20A.asp.
76. Baumans V, Augustsson H, Perretta G. Animal needs and environmental refinement. In: Howard BR, Nevalainen T, Perretta G, editors. *The cost manual of laboratory animal care and use: Refinement, reduction, and research*. USA: CRC-Press, Taylor & Francis Group; 2011. [Chapter 4].
77. Forbes D, Blom H, Kostomitsopoulos N, Moore G, Perretta G. *Euroguide on the accommodation and care of animals used for experimental and other scientific purposes*. London, UK: Published by Royal Society of Medicine Press Limited; 2007. ISBN: 978-1-85315-751-6.
78. Reed B, Jennings M. *Guidance on the housing and care of Zebrafish (Danio rerio)*. Southwater, Horsham, UK: RSPCA; 2010.
79. Committee for the Update on the Guide for the Care and Use of Laboratory Animals. *Guide for the care and use of laboratory animals*. 8th ed. Washington, DC, USA: Institute for Laboratory. Animal Research, National Research Council. The National Academies Press; 2011.
80. Voipio HM, Baneux P, Gomez de Segura IA, Hau J, Wolfensohn S. Guidelines for the veterinary care of laboratory animals: report of the FELASA/ECLAM/ESLAV joint working group on veterinary care. *Lab Anim* January 2008;**42**(1):1–11.
81. Unites States Public Health Service. *Public health service policy on humane care and use of laboratory animals*. 2002. Available from: http://grants.nih.gov/grants/olaw/references/phspolicylabanimals.pdf.
82. https://www.humane-endpoints.info/en.
83. https://www.gov.uk/government/uploads/system/uploads/attachment_data/file/276014/NotesActualSeverityReporting.pdf.
84. Antunes L. Euthanasia. In: Howard B, Nevalainen T, Perretta G, editors. *Chapter 8 of the cost manual of laboratory animal care and use: Refinement, reduction and research*. Boca Raton, FL, USA: CRC Press, Taylor and Francis Group; 2011.
85. Close B, Banister K, Baumans V, Bernoth E-M, Bromage N, Bunyan J, et al. Recommendations for euthanasia of experimental animals: Part 1. *Lab Anim* 1996;**30**:293–316.
86. Close B, Banister K, Baumans V, Bernoth E-M, Bromage N, Bunyan J, et al. Recommendations for euthanasia of experimental animals: Part 2. *Lab Anim* 1997;**31**:1–32.
87. Hawkins P, Prescott MJ, Carbone L, Dennison N, Johnson C, Makowska IJ, Marquadt N, Readman G, Weary DM, Golledge HD. A good death? Report of the second newcastle meeting on laboratory animal euthanasia. *Animals* 2016;**6**(9).

Chapter 6

Israeli Legislation and Regulation on the Use of Animals in Biological and Medical Research

Rony Kalman[1], Alon Harmelin[2], Ehud Ziv[3], Yacov Fischer[4]
[1]*Hebrew University, Jerusalem, Israel;* [2]*The Weizmann Institute of Science, Rehovot, Israel;* [3]*Hadassah University Hospital, Jerusalem, Israel;* [4]*Israeli Council for Animal Experimentation, Ministry of Health, Jerusalem, Israel*

GENERAL FRAMEWORK

Animal care and use is fully regulated in Israel since 1994 when the Animal Welfare Law—Animal Experiments[1] (Law) was issued. Until then, research animal care and use was carried out on a voluntary basis according to the National Research Council *Guide for the care and Use of Laboratory animals*[2] (*Guide*) standards. The Law incorporates the internationally recognized principles of the Three Rs. At the heart of the Law is the establishment of a National Council (Council) (Fig. 6.1). The 23-member Council represents different stake holders including government, professional medical and veterinary organizations, industries, animal welfare groups, and the national academy of sciences. The Council operates in the Ministry of Health, and all members are appointed for a 4-year term by the Minister of Health. The Law regards the Council as a professional body, and the chairman is appointed by the Minister of Health from Council members representing the National Academy of Sciences (usually from life science and medicine disciplines). The Council has all legal authority regarding animal care and use and is entrusted with regulation and supervision duties.

The essence of the Law is the establishment of an Institutional Animal Care and Use Program (Program) in each and every institution that wishes to carry out research with animals. Each institution must be authorized by the Council prior to the introduction of animals to the facility or obtaining research permits by its researchers. Council authorization indicates the existence of a Program including an appropriate infrastructure, the appointment of an attending veterinarian (AV), the existence of an Institutional Animal Care and Use Committee (IACUC), and a written Program manual. Animals cannot be introduced to the

Laboratory Animals. http://dx.doi.org/10.1016/B978-0-12-849880-4.00006-4
Copyright © 2018 Elsevier Inc. All rights reserved.

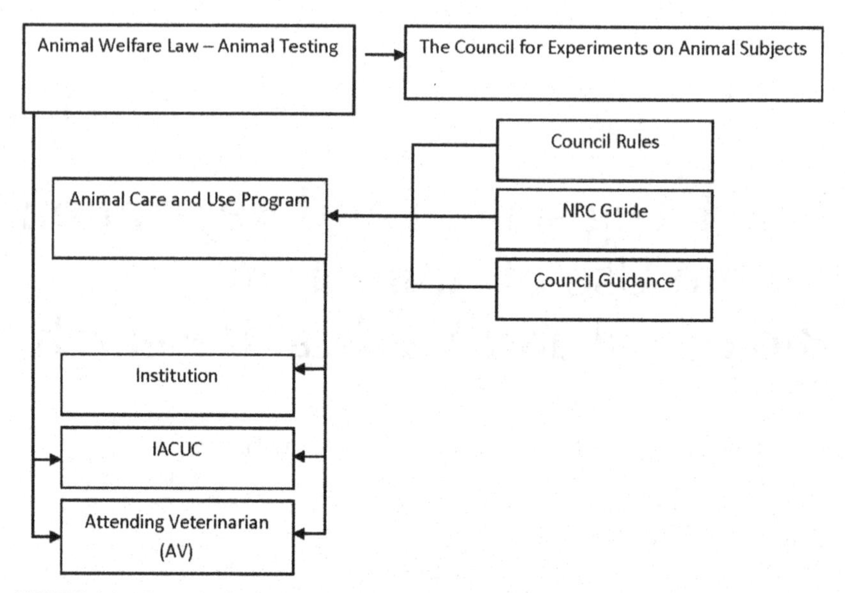

FIGURE 6.1 General schema of the legislative and supervising bodies involved in the regulation of the use of animals for biological and medical research in Israel.

animal facilities without a valid Animal Experiment Permit (Permit), which can be obtained from a designated (by the Council) committee. Not all IACUCs are authorized by the Council to issue Permits. Therefore Permits can be obtained either from IACUCs (in institutions where the IACUC is authorized to issue Permits) or from the National Permit Committee (NPC, composed from Council members). The NPC currently issues 25% of all approved permit applications nationwide, while IACUCs issue the remaining 75%. Permit applications (Applications) can be submitted only by principal investigators (PIs) who have passed appropriate training for the purpose of the Law and have been certified by the institutional manager to conduct such research in the institution. Training performed in one institution is valid in all the institutions, while the individual certification issued in a certain institute is valid only in that institute. Thus, PIs who transfer to a different institution will need to be recertified but will not need to retrain.

Permit-issuing activity in an IACUC can be established only following an authorization by the Council. Each IACUC that issues Permits is independent in its deliberations, and the Applications are submitted on a Council-approved National Standard Form, which applies for the entire country and includes all aspects of animal care and use.

Applications can be submitted by certified PIs. One of the parameters to be considered for PIs certification is the training they have received. The Council places a particular importance to the issue of training, and each of the training courses is approved by the Council. All courses are built on a unified national framework of two levels: a basic training and a specific supplementary course for each animal species not covered by the basic course. PIs have to be qualified for the particular animal species they intend to research, as well as all research stuff that is intended to perform the procedures.

The institutional Program is built on the commonly accepted three arms system: the institution, the IACUC, and the AV. Council guidance on the establishment of the Program comprises Council rules, Council guidance, and the instructions of the latest edition of the *Guide*, which has an official status in Israel. The Council requires all institutions to fulfill all the functions and responsibilities of these three arms (as stated by the *Guide*), irrespective of the institution having a Permit-issuing activity authorized.

The Law places special importance on the role of the AV, whose post is a legal requirement, which also gives the AV authority and duties. The AV is responsible for the health and welfare of animals used in research and entrusted among other things with supervisory duties at the institution. Moreover, the AV has to be specifically authorized for this duty by the National Director of Veterinary Services and Animal Health—Ministry of Agriculture or be a diplomat in Laboratory Animal Medicine (LAM), and the institution has to employ them according to a specific Council guidance.[3] The guidance indicates that by 2020 all the AVs should be diplomats in LAM (this decision is under review due to the relatively long periods of time and the difficulty in obtaining the diplomat status). Furthermore, Council guidance on the employment of AVs includes a minimal part of position or working hours that the AV must be employed,[3] depending on animal species diversity, span of activity, and sites location in the institution. As a minimum, a 25% FTE (full-time employment) veterinarian position (LAM diplomat) is required for an institution with minimal activity. That percentage can get up to 200% FTE veterinarian position (LAM diplomat) for an institution with extensive activity (as defined in the decision).

Institutions are required to report to the Council on all approved permits, on the actual progress of research projects, and on the general veterinary conditions. Institutions are also supervised by Council members and officers of the Council (veterinary and scientific) who have the legal authority to inspect and implement all needed measures, and even to suspend permits or order the removal of animals from an experiment when severe irregularities are identified.

THE PRINCIPLES

The principles by which animal care and use are regulated in Israel are based on the Three Rs and are explicitly written in paragraphs 8 (regarding reduction and refinement) and 9 (regarding replacement) of the Law[1]:

- The number of animals in any experiment shall be limited to the minimal number required for the performance of the experiment.

- Experiments on animals shall be carried out with close attention to minimizing the pain and suffering caused to them.
- No permit shall be granted for the conduct of any experiment on animals, if the objective of the experiment can be attained by reasonable alternate means.

In addition to this the Animal Welfare Rules—on animal experiments[4] (Rules) published in 2001 give a legal status to the *Guide*[2] in Israel, which also follows the same principles.

The above legal instructions find their expression in each and every activity that follows the legislative act. The Rules state the necessary elements that need to be part of the training process for prospective animal users. These include detailed rules on the use of animals and the contribution to scientific research, the fundamental principles of minimizing the number of animals in an experiment, the search for alternatives to experiments with animals, the moderation of experimental methods, the implementation of legal principles, and the functioning of the ethical system in the country and in the institution.

SCOPE/APPLICABILITY

The Law[1] applies to the use of live vertebrate animals only (invertebrates and fetuses are not covered) for research purposes to achieve one of the following goals:

1. the advancement of health and of medicine and the prevention of suffering;
2. the advancement of scientific research;
3. testing or producing materials or objects; and
4. education and teaching (this is further discussed under "Education and Teaching With Animals" section);

Achieving the above goals must be done, only after careful deliberations and taking into account numerous limitations and restrictions, including the Three Rs.

The bodies responsible for implementing the Law in Israel are the Council and IACUCs empowered by the Council.

All the activities related to animal experimentation in Israel are under the Council's authority. The Council grants authorization to institutes to perform experiments and to establish IACUCs and training courses. Additionally, the Council receives reports from the IACUCs, from Council's members,[5] and from the Council's supervising veterinarians, as to the activities in these institutes.

An important role in the Council's activity is performed by the Council's supervising veterinarians whose role and authorities are specified in paragraph 15 of the Law.

The Council oversees all research activities with animals in Israel including but not limited to academic institutes—public or private, commercial institutions, public health institutions, and governmental agencies.

Requirements for institutional permits are described in paragraph 12 of the Law. Council authorization confirms the existence of a Program, the existence of appropriate infrastructure, the appointment of an AV, and the appointment of an IACUC.

NONHUMAN PRIMATES

According to the Law,[1] the care and use of nonhuman primates (NHPs) must follow the *Guide*[2] recommendations. In addition, since the large majority of NHP experiments in Israel are electrophysiological studies, and because these studies are long termed (usually 3–4 years) and involve specific animal welfare aspects, the Council published special guidelines on harmonization of electrophysiological studies with NHP in Israeli institutions.[6] The purpose of these guidelines is to update users on available means of conducting the research and to harmonize the way electrophysiological studies with NHP are performed.

These guidelines emphasize the importance of specific aspects related to studies with NHP such as veterinary care, behavioral studies, environmental enrichment and behavioral welfare, task learning by the animals, requirements for personnel, and the training of the research team at all levels. All the points listed above must be addressed in each new application for an experiment with NHP before approval is granted. To meet the goals of this harmonization, each institute performing this type of research is expected to acquire knowledge on all the subjects listed below, including the help of behavioral research groups, either locally or abroad. One of the innovations of the harmonization guidelines is the requirement for institutions that conduct research with NHP to establish an institutional animal welfare position.

The guidelines apply to the following:

1. The number of animals allocated to the project. When examining the number of NHP used in the research project, the IACUC must evaluate, among other things, if the minimum possible number of animals is allocated to the study while considering:
   a. The study objectives, taking into account the validity and significance of the potential findings.
   b. Potential loss of animals due to health problems and/or failure of training.
   c. The social needs of NHP, who should be housed in direct proximity with at least one more individual when they are not involved in experimental procedures.
2. Personnel. The IACUC must condition the approval of an application to the constant presence of appropriate personnel, in quality and quantity, to ensure animal suffering is minimized and animals are maintained in the study for the minimum period of time necessary to achieve the study objectives. The IACUC must ensure that:
   a. There are enough personnel to fulfill the tasks outlined in the planed protocol of the experiment at all times.

   b. The absence of the PI or other researchers will be backed up by suitable replacements so that the experiment will not be unnecessarily delayed.
   c. Any invasive and/or surgical procedure that has to be carried out under anesthesia requires the presence of a trained veterinarian, and a person competent in performing the procedure.
   d. Institutions must have an institutional behavioral animal welfare position (person) that addresses the NHP social and behavioral needs, establishes learning paradigms and the environmental enrichment program, as well as addresses the teaching and training of the research team on the use of NHP.
3. NHP training. The IACUC must verify that the research team has the appropriate knowledge regarding the training of NHP (or uses an external expert for this purpose) and that a training program is in place. This training program will address the following aspects:
   a. The researchers must have thorough acquaintance with the NHP.
   b. To allow NHP to appropriately perform the required experimental procedures, nonaggressive training methods shall be preferentially selected using positive reinforcement paradigms. Negative reinforcement paradigms of aggressive measures should be avoided as much as possible.
   c. If the training paradigms are not successful in the preselected NHP, the paradigms shall be revised and adapted.
   d. All members of the research team (including supporting staff) must be updated on the training objectives.
4. Education. IACUC approval will be granted only after verification that all members of the research team are qualified to work with NHP and have followed and acquired the relevant theoretical and practical training and knowledge. This training must be implemented under the guidance of a professional and experienced instructor in the field.
5. NHP not in experiment. The IACUC needs to verify the existence of an environmental enrichment program appropriate to the NHP daily routines and needs, especially for periods prior to the experiment, between the various parts of the study, and until the NHP leaves the facility on the completion of the study.

GENETICALLY ALTERED ANIMALS

The Israeli legislation does not have special regulations and requirements regarding genetically modified animals (GMAs). As GMAs are becoming an increasingly large proportion of the animals used in research in Israel, the IACUC and the AV have the mutual responsibility of assessing and monitoring the protocol and the experiments.

GMAs require special genetic management procedures including maintenance of detailed pedigree records and genetic monitoring, which is the responsibility of the scientist. However, generation of animals with multiple genetic alterations often involves crossing different GMA lines and can lead to unexpected phenotypes. Any new phenotypes that negatively affect the animal's well-being should be reported to the IACUC and managed in a manner that ensures the animals' health and well-being. The issue is further addressed in the mandatory section of Humane End Points of the research permit application form.

A growing concern in the use of GMA, particularly mice, is the maintenance and preservations of colonies. Institutions are encouraged to maximize the use of cryopreservation for GMA strains whenever possible, to promote reduction (Three Rs) and make the research more efficient.

INSTITUTIONAL AND DESIGNATED PERSONNEL RESPONSIBILITIES

Several designated institutional bodies and professionals are specifically mentioned in the Law[1] and in Council's regulations. These have a major role in approving and supervising animal experimentation, to guarantee their adherence to the legal requirements.

The Institutional Animal Care and Use Committee

All institutions are required to establish an IACUC. IACUCs are required to fulfill all responsibilities as stated in the *Guide*,[2] with exception to issue Permits, which requires Council authorization. According to the Law, Permits can be obtained from an authorized committee, either from one acting as a Council-authorized IACUC or from the NPC formed by members of the Council. Issuing Permits by an IACUC is subject to Council authorization. Institutions that are not authorized to conduct such activity within their own IACUC will receive Permits from the NPC. Currently, only about half of authorized institutions are authorized to issue Permits by their IACUC.

The IACUC's roles as outlined in the Law (paragraph 14) indicate that the IACUC in authorized institutes must include at least three members: a veterinarian (according to the definition of veterinarian in the Law), a researcher from the life sciences or medicine disciplines, and a member who is not from the disciplines of the life sciences or medicine. The legal obligation to include at least one public member to represent general community interests in the proper care and use of animals in IACUCs with authorized permit committee activity is currently under review. Nevertheless, in practice, most IACUCs in Israel do include already a public member.

The IACUC's activities include the following:

- The examination and evaluation of all the animal-related activities in the institution.
- The approval of protocol applications and amendments to already approved protocols.

- The control of all the animal facilities and the research activities.
- Reporting of the regular activities and any irregularities to the Council. Regular activities include the following:
  - Online reporting of all the approved applications.
  - Annual reporting on the project progress for each approved protocol.
  - Annual summary of institutional animals used by species.
  - Details of the program and facilities, including personal details of the AV.
- Approval of the syllabus of the institutional courses for personal accreditation of researchers, students, and technicians.
- Supervision of research permits, the progress of research projects, and the allocation of animals and their fate.
- Processing complaints from institutional personnel and the public about concerns regarding the use of animals in research.

The Attending Veterinarian

To become the AV (or as described in the Law "the veterinarian") a veterinarian must, as stated in the Law, be either a diplomat in LAM (by the Israeli college or any other recognized college) or personally accredited by the National Director of Veterinary Services and Animal Health—Ministry of Agriculture.

In September 2008 the Council instructed the minimum time dedication for AVs at institutions.[3] The percentage of dedicated time depends on the size and complexity of the animal research at the specific institution:

- Facilities housing over 1000 rodent cages and also other species different than rodents—at least two full-time job veterinarians.
- Facilities housing over 1000 rodent cages and without other species—at least one full-time job veterinarian.
- Facilities housing 500–1000 rodent cages and also other species different than rodents—at least one full-time job veterinarian.
- Facilities housing 500–1000 rodent cages and without other species—at least 50% part-time veterinarian (at least three weekly visits).
- Facilities housing less than 500 rodent cages and without other species—at least 25% part-time veterinarian (at least two weekly visits).
- Facilities housing up to 50 "large animals" (nonrodent species that are of a higher phylogenetic level)—at least 50% job veterinarian (at least three weekly visits).

In addition, the Council passed a resolution[3] that as of 2015 all newly appointed AVs will be diplomats in LAM and that by 2020 all AVs will be diplomats in LAM.

The Law (paragraph 12) describes in general terms the roles of the AV.

In practice, AVs in Israel have many additional roles as indicated in the *Guide*, and therefore the AVs authorities are broad, according to the ones written in the *Guide*. These roles include membership in IACUCs;

conduction of courses for new researchers, students, and technicians; advice on technical methods (i.e., analgesia and anesthesia) in research models; purchase and breeding of animals, and defining environmental enrichment programs.

OVERSIGHT AND ETHICAL REVIEW PROCESS

The IACUC has a dual responsibility. It is the highest authority regarding the care and use of animals in research and at the same time it is the delegate of the Council and as such responsible for the application of the Law[1] and National Regulations at institutional level.

Animal experiment permit applications are reviewed either by the IACUC (in institutions where the IACUC has been granted this authority by the Council) or by the NPC (in institutions where the IACUC has not been given the authority to issue Permits). IACUCs conduct the review process according to the principles of the *Guide*,[2] and applications are submitted on a standard national Council approved form, which requires the following information:

- Scientific abstract of the research
- Duration of the experiment
- Personnel authorized to perform the experiment
- The purpose of the experiment (according to the Law)
- Justification for the use of animals in the experiment
- Number of animals required and justification for the species, strain and number
- Description of the procedures to be performed, including a clear time course for the experiment in individual animals
- Anesthetics and analgesics to be used
- Follow up on the animals and humane end points
- Fate of the animals at the end of the experiment and the method of euthanasia if performed
- Severity level (these are described in "Conduct of Experimental Procedures" section)
- Signed declaration of the PI acknowledging awareness to the Law

There are several exceptions where specific guidance and preruling are required from the relevant authorities before IACUC deliberations can proceed.

- Applications that propose the use of LD50 protocols have to gain specific Council authorization for each suggested use before IACUC deliberations occur.
- Animal testing for cosmetic substances is prohibited in Israel, unless it is intended for medicinal purposes. Therefore any application with such intent has to be cleared by the regulatory authority for pharmaceuticals as one that satisfies the regulatory definition for cosmetics for medicinal purposes, which is then communicated to the IACUC by the Council.

- In applications intended for the purpose of teaching, the Council requires that the evaluation of alternative methods will not be the sole responsibility of the PI and asks for additional prereview by the institutional teaching committee.[7] The approval by this committee regarding the justification of using animals for the purpose of teaching is required before the IACUC commences its independent deliberations. Additionally, permits for teaching purposes are limited in time for 1 year only.

Each Permit is submitted on IACUC approval to the Council, where they have a legally protected status for the purpose of supervision and statistical requirements of the Council. The representatives of the Council (professional civil servants) will conduct a review of the permits. In some cases this will lead to further exchange between the Council and the IACUC for discussion and clarifications, which may result in protocol modifications. In extreme cases, where common ground cannot be found, or when there is a significant deviation from requirements, the Council representatives may instruct to cancel the permit.

The IACUC oversight does not terminate in the approval of applications. The IACUC is the highest authority at institutional level concerning the use of animals for research. As such, the IACUC supervises and controls all the animal-related activities such as animal facilities, research laboratories, and training programs.

VETERINARY CARE

According to the Law,[1] the *Guide*[2] standards (pages 106–124) apply with regard to the veterinary care of laboratory animals in Israel.

The AV roles according to the Law are indicated in the *Guide*, which broadens the AV's authority, making him the final voice concerning the animal's health and fate. The AV is required to oversee animal health and welfare; provide medical care; prevent diseases; reduce the suffering of animals before, during, and after the experiments; euthanize animals where necessary; and instruct staff members on these issues.

The AV must be a diplomat in LAM or a veterinarian authorized by the National Director of Veterinary Services and Animal Health—Ministry of Agriculture for the purposes of the Law. The Council, in collaboration with the National Director of Veterinary Services and Animal Health—Ministry of Agriculture, set accreditation criteria for veterinarians who are not diplomats in LAM and decided that this authorization shall be limited for a period not exceeding 3 years from the first approval. After such period, renewal applications shall be submitted to the accreditation committee.

REUSE

The reuse of animals is a delicate issue that demonstrates the possible tension between the principles of reduction and refinement. The Council has given its attention to this point in the Rules[4] Article 6 indicating that a reuse can be considered by the IACUC only if one of the following conditions exist:

- The Committee is convinced that the experiment already carried out involved little suffering, although no association to a defined severity level is required.
- The animal will be anesthetized at the beginning of the additional experiment and will be euthanized at its conclusion, without regaining consciousness in its course.

The reuse may be approved only following a specific request of the PI to the IACUC. Such request should include information regarding both the experiment that was already performed as well as the requested additional experiment. All these approvals must be specifically marked by the IACUC and receive special attention by the Council.

In addition, the Rules clearly state that reuse of animals cannot be justified for economic reasons. There must be a scientific (or other) justification.

SETTING FREE/REHOMING

Research applications in Israel must include a paragraph describing the fate of the animal at the end of the experiment. Although there is no legal requirement to set free or rehome animals, it is the common understanding that farm animals, dogs, cats, and NHP should not be euthanized at the end of the experiment, unless there is a scientific justification. If there is the need for euthanasia, then the process needs to be carried out according to the American Veterinary Medical Association (AVMA) guidelines on euthanasia.[8] The Council supports the rehoming of cats and dogs, the return to nature of appropriate wild animals, the return to the herd of farm animals, and the rehabilitation of NHP. The rehoming of farm animals and NHPs and setting free of wild animals are significant, and annual statistics are available.

The setting free of NHP is currently a well-established solution. Research institutes can set free NHP at the end of the experiment to a specific NHP shelter. This shelter is supported and controlled by the Ministry of Environmental Protection. According to the Council's data, over 83% of NHP that ended their participation in experiments in the past 5 years were set free.

OCCUPATIONAL HEALTH AND SAFETY

Legally required institutional safety units monitor the health of faculty, staff, students, and any other persons on campus exposed to hazardous materials and situations, in accordance with Israeli laws and institutions' policies. All new employees must undergo a health evaluation, and persons whose work involves hazardous substances or procedures are required to undergo special medical tests at regular intervals, in accordance with the Ministry of Labor guidelines.[9] Special medical examinations are required if a significant risk of excessive exposure to a hazard exists and for work-related health problem. The supervisor and the institutional safety officer, in agreement with an Industrial-Hygiene Medical Doctor, decide on the specific medical examinations. Any person suspecting adverse health effects from a substance or procedure in the work environment is entitled to consult the Doctor, who will decide if further testing is required. Noncompliance with the medical examinations is a disciplinary transgression and can bring disciplinary punishment.

The institutional safety unit maintains health records and monitors the medical surveillance program. The individuals are summoned to the examinations by the personnel department or the safety officer, who coordinate the process with the health authorities. The process may include any measurements taken to monitor exposures such as questionnaires, medical consultations, or physical examinations. The Industrial-Hygiene Medical Doctor enters the results to the workers' medical file and deals with any cases requiring a second test, medical treatment, or change in work conditions. The employee may ask to see and receive a copy of their personal records.

Experiments with special agents such as biological hazards and chemicals require biohazard specialized facilities or other adequate methods of containment.

The following are principles of biosafety, which are used in Israel[10]:

The term containment describes safe methods or properly managing infectious agents in the laboratory environment where they are being handled or maintained. The purpose of containment is to reduce or eliminate the exposure to potentially hazardous agents of laboratory workers, other persons, and the outside environment.

Primary containment, which is the protection of personnel and the immediate laboratory environment from exposure to infectious agents, is provided by both good microbiological technique and the use of appropriate safety equipment. The use of vaccines may provide an increased level of personal protection.

Secondary containment, which is the protection of the environment external to the laboratory from exposure to infectious materials, is provided by a combination of facility design and operational practices.

The three elements of containment include the following:

- laboratory practice and technique;
- safety equipment; and
- facility design.

The risk assessment of the work to be done with a specific agent will determine the appropriate combination of these elements.

All regulations, publications, and guidelines of the Israeli Institute for Occupational Safety and Hygiene, Ministry of Labor are online.[11]

EDUCATION, TRAINING, AND COMPETENCE OF PERSONNEL

The issue of education and training has been extensively dealt with by the Law,[1] Rules,[4] and Council decisions.[8,12] Instructions detail both the requirement on how to obtain a personal license as well as the relevant content of the accreditation courses to be taken for this purpose.

To conduct any experiment on animals, the PI and all the personnel who handle the animals (students, technicians, and animal caretakers) must obtain prior to the experiment a personal license as indicted in the Law (paragraph 11).

In the 2001 Rules[4] (paragraph 9) the Council issued the general outline for such accreditation courses including topics to be covered, minimal content, as well as the qualification and supervision methods to be taken by the Council to ensure the academic level of such courses.

Nevertheless, as courses are not available continuously, and to enable students to start working under the supervision of qualified researchers even before they take the courses, the Council, in its 2001 Rules, enabled to obtain a one-time temporary-supervised permit for up to 6 months. Such temporary permits are supervised by the IACUC and must be followed by a permanent qualification to continue and conduct experiments with animals. The training should include at least the following:

- Explanations on the correct handling of laboratory animals, the use of alternatives, and minimization of the suffering of animals used in experiments.
- A practical demonstration of the experiment on animals in which the employee will engage.
- The conduct of at least two of those experiments on animals by a qualified researcher in the employee's presence.

In its circular dated September 2006,[12] the Council issued further instructions regarding qualification courses. The circular indicated that qualification courses must open with a basic theoretical part common to all the species in research. Participating in this basic part is a prerequisite to participate in the second practical part. The practical part is specific for each species, and each person will be qualified only to work with the species he or she has passed in the practical part. The courses in Israel which syllabus has been approved by the Council are listed in the circular dated September 2006.[12]

EDUCATION AND TEACHING WITH ANIMALS

Animals can be used for two main teaching purposes: personnel qualification and academic education. These purposes were regulated by the Council in its circular dated October 2007.[7] The purposes are as follows:

- Qualification courses—(see Education, Training, and Competence of Personnel section)
- Courses for teaching students known biological principles. The regulations for such courses include the following:
  - Each course has to be approved by the institutional academic teaching committee, which has to evaluate the need to use live animals—and not alternatives—for that specific course, prior to being finally reviewed by the IACUC.

- The approval has to be renewed every academic year.
- The use of animals in the course must be known to the students during their registration.

TRANSPORT

Transportation of animals is governed by a number of Israeli regulatory agencies and international bodies. The International Air Transport Association (IATA) updates the Live Animals Regulations annually, and IATA member airlines and many countries agree to comply with these regulations to ensure the safe and humane transport of animals by air.[13] The Ministry of Agriculture with its veterinary services regulates the permit and vaccination programs if needed for the importation and exportation of animals; enforces regulations to prevent the introduction, transmission, or spread of communicable diseases; and regulates the importation of any animal or animal product capable of carrying a zoonotic disease. The Ministry of Environmental Protection regulates importation/exportation of wild vertebrate and invertebrate animals and their tissues through the Wild Life Protection Law[14] and the national authority of the Convention on International Trade in Endangered Species of Wild Fauna and Flora.[15] The National Parks Service, operating under the Minister of Environmental Protection, regulates movement of the Convention on International Trade in Endangered Species of Wild Fauna and Flora (CITES)-listed species that are captive bred, including NHP. Institutions should contact appropriate authorities to ensure compliance with any relevant statutes and other animal transportation requirements that must be met for animals to cross international boundaries.

Animal transportation may be intrainstitutional, interinstitutional, or between a commercial or noncommercial source and a research facility. For wildlife, transportation may occur between the capture site and field holding facilities.

According to the Law[1] and Rules, the *Guide*[2] standards (pages 107–109) apply to transport of animals to and from Israel.

HOUSING AND ENRICHMENT

According to the Law[1] and Rules, the *Guide*[2] standards (pages 42–88) apply to the housing and environmental enrichment of laboratory animals in Israel. This includes the minimum cage sizes for all species except rabbits. For rabbits, the Council made an exception and decided they should be housed according to the European Convention ETS 123 Appendix A.[16]

HUSBANDRY AND ENVIRONMENT

All the parameters related to husbandry and environment are equivalent to those indicated in the current edition of the *Guide*.[2] The 2001 Rules[4] specifically indicate that the leading document in Israel regarding environment and animal management is the *Guide* (the current updated version applies).

CONDUCT OF EXPERIMENTAL PROCEDURES

The need for refining experimental procedures is one of the basic principles of conducting animal procedures. It is referred in the Law[1] and in all the Council's publications.

In 2006 the Council instructed that all research projects must be assigned a severity level ranging from 1 to 5.[8] The severity level must be indicated in the research protocol and reported to the Council. The instructions were updated by the Council in 2015.[8]

The five severity levels are as follows:

1. Collection of organs from animals that did not go under any experimental procedure and were euthanized using a method acceptable for organ collection.
2. Experiments that cause slight temporary discomfort or stress. Examples: intravenous, intramuscular, intraperitoneal, and subcutaneous injections; behavioral experiments that do not cause stress (but not including water maze or predator experiments); infliction of slight pain that the animal can avoid; withdrawing blood from peripheral vessels up to the quantity of blood that does not require anesthesia (10% of the total blood volume or 1% of the body weight); feeding experiments that do not cause clinical manifestations; tail tip sampling.
3. Experiments that cause slight stress or short-term pain. Such experiments should not cause significant changes to the animal's appearance, to physiologic parameters such as heart rate or respiratory rate, or to social behavior. During and after such experiments, animals should not exhibit signs of self-injury, anorexia, dehydration, anxiety, excessive recumbence, vocalization, overaggressiveness, or tendency for isolation. Examples: nonsurvival major surgery; cannulation; minor survival surgery; blood withdrawal under anesthesia from the retroorbital sinus or from the heart; restraint for short periods; water or food restriction for less than 12 h a day.
4. Experiments that cause medium pain or distress that is alleviated by analgesics. Examples: major survival surgeries where animals receive analgesics; local nonmetastatic tumors where animals receive analgesics; restraining animals for over 60 min; restriction of water or food for over 12 h during the animal's activity phase; significant changes in environmental parameters (temperature, lighting); procedures that cause sensory or motor damage or severe and constant anatomical and/or physiological changes; use of complete Freund's adjuvant.
5. Experiments that cause severe and lasting pain or distress that are not alleviated by analgesics. Examples are metastatic tumors or experiments in which the endpoint is death. In all such experiments the researcher is requested to justify why analgesics are not used.

EUTHANASIA

Both the Law[1] and the Rules[4] indicate the need to include in the qualification courses the adequate euthanasia methods and materials for each species.

The adequate methods and materials required by the Israeli regulation are the ones detailed in the AVMA guidelines on euthanasia.[17]

EQUIPMENT AND FACILITIES

According to the Law[1] and Rules,[4] the *Guide*[2] standards (pages 133–151) apply with regard to equipment and facilities for laboratory animals in Israel.

AUTHORIZATION AND INSPECTION

The Council must approve facilities in each institution before animals can be housed and before any experiment can take place. The Council's supervising veterinarian must approve that the facility of an institution complies with the *Guide*[2] standards. Once the facility is approved, periodic monitoring by the Council veterinarians is performed. Any change in the facility infrastructure and any addition in the variety of animal species must be approved before it can be used. The Council's supervising veterinarians and the Council's scientific advisor may enter at any time any institution and any animal facility, on the condition that they take the necessary steps to prevent interference with an experiment. In addition, any Council member can enter any institution after informing the Council chair of their intent to visit, and after receiving the Chair's approval. Such person may read any document to check whether the provisions of the Law are complied with. Reports to the Council on the state of the veterinary program at the institutions are submitted on a quarterly basis, and a more detailed report is submitted annually. The Annual Report provides description of any changes in the facilities and the veterinary program in the institution and should report among other things special problem or irregularities that occurred.

REFERENCES

1. *Animal Welfare Law – prevention of cruelty to animals law (experiments on animals) 5754.* 1994. Available from: http://www.weizmann.ac.il/vet/Law.htm.
2. National Research Council. *Guide for the care and use of laboratory animals.* Washington: National Academies Press; 2011. Available from: http://www.aaalac.org/resources/Guide_2011.pdf.
3. *Veterinarian role, council circular.* September 14, 2008. Available from: http://www.health.gov.il/Services/Committee/animax/Documents/D1_2008.pdf.
4. *Animal Welfare Rules – prevention of cruelty to animals rules (experiments on animals) 5761.* 2001. Available from: http://www.weizmann.ac.il/vet/Law.htm.
5. *Visits by council members, council circular.* December 26, 2004. Available from: http://www.health.gov.il/Services/Committee/animax/Documents/D2_2004.pdf.

6. *Council's report on harmonization of electrophysiological studies with NHP in Israeli Institutions*. February 2011. p. 19–21. Available from: http://www.health.gov.il/PublicationsFiles/electro_monkey_feb2011.pdf.
7. *Use of animals for teaching, council circular*. October 23, 2007. Available from: http://www.health.gov.il/Services/Committee/animax/Documents/D2_2007.pdf.
8. *Classification of severity levels in animal experimentation, council circular*. December 29, 2015. Available from: http://www.health.gov.il/Services/Committee/animax/Documents/D2_2015.pdf.
9. *Ministry of labor guidelines for workers with hazardous materials*. 2011. https://www.nevo.co.il/law_html/Law01/500_552.htm.
10. *Israeli regulations for safe work with biohazard agents*. 2001. Available from: http://www.osh.org.il/uploadfiles/takanot_gehut_maabadot.pdf.
11. Israel Institute for Occupational Safety and Hygiene Website. Available from: http://osh.org.il/site/english_main.html.
12. *Format of qualification courses, council circular*. September 4, 2006. Available from: http://www.health.gov.il/Services/Committee/animax/Documents/D2_2006.pdf.
13. IATA Live Animals Regulation. Available from: http://www.iata.org/publications/store/Pages/live-animals-regulation.aspx.
14. *Israeli Wild Life Protection Law 5715*. 1955. Available from: http://www.sviva.gov.il/English/Legislation/Documents/Wildlife%20Protection%20Laws%20and%20Regulations/WildlifeProtectionLaw1955.pdf.
15. CITES – convention on international trade in endangered species of wild fauna and flora. Available from: https://www.cites.org/eng/disc/text.php.
16. Appendix A of the European convention ETS 123. Available from: http://www.coe.int/en/web/conventions/full-list/-/conventions/treaty/123.
17. AVMA. *AVMA guidelines on euthanasia*. 2013. Available from: https://www.avma.org/KB/Policies/Documents/euthanasia.pdf.

Chapter 7

Animal Experimentation in Africa: Legislation and Guidelines

Prospects for Continued Progress

Amanda R. Hau[1], Faisal A. Guhad[2], Margaret Cooper[3], Idle O. Farah[4], Bert J. Mohr[5], Ouajdi Souilem[6], Jann Hau[7]

[1]Administrative Court, Malmo, Sweden; [2]Jigjiga Export Slaughter House (JESH), Jigjiga, Ethiopia; [3]The University of Kent, Canterbury, United Kingdom; [4]Institute of Primate Research, Karen, Kenya; [5]University of Cape Town, Cape Town, South Africa; [6]National School of Veterinary Medicine, Sidi Thabet, Tunisia; [7]University of Copenhguen, Copenhaguen, Denmark

INTRODUCTION

Animal experimentation is common in Africa, a region that according to Kimwele, Matheka, and Ferdowsian accords little priority on animal protection in comparison to economic and social development.[1] Animal welfare has become a major issue worldwide, and it is of interest to examine the state of the situation in a region of the world where animal experimentation is common, but little or no legislation exits.

The aim of this chapter is to investigate and review shortfalls in the legislation and guidelines concerning animal experimentation in Africa. The investigation can be recapped into three main areas of research:

- What is the present situation?
- What are the main issues?
- Prospects for continued progress?

The method used in the process of writing this chapter was a study of various documents found on the subject: Acts of Parliament, ordinances and proclamations, national standards, university guidelines, and scientific papers. Kimwele, Matheka, and Ferdowsian's paper, "A Kenyan perspective on the use of animals in science education and scientific research in Africa and prospects for improvement," and the Office International des Epizooties (OIE) also known as the

Laboratory Animals. http://dx.doi.org/10.1016/B978-0-12-849880-4.00007-6
Copyright © 2018 Elsevier Inc. All rights reserved.

World Organisation for Animal Health (WOAH) summaries of animal welfare in OIE member countries and territories in the Southern African Development Community (SADC) region have been very helpful. Summaries and statements from nongovernmental organizations (NGOs) and other organizations were also reviewed. Correspondence with experts in the field has also been helpful throughout the process.

After our initial expectation of studying legislation and guidelines from as many African countries as possible, it soon became apparent that due to the scarceness of legislation in most countries, the main focus of the chapter would be on Kenya, Tanzania, and South Africa. These countries have produced the four most comprehensive pieces of legislation, namely, The Animal Welfare Act of Tanzania,[2] the Prevention of Cruelty to Animals Act, Cap 360 of the Laws of Kenya,[3] and the three main laws regulating the welfare of animals in South Africa, the Animals Protection Act 71 of 1962,[4] the Performing Animals Protection Act 24 of 1935,[5] and South African Medical Research Council Act, No. 58, 1991.[6]

THE PRESENT SITUATION

Empirical data regarding the existence or adequacy of national or institutional policies and guidelines on the use of animals in research in Africa are limited. However, the little evidence available indicates that most African countries lack relevant legislation and guidelines.[7]

General Framework

Masiga and Munyua[8] report "currently in West, East and Southern Africa, animal welfare issues are addressed under several Acts of Parliament, ordinances and proclamations including acts with the following titles (the titles of these acts vary from country to country, so some of the alternative names have been provided in brackets):

1. the Prevention of Cruelty to Animals Act (Animal Welfare Act; Animals Protection Act);
2. the Branding Act (Animal Identification Act);
3. the Animal Diseases and Pest Control Act;
4. the Wildlife Conservation and Management Act (Wildlife Act);
5. the Meat Control Act (Meat Inspection Act, Veterinary Public Health Act);
6. the Veterinary Surgeons Act (Veterinary and Para-Veterinary Professions Act);
7. the Animal (Control of Experiments) Act; and
8. the Scientific Experiments on Animals Act."[9]

However, specific legislation concerning the welfare of animals *used in research and education* has in this study only been found in a limited number of cases, namely, in the animal welfare and related laws of Kenya, Tanzania, the Seychelles, Zimbabwe, and South Africa.

Britain's Protection of Animals Act, which was introduced in 1911, has been the model for most of the legislation in African countries that were under British influence. Some nations such as Kenya, South Africa, and Tanzania have enacted their own animal welfare legislation,[2–6] but even in these cases, implementation and enforcement mechanisms have according to Kimwele, Matheka, and Ferdowsian been largely ineffective.[1]

Kenya's Prevention of Cruelty to Animals Act[3,10] is a comprehensive piece of legislation regarding animal welfare issues in general. The Animal Welfare Act of Tanzania[2] also contains comprehensive legislation.

South Africa has a comprehensive governance framework for the care and use of animals for scientific purposes, although the legislation is fragmented and requires review.[11] The two main laws regulating the welfare of animals in South Africa, the Animals Protection Act 71 of 1962[4] and the Performing Animals Protection Act 24 of 1935[5] contain comprehensive legislation. The use of animals in research explicitly is covered in South African legislation in the rules promulgated under the Veterinary and Para-Veterinary Professions Act 19 of 1982[12] (i.e., Rule 32 Minimum Standards for Research Animal Facilities),[13] the National Health Act 61 of 2003,[14] and the South African Medical Research Council Act No. 58 of 1991.[6] The South African government has initiated the drafting of a new Animal Welfare Act. This Act is expected to specifically include animal use for scientific purposes, thus addressing both the fragmented nature of current legislation and the incomplete coverage of certain aspects of animal care and use for scientific purposes.

The Mauritius Animal Welfare Act of 2013 contains a dedicated section on animal experiments.[15]

Other pieces of applicable legislation are Uganda's Animals (Prevention of Cruelty) Act, Chapter 39 (consolidated 2000)[16]; Nigeria's Criminal Code 1990 Chapter 50[17]; Niger Law No. 2004-048 of June 30, 2004, Article 3 and 5[18]; and Zambia's Prevention of Cruelty to Animals Act.[19] There is also a draft animal welfare law in Morocco titled Projet de Loi 122-2 produced in 2013 where Article 14 concerns maltreatment and Article 15 addresses experiments on animals.[20]

Angola, DR Congo, Madagascar, and Mozambique are reported to have literally nothing significant on animal welfare legislation.[21]

Comprehensive and modern guidelines, very much in agreement with the National Research Council (NRC) Guide for the Care and Use of Laboratory Animals[22] and the European Directive 2010/63/EU on the Protection of Animals used for Scientific Purposes,[23] have now been developed in Kenya. These National Guidelines for the Care and Use of Animals in Research and Education in Kenya[24] (produced by the Consortium for National Health Research) provide excellent guidance on all aspects of an animal care and use program and compliance with these guidelines would assure standards similar to those of quality programs in the United States (US) and Europe.

Modern standards for animal research, based on international recommendations, are encapsulated in the South African National Standard (SANS 10386:2008) for the Care and Use of Animals for Scientific Purposes.[25] The SANS 10386 is currently being revised based on the 2013 Australian Code for the Care and Use of Animals for Scientific Purposes[26] and Directive 2010/63/EU[23] on the protection of animals used for scientific purposes. The purpose of the SANS 10386 is to ensure the ethical and humane care and use of all animals involved in scientific activities. The document establishes minimum standards for animal care and use and defines the responsibilities of researchers, teachers, and institutions that use animals, including the functioning of Institutional Animal Ethics Committees (i.e., animal care and use committees).

The authors have reviewed all relevant documents accessible, but it is important to bear in mind the possibility that guidelines and other regulations may exist in some African countries that are not available in the public domain.[1]

The Principles

There are certain principles generally accepted by the scientific community that are followed when using animals in research. One of the most well known and internationally accepted is the principle of the Three Rs (3Rs) (Replacement, Reduction, and Refinement); a guiding principle for the use of research animals in many countries, and now also explicitly endorsed in the text of the European Directive.[23]

The 3Rs were first described by Russell and Burch in 1959[27] and represent a practical strategy for researchers to apply when considering experiments with laboratory animals. *Replacement*, the first "R", refers to striving to avoid using nonhuman animals completely if possible. Replacement also encompasses replacing animals high on the phylogenetic scale with animals lower on the phylogenetic scale, although this may be challenging in terms of defining differential levels of sentience. *Refinement* concerns the objective to use research methods that alleviate or minimize potential pain, distress, suffering, or lasting harm of the animals still used. The enhancement of animal well-being is also included. *Reduction* refers to strategies to allow researchers to obtain valid scientific information from the use of as few animals as possible, or to obtain more information from the same number of animals.[27]

The Five Freedoms[28] are another set of guiding principles used by, among others, animal researchers and endorsed by numerous international guidelines on animal welfare. They have been the basis of much legislation in many parts of the world, and were originally developed in 1965 in a UK Government report on livestock husbandry. The freedoms entail the freedom from hunger, thirst, and malnutrition; freedom from fear and distress; freedom from physical discomfort; freedom from pain, injury, and disease; and freedom to express normal patterns of behavior. The freedoms define ideal states rather than standards for acceptable animal welfare, but form a logical and comprehensive framework for study of welfare in almost all situations.[28]

The Universal Declaration on Animal Welfare[29] is another source of guiding principles in the animal welfare field. It is a draft agreement built on feedback from the United Nations (UN) member states, international organizations, and NGOs. If endorsed by the United Nations, the declaration would encompass a nonbinding set of animal welfare principles affirming, among other things, that animals are sentient beings and acknowledging that their welfare must be respected. The declaration would provide a solid uniform basis for member states to improve already existing animal welfare legislation or introduce new national animal welfare legislation built on the agreed upon principles.[29]

Principles of animal welfare can be found both implicitly and explicitly in various relevant documents in African legislation. The most obvious reference to fundamental animal welfare principles can be found in the Animal Welfare Act, 2008, of the Laws of Tanzania.[2] According to Part I section 4 (a) in aforesaid Act, "every person exercising powers under, applying or interpreting this Act shall have regard to-ensuring that animals are cared for according to their universally adopted five freedoms." Additional fundamental principles of animal welfare to be regarded are listed in Part I section 4 (b), "recognizing that- and animal is a sentient being; animal welfare is an important aspect of any developed society which reflects the degree of moral and cultural maturity of that society; animal welfare enhances livestock productivity and that a human being has a moral obligation to care, respect and protect an animal." The concept of the 3Rs is mentioned explicitly in the Act, Part I section 4 (c), "ensuring that animal experimentation promotes the reduction in number of animals used, refinement of experimental methods and replacement of animal with non-animal use techniques."

Principles regarding the ethics of animal experimentation are also found in a less formal form, namely, in the guidelines and codes of conduct governing the use of animals in research and teaching at universities in Africa. A telling example is the guide presented by the Animal Ethics Committee of the University of KwaZulu-Natal, South Africa.[30] The guide describes the concept of the 3Rs and also lists principles concerning the humane treatment of animals by referring to principles described by the *Canadian Council on Animal Care*.[31] The University of Cape Town similarly recognizes animal sentience as the cornerstone for the moral obligation to respect animals and to diligently apply the 3Rs in animal care and use.[32]

Although there are no legislative requirements for ethical review prior to animal experimentation, in Kenya, initiatives to establish ethical review systems have been made by leading research institutions like the National Museums of Kenya's Institute of Primate Research (IPR) in Nairobi (Institute of Primate Research, Kenya).[33]

Nonhuman Primates and Genetically Altered Animals

The use of nonhuman primates in scientific research is a controversial subject and therefore a part of animal welfare that often entails additional attention. Internationally, the use of nonhuman primates and genetically altered animals in research regularly require special considerations and regulations. In the European Union, for example, several member states such as Austria, United Kingdom, the Netherlands, and Sweden introduced special regulations and/or some form of prohibition on the use of nonhuman primates in research, before the revised directive banned experimentation on apes and restricted the use of monkeys to studies of debilitating human diseases.[23] In Africa, legislation covering the subjects is generally lacking.

Nonhuman primate models are used in several research institutions in Africa. For instance, the National Museums of Kenya's Institute of Primate Research (IPR), which is located in Nairobi, Kenya, is a well-known research center focused on primatology. Animal models, in particular vervet monkeys and baboons, are used in studies of human reproductive disorders and tropical infectious diseases, and many scientific publications have been published from IPR.

Research ethics and animal welfare issues are taken into account at IPR, and according to their webpage "Ethical and animal welfare concerns form a strong component of the Department's animal husbandry and research activities." Before any experimental procedure is carried out, review committees assess all the proposals for scientific merit and welfare concerns.[33] Nyika reports that to ensure that animals are used humanely, the institute has an Animal Resources Department equipped with modern diagnostic, therapeutic and surgical facilities, which takes care of all animals used in research at the institution.[7] The Institute of Primate Research is currently leading a multi-institutional effort to develop comprehensive guidelines for laboratory animals used in research and education. IPR has partnered with the National Council for Science and Technology, Kenya, and the Consortium for National Health Research (a local funding agency for Health Research) for this task.

Research using nonhuman primates as animal models is also conducted in other African countries including Ethiopia where vervet monkeys have been used.[34]

The SANS 10386:2008 for the Care and Use of Animals for Scientific Purposes includes specific recommendations of the care and management of nonhuman primates.[25] The University of Cape Town has a specific policy to protect the interests of nonhuman primates used for scientific purposes.[35]

Special considerations and restrictions regarding genetically altered animals in the reviewed documents are also limited. However, according to the Tanzanian Animal Welfare Act, a special permit is necessary for practices involving "the alteration of the genetic material of an animal in a manner which infringes the natural barriers of sexual reproduction and of recombination; the application of biotechnical technology or the administration of substances to animals or embryos that modify genetic make up of animals."[36]

In South Africa, there is a specific Act that addresses genetically altered animals, namely, the Genetically Modified Organisms Act, No. 15 of 1997.[37] The objective of the Act is to provide for measures to promote the responsible development, production, use, and application of genetically modified organisms. This is implemented by the establishment of a council to advise the Minister on all aspects concerning genetically altered organisms, and to ensure that all activities with regard to the development, production, use, and application of genetically modified organisms are performed in accordance to the Act.[37]

Transport, Housing, and Enrichment

The transport of laboratory animals, housing of laboratory animals, and the enrichment of the animal's environment are areas of animal welfare typically covered in legislation in other parts of the world.

Among all the reviewed documents, the Animal Welfare Act of Tanzania[36] alone covers transportation of animals, and even in the Tanzanian Act, it is not apparent if the transportation of laboratory animals specifically is included. The Act deals with the transportation of animals, animals unfit for transportation, revocation of movement permits, and certificates and documents required for transporting animals.[38]

There have been attempts to implement draft legislation concerning animal welfare in North Africa. An interesting example of this was worked on by the Society for the Protection of Animals and Nature (SPANA) in Morocco in 1997. The draft legislation did not relate to laboratory animals, but working animals—more specifically tourist carriage and cart horses. However, the considerations made when putting the framework for draft legislation together concerning animal welfare, in general, provide an interesting example as to how legislation regarding laboratory animals explicitly, could similarly be designed. The framework of SPANA's draft included sections such as enhancing animal protection, working instructions, formulation of provisions/consultations, enforcement, and ethical/advisory committees for animal protection.[39]

Although actual legislation in the area of animal transportation may be limited, guidelines within the various institutions can be found. One example is the guide to the care and use of animals in research and teaching used at the University of KwaZulu-Natal in South Africa, a guide containing standards to be met when transporting animals.[30]

The OIE has also set standards for the transport of animals.[40]

Husbandry and environment is covered only on a general level in the reviewed documents and no specific requirements for cage sizes, ventilation, temperature, etc., can be found. Confining an animal in a manner which causes that animal unnecessary suffering and starving, and underfeeding or denying water to an animal are considered offenses according to the animal welfare acts of both Kenya and South Africa. In this context, it is interesting to consider if the aforesaid regulations have their basis in the principle of the Five Freedoms, mentioned in The Principles section.

The SANS 10386:2008 for the Care and Use of Animals for Scientific Purposes includes species-specific minimum standards for the housing and enrichment of animals used for scientific purposes.[25]

Euthanasia

Euthanasia is another example of a frequently specifically regulated procedure. Euthanasia is the act of inducing humane death in an animal by a method that induces rapid loss of consciousness and death with a minimum of pain, discomfort, or distress.

Material and information concerning euthanasia of research animals in the various African countries was very difficult to come by. No approved methods are stated, and only general guiding principles regarding slaughter such as "A person who, whether in a slaughterhouse or abattoir or in any other place, and whether for human consumption or not, slaughters and animal-in such a manner as to cause it more suffering than is necessary (…) shall be guilty of an offence (…)"[41] can be found in the reviewed documents. However, in Kenya, reported methods of euthanasia have according to Kimwele, Matheka, and Ferdowsian raised serious welfare concerns. For example, the use of chloroform has been criticized based on concerns about human and animal welfare. Various methods of euthanasia have been reported being used in Kenyan institutions. Inhalant chloroform and diethylether were most commonly used, indicating a need to update guidelines introducing more modern and internationally recommended methods. Other reported methods included cervical dislocation, barbiturate overdose, and slaughter.[1]

In South Africa, the euthanasia of experimental animals is a regulated procedure, as per requirements of the Veterinary and Para-Veterinary Professions Act[12] and the Rules made thereunder.[13,42] As for other clinical or experimental procedures on animals, persons who perform euthanasia need to be confirmed competent by a registered veterinarian, to be authorized by the South African Veterinary Council (i.e., the national competent authority) to be legally allowed to conduct the procedure; authorized persons then work under the supervision of a registered veterinary professional to maintain competence.

Education and Awareness

Levels of public awareness on animal welfare are according to the OIE generally low in most of the African countries. An exception is South Africa, where awareness is high in some sectors of the society. However, according to OIE's paper on Animal Welfare in OIE Member Countries & Territories in the SADC

Region, there are "considerable activities being carried out on humane education and or animal welfare classes in schools by NGOs such as in, Tanzania, South Africa and the Seychelles." NGOs in North Africa are also reported to play key roles supporting research institutions and researchers in the area.

Several international training programs have been organized in Tunisia by the International Council for Laboratory Animal Science (ICLAS) for North and West African scientists. Courses outlined after the guidelines for Federation of European Laboratory Animal Science Associations (FELASA) category C have also been organized on several occasions in countries including Tunisia, Kenya, and South Africa.[43] The OIE also reports that "In some countries, humane education is included in the national environmental education programs (NEEPs) and is widely defused and resourced with assistance from civil societies and animals welfare networks. South Africa, Tanzania and Zimbabwe have included animal welfare in their veterinary curricula." Very few veterinary authorities seem to be conducting joint activities on animal welfare education and awareness.[21]

In some countries, e.g., Botswana, veterinarians are trained outside the country, as there is no university-level veterinary educational establishment in the country. In Botswana, the veterinarians are trained at the University of the West Indies, in Trinidad. In other countries, e.g., Malawi, there is reportedly a complete lack of training in professional and occupational courses in veterinary education. However, in countries such as South Africa, Kenya, Tanzania, Uganda, Zambia, and Zimbabwe, Faculties of Veterinary Science or veterinary educational establishments exist whose mission is to provide undergraduate and postgraduate professional veterinary training of an international standard.[21] Egypt, Nigeria, and Ghana are also reported to have veterinary schools.[44]

Acts of Parliament regarding the registration of veterinarians exist in a very similarly worded fashion in Kenya, Tanzania, Uganda, Zambia, Zimbabwe, Nigeria, and Ghana.[45]

The training of animal researchers in South Africa is conducted at institutional level, with some institutions offering in-house training. One well-established example is the University of Cape Town's Introductory Course in Laboratory Animal Science,[35] a 5-day modular course including theoretical and practical components, as based on the FELASA accreditation system for persons who carry out procedures on animals.

ACURET (animal care and use in research, education, and training: www.acuret.org) is a chartered multinational, interdisciplinary NGO, incorporated for the purpose of promoting humane animal care and use in developing countries, which is taking an active part in education and training in Africa. ACURET has organized several training courses in countries such as Nigeria and Ghana. ACURET is promoting the establishment of institutional animal care and use committees in African institutions and is focusing on the training of their members.

Inspection, Oversight, and Enforcement

There is no uniform approach for dealing with inspection, oversight, and enforcement in the African countries.

In Tanzania, Kimwele, Matheka, and Ferdowsian report "Researchers must obtain a permit issued by the relevant minister or by the Director of Animal Welfare under advice of the Animal Welfare Council. Some guidelines are also provided, such as the need for veterinary supervision. However, implementation has been delayed, and the various statutory organs stipulated in the Animal Welfare Act have yet to be established."[1]

In Kenya, a license to perform animal experimentation may be granted only to a person who is registered under the Veterinary Surgeons Act or the Medical Practitioners and Dentists Act (or under the supervision of a person duly registered under one or other of those Acts).[46] A majority of the institutions in Kenya do not have established committees to review experimental protocols or to provide oversight regarding the use of animals in education and research. Kimwele, Matheka, and Ferdowsian account that "Only two of the institutions with an established animal care and use committee refer to documented guidelines, one of which makes reference to external guidelines, specifically the United States National Research Council's Guide for the Care and Use of Laboratory Animals and the United Kingdom Animals (Scientific Procedures) Act 1986."[1] There are no requirements by law for protocol review by an ethics committee.[46]

In the Seychelles, only persons licensed by the Minister under the Animals (Control of Experiments) Act may carry out experiments on animals. The Minister may grant a license for animal experimentation to any person to perform any experiment for any purpose specified in such license during such period subject to such conditions in addition to the considerations specified in Animals (Control of Experiments) Act, as he may think fit. The Minister may grant a teaching permit to perform specified experiments for teaching purposes and may authorize experiments on animals without the use of anesthetic.[47]

In South Africa, animal research standards are guided by the SANS 10386:2008 for the Care and Use of Animals for Scientific Purposes.[25] Specifically, each Institutional Animal Ethics Committee is responsible for inspecting all animal areas regularly to ensure compliance with the SANS 10386 and to perform postapproval monitoring of animal research protocols. The NSPCA (National Council of Societies for the Prevention of Cruelty to Animals) Animal Ethics Unit, established in 2001, performs routine welfare inspections of animal facilities under the Societies for the Prevention of Cruelty to Animals (SPCA) Act 169 of 1993.[48] The South African Veterinary Council audits research animal facilities for compliance with required minimum standards.[13]

Many African countries have according to Masiga and Munyua adopted initiatives that support research and development work; however, legislation in place to support the initiatives is often lacking. In former British colonies such as Kenya, the legal system criminalized animal abuse without any provisions for community awareness or education about animal welfare. This is not unique for simply Africa, as few countries at that time did more than make animal cruelty and unnecessary suffering an offence, but is an issue nonetheless. Furthermore, Masiga and Munyua consider that "internal committees at the various research institutions, which are often subservient to the directorate, do not have the capacity and/or authority to perform self-inspections or enforce regulations, which makes them essentially powerless." For the African countries that are affected by this situation, it would be beneficial if provisions for regular external monitoring and evaluation of the current systems were incorporated in legislation so a legal framework exists to support the new initiatives.[8]

DISCUSSION

Compared to most of the rest of the world, policies, legislative frameworks, and guidelines on the use of animals in experimentation are generally poorly developed in Africa.

Main Issues

The reasons as to why legislation and guidelines concerning animal experimentation are poorly developed in Africa are numerous.

The first issue encountered when reviewing relevant material for this chapter, was that although many of the African countries had laws aiming at protecting the welfare of animals, most of these laws covered the welfare of animals in general, without specifically addressing animals used in research or teaching. This is a major problem, since there is a large difference between regulating the welfare of animals on a general level (animal abuse, animal hunting, etc.) and on the rather complex and multifaceted level of animal experimentation. Laws and guidelines covering the various sectors—including universities, breeding facilities, wildlife research, the pharmaceutical and chemical industry, and the various producer groups— should preferably be specialized and detailed. To regulate and legislate animal welfare on such a general level, as is commonplace in most African countries, will not suffice in protecting animals in research and education.

The second obstacle the authors faced during this process was simply retrieving relevant information on the subject of animal experimentation. The inaccessibility of legislation is an issue in Africa; and not only concerning animal experimentation and welfare. It is possible that some African countries may have developed guidelines that are not available in the public domain and are therefore not known by experts in the field, and that are generally unknown to the larger scientific community. Animal welfare is a worldwide issue, and the dissemination and inaccessibility of relevant information to potential users is a major problem since the possible efforts made to live up to international standards subsequently will not serve the purpose for which they were developed.

Enforcement is another important issue. Many of the reviewed acts lacked enforcement mechanisms. The success of animal experimentation legislation depends strongly on ensuring effective mechanisms for enforcement, since all efforts to regulate the welfare of research animals become to a large extent meaningless if there is no authority to enforce regulations. The absence of legal and ethical frameworks and committees to review protocols that involve animals in research and teaching leaves major gaps in the protection of the animals involved. It is important to consider that such inadequacies in the levels of oversight of animal care and use standards are often associated with perceived inadequacies of scientific quality-control standards, thus significantly hampering the potential for scientific publication of the research.

It is perhaps not surprising that legislation regulating animal experimentation is not a top priority in many African countries. Africa is the world's poorest and most underdeveloped continent, and political instability, civil unrest, oppression, poverty, illiteracy, malnutrition, and poor health affect a large proportion of the people who reside in the continent. Animal welfare involves the physical and psychological well-being of animals and the human concern for the welfare of them. It is neither surprising nor unexpected that in regions where the prevalence of respect for even basic human rights is questionable, more important issues relating to human welfare overshadow the concern for the welfare of animals. In countries with considerable social welfare challenges, meeting animal welfare standards may not be a priority nor, in some cases, a possibility.

Prospects for Continued Progress

Policies, guidelines, and legislation regarding the use of animals in research and teaching need to be established and enforced. Improvements must be made.

The first issue considered was the fact that many of the African countries have laws aiming at protecting the general welfare of animals, without specifically addressing animals used in scientific research or education. This existing legislation urgently needs to be adapted and developed to include codes on animal experimentation specifically. The needed regulations can initially be implemented within the current legislative framework (covering the welfare of livestock, transport of game, etc.), while more effective legislation is being drafted.

The problem of dissemination and accessibility of relevant codes and regulations also needs to be remedied. One approach to this issue would be to either create a regional organization dealing with the assembly of legislation, or developing already existing bodies for that purpose. It is naturally problematic to decide which body would be an appropriate candidate to place the responsibility of assembling the legislation and guidelines upon, but there does exist various government and nongovernment organizations both locally and internationally in Africa that might be up to the task. The establishment of an online repository for African laws, standards, education and training courses, and ethical review systems would be integral toward establishing the true state of affairs of animal welfare standards in the scientific arena in Africa.

Improving enforcement is vital. Mechanisms should be incorporated in the animal welfare acts. However, the addition to relevant laws is not enough to ensure enforcement of regulations; ethical review committees and oversight boards should be established to oversee and exercise control over research involving animals.

The main prospect for improvement begins with changing and increasing the education and awareness on the subject. Levels of public awareness on animal welfare are generally low, and although awareness has risen over the years, humane education on animal welfare is lacking at most of institutions of higher learning. This is unacceptable by international standards, considering how commonly animals are used in academic research in Africa. It should be mandatory that all persons who perform procedures on animals, design research protocols or procedures, euthanize animals, care for animals, or serve on Animal Ethics Committees (i.e., animal care and use committees) should undergo appropriate training to ensure competence in the tasks, with continuing education aimed at ensuring maintenance of competency. Education concerning the welfare of animals in research should be a major part of every veterinary curriculum, and the simplest way to implement an increased awareness and education would be to adopt international standards and incorporate them into national legislation.

Internationally, the accepted guiding principles include the universally recognized "Five Freedoms," the "Three Rs," and the fact that the use of sentient animals carries with it a duty to ensure the welfare of such animals to the greatest extent achievable. These principles need to be distinctly harmonized into all African legislation concerning animal experimentation. Potentially appropriate legislation and guidelines from Europe, the United States, Canada, New Zealand, Australia, and relevant African countries could be reviewed, edited, and adapted, along with the inputs from international animal welfare organizations and the draft Universal Declaration on Animal Welfare[29] (reference needed for Declaration here). Other widely accepted international principles for the care and use of animals for scientific purposes, which may be broadly applicable to many African countries and thus form the base for policy development, include The World Organisation for Animal Health's *Terrestrial Animal Health Code for the Use of Animals in Research and Education*[49]; The *International Guiding Principles for Biomedical Research Involving Animals* [Council for International Organizations of Medical Sciences (CIOMS) and the ICLAS, 2012][50]; and the *Guide for the Care and Use of Laboratory Animals*, produced by the Institute for Laboratory Animal Research, also known as "The ILAR Guide."[22]

However, it is of paramount importance to honor the sovereignty of African nations and to ensure that each country remains able to maintain a system of oversight that reflects its own cultures, traditions, and religions. Harmonization of principles, with incorporation of full stakeholder engagement—rather than a case of generic simple standardization—should thus be the ultimate goal for sustainable positive change.

CONCLUSION

African legislation on animal experimentation is in urgent need of review. In order for the relevant legislation to harmonize with international standards, significant improvements must be made. A much greater awareness and a vast improvement in animal welfare education are necessary.

One approach to resolve this major problem is to immediately adapt current legislation concerning animal welfare in general, to also cover the welfare of animals in research specifically. This would constitute a short-term solution; a provisional solution while new, comprehensive, and effective legislation is being drafted. In order for new legislation to be efficient and to suit the right purpose, it is important that international standards are clearly harmonized into the statutes. Appropriate legislation and standards from other parts of the world could be used as a basis for consultation. However, regardless of how comprehensive and inclusive the novel legislation would be, the acts and ordinances will be of limited usefulness without effective enforcement mechanisms. New legislation must consequently be created hand-in-hand with new methods of enforcement and inspection. Provisions must be made in the statutes for regular external monitoring and evaluation of the current systems. Systems must be developed to inspect animal facilities and review research practices and personnel competence to ensure that animal welfare issues are addressed adequately in all institutions and facilities dealing with animals. Ethics committees and inspection agents should be given the power to execute the provisions stated in the legislation.

REFERENCES

1. Kimwele C, Matheka D, Ferdowsian HA. Kenyan perspective on the use of animals in science education and scientific research in Africa and prospects for improvement. *Pan Afr Med J* 2011;**9**:45.
2. *Tanzania Animal Welfare Act, Part V*. 2008.
3. *Kenya Prevention of Cruelty to Animals Act*. 1983.
4. *South Africa Animals Protection Act, No. 71*. 1962.
5. *Performing Animals Protection Act 24*. 1935.
6. *South African Medical Research Council Act, No. 58*. 1991.
7. Nyika A. Animal research ethics in Africa: an overview. *Acta Trop* 2009:112.
8. Masiga WN, Munyua SJM. Global perspectives on animal welfare: Africa. *Rev Sci Tech Off Int Epiz* 2005;**24**(2):579–86.
9. Scientific Experiments on Animals Act, Chapter 19:12, Zimbabwe and Animals (Control of Experiments) Act, (Cap 8), Seychelles.
10. Cap 360 of the Laws of Kenya.
11. Mohr B. The current Status of laboratory animal ethics in South Africa. *Altern Lab Animals* 2013;**41**:48–51.

12. Department of Agriculture, Forestry and Fisheries. *Veterinary and Para-Veterinary Professions Act. South Africa. Act No 19 of 1982*. 1982.
13. Department of Agriculture, Forestry and Fisheries. Rules relating to the practising of veterinary professions. Government notice no R1082 of 2015. South Africa. *Gov Gazette No 39380 9 November 2015* 2015.
14. Department of Health. National Health Act. South Africa. *Gov Gazette No 26595 23 July 2004* 2003. Act No 61 of 2003.
15. Parliament of Mauritius. Animal Welfare Act 2013, Act No 19 of 2013. *Gov Gazette Maurit No 63 18 July 2013* 2013.
16. Uganda's Animals (Prevention of Cruelty) Act, Chapter 39 (consolidated 2000).
17. *Nigeria's criminal code*. 1990. [Chapter 50].
18. Niger Law No 2004-048 of 30 June 2004, Articles 3 and 5.
19. Zambia's Prevention of Cruelty to Animals Act.
20. *Draft Animal Welfare Law of Morocco: Projet de Loi 122-2*. 2013.
21. World Organisation for Animal Health, Animal Welfare in OIE member countries & territories in the SADC region, Gaborone, Botswana, April 2011.
22. Committee for the Update of the Guide for the Care and Use of Laboratory Animals, National Research Council, Institute for Laboratory Animal Research. *Guide for the care and use of laboratory animals*. Washington, DC: National Academies Press; 2011.
23. EU Directive 2010/63/EU of the European Parliament and of the Council of 22 September 2010 on the protection of animals used for scientific purposes.
24. Consortium for National Health Research (CNHR). *National guidelines for the care & use of animals in research & education in Kenya*. 2015.
25. South African Bureau of Standards. *South African national standard: the care and use of animals for scientific purposes (SANS 10386:2008)*. Groenkloof, Pretoria: SABS Standards Division; 2008.
26. Australian code for the care and use of animals for scientific purposes.
27. Russell WM, Burch RL. *The principles of humane experimental technique*. Springfield, II: Charles C Thomas; 1959.
28. Farm Animal Welfare Committee. *Five freedoms*. Available from: http://www.defra.gov.uk/fawc/about/five-freedoms/.
29. Universal Declaration on Animal Welfare.
30. University of KwaZulu-Natal. Guide to the care & use of animals in research & teaching.
31. Canadian Council on Animal Care. *Guide to the care and use of experimental animals*, vol. 1. 1984. Available from: http://www.caresa.uct.ac.za.
32. University of Cape Town. *Code of ethics and procedures for the use of animals in teaching and research*. 2016. Available from: http://www.researchsupport.uct.ac.za/sites/default/files/image_tool/images/362/Documents/Integrity/animalresearch_ethics_policy.pdf.
33. Institute of Primate Research, Kenya. Available from: http://www.primateresearch.org//index.php?option=com_content&task=view&id=39&Itemid=41.
34. Hailu A, Negesse Y, Abraham I. *Leishmania aethiopica*: experimental infections in non-human primates. *Acta Trop* 1995;**59**(3):243–50.
35. University of Cape Town. *Introductory course in laboratory animal science – the ethics, care and use of animals for scientific purposes*. 2016. Available from: http://www.ceuhealth.uct.ac.za/introductory-course-to-laboratory-animal-science.
36. The Animal Welfare Act, 2008, Tanzania, Part V section. 39(1)(a-b).
37. Genetically Modified Organisms Act, No 15 of 1997, South Africa.
38. Animal Welfare Act, 2008, Tanzania, Part III, section. 22–25.

39. Jones KE. Transport animal welfare legislation and inspection: how to progress and succeed. In: *World association for transport animal welfare and studies (TAWS), workshop*. UK: Silsoe Research Institute; April 24, 2003.
40. Available from: http://www.oie.int/animal-welfare/animal-welfare-key-themes/.
41. Kenya Prevention of Cruelty to Animals Act (1983), Part I, section. 8.
42. Department of Agriculture, Forestry and Fisheries. Rules relating to the practising of the para-veterinary profession of laboratory animal technologist. South Africa. *Gov Notice No 1445 3 October 1997* 1997.
43. Wilson MS, Berge E, Maess J, Mahouy G, Natoff I, Nevalainen T, van Zuphten LFM, Zanielli P, Fosse RT, Bonnod J, Hau J, Leyten R, O'Donoghue PN, Sabourdy M, Schwartz K, Thomann P. FELASA recommendations on the education and training of persons working with laboratory animals: category A and category C. *Lab Anim* 1995;**29**. 121±31.
44. World Veterinary Association. Available from: http://www.worldvet.org/about.php?sp=members®io=Africa.
45. Veterinary Surgeons Act, Cap 366, Kenya; Veterinary Act 2003, Tanzania; Veterinary Surgeons Act, Chapter 243, Zambia; Veterinary Surgeons Act, Chapter 27:15, Zimbabwe, Veterinary Surgeons Act Nigeria 2004, section 8 and Veterinary Surgeons Law Ghana 1992, sections 13–21.
46. Prevention of Cruelty to Animals Act, Cap 360, Kenya, Part III section. 16.
47. Animals (Control of Experiments) Act, Seychelles, section 7.
48. Department of Agriculture, Forestry and Fisheries. *Societies for the Prevention of Cruelty to Animals Act. South Africa. Act No 169 of 1993*. 1993.
49. World Organisation for Animal Health (OIE). Terrestrial animal health code. *Use of animals in research and education*2015. [Chapter 7.8].
50. Council for International Organizations of Medical Sciences (CIOMS), The International Council for Laboratory Animal Science (ICLAS). *International guiding principles for biomedical research involving animals*. 2012. p. 1–4.

Chapter 8

Laws, Regulations, and Guidelines Governing Research Animal Care and Use in India

Syed S.Y.H. Qadri[1], Subbaraya G. Ramachandra[2]
[1]*National Institute of Nutrition (ICMR), Hyderabad, India;* [2]*Indian Institute of Science, Bangalore, India*

GENERAL FRAMEWORK AND PRINCIPLES

The Indian culture traces its roots to the Indus valley civilization with a history of over 5000 years. Animals have always been an ingrained part of the ethos of this culture with several species both domesticated and wild being revered and worshiped. For many in other parts of the world, worship of animals would have a different connotation but, for those who understand the Indian culture and its philosophy, this considers all animals as creations of "The Almighty," and the deep sense of reverence, respect, and love of "The Almighty" is extended to all "His" creations including trees, forests, rivers, mountains, etc. The universal concept of harmony between humankind, animals, and the environment over the ages developed into a perfect symphony so deeply embedded in the Indian culture and it has become an ingrained part of the spiritual life of its people. These cultural concepts based on sound reasoning also had a profound effect on the subsequent invading cultures compelling them to adopt them into their own cultures. The same culture is reflected in the Constitution of India Article 48A, which directs the state to protect and improve the environment and to safeguard the forests and wildlife of the country. In Part IV A, under Fundamental Duties, a citizen is duty bound to protect and improve the natural environment including forests, lakes, rivers, and wildlife, and to have compassion for all living creatures.[1]

People in the new government, under the administration of Honorable Prime Minister Mr. Narendra Modi, have rediscovered the fervor and enthusiasm that has led to an awakening and realization of the unused potential. This has had a significant impact on the people; especially on the awareness levels about environmental issues that includes animal protection and welfare. The positive outcome has been the willingness of the people to address issues that earlier were most often ignored.

Laboratory Animals. http://dx.doi.org/10.1016/B978-0-12-849880-4.00008-8
Copyright © 2018 Elsevier Inc. All rights reserved.

The Constitution of India (Part IV) in the Directive Principles of State Policy; Article 48, says that the State shall endeavour to organize agriculture and animal husbandry on modern and scientific lines and shall take steps to preserve and improve breeds, prohibit slaughter of cows and calves and other milch and draft cattle. Article 48A directs the state to protect and improve the environment and to safeguard the forests and wildlife of the country. The Indian Penal Code, 1860 (45 of 1860), Chapter II, Section 47, defines the word "Animal" as any living creature, other than a human being. Chapter III of the code deals with the punishments for mischief by killing or maiming animals.[2]

The use of animals in transportation as draught animals to pull carts and plow fields in the agricultural sector and in other field operations has been the backbone of the Indian economy since ages. Similarly, the use of other domestic animals such as buffaloes, horses, ponies, donkeys, mules, dogs, cats, sheep, goats, pigs, camels, elephants, yaks, and other animals since the beginning of the Indian civilization has had a profound effect not only in agriculture but also on the cultural, religion, and social practices. This has led to the development of bonding and interrelationship between man and animal to the extent that several animals are revered and worshiped. Therefore, animal welfare and compassion in the Indian context is more a part of the Indian genetic make-up. There are several reports and instances where animals such as cows and bulls were sheltered by the farmers within their huts during the rainy season. However, there were also instances where animals were being abused or overused more out of ignorance and some due to human greed. The Prevention of Cruelty to Animals Act (PCA) 1960[3] was promulgated to protect these animals from human exploitation and misuse.

The use of Laboratory animals (especially rodents) in research started in India early in 1876 at the Madras Veterinary College (presently the Tamil Nadu University of Veterinary and Animal Sciences) during the rule of the British followed by the Indian Veterinary Research Institute and the Haffkine's Institute in 1889. The first laws that governed animal welfare were passed by the Indian Parliament, which was called as the PCA Act 1960 (Act 59 of 1960).[3] This particular act, under definitions, describes an "animal" as any living creature other than a human being. Chapter II of this Act deals with the establishment of an Animal Welfare Board of India (AWBI), functioning to promote animal welfare and to protect animals from being subjected to unnecessary pain or suffering. Chapter III of the Act lists and describes cruelties, penalties, and the procedure to destroy suffering animals.

Chapter IV deals entirely with Experiments on animals, the constitution of the Committee for the Purpose of Control and Supervision of Experiments on Animals (CPCSEA), and its functioning. Chapter V deals with performing animals, and Chapter VI deals with miscellaneous issues of cognizability of offenses, powers of search, seizures, treatment of animals, and other rules. The general understanding of animal experimenters has been that experimentation on animals meant the use of only laboratory animals such as mice, rats, guinea pigs, hamsters, rabbits, cats, and dogs. Even to this day, there are quite a few who still have not realized that this Chapter IV covers all animals including invertebrates, amphibians, crustaceans, fishes, reptiles to larger mammals such as horses, cattle, and other animals. Therefore, the presumption that only Chapter IV of the Act is applicable on experimentation on animals would not be true in its entirety as the other chapters would also be applicable especially in cases of larger animals such as horses, ponies, and mules used in antiserum production, and cows and buffaloes used in nutrition and feeding trials. In instances where large animals are being exploited/misused in the name of experimentation, it would entirely be the prerogative of the prosecution to identify the cruelty meted out to the animal and charge the offender under the relevant section of the law.

Several organizations in India are in the business of producing antibodies against snake, scorpion, and spider venoms and also against rabies, tetanus, and diphtheria using horses. The CPCSEA has been regulating all these organizations. There have been concerns raised by animal rights activists regarding the welfare of these animals. The matter is under consideration with the CPCSEA. A spurt in the CPCSEA registrations between 2011 and 2014 were observed due to several colleges, especially pharmacy colleges having postgraduate and PhD programs, started registering laboratory animal facilities as mandated by the Pharmacy Council of India. However, since 2015, the demand for pharmacy as a profession declined with fewer students taking up the profession, most of the Institutional Animal Ethics Committees (IAECs) in these institutions have become nonfunctional. Several pharmacy colleges have not renewed their CPCSEA registration. The dramatic decline in the number of pharmacy colleges as well as the decline in number of students opting for pharmacy as a profession could possibly be related partly to the slump in the pharma sector (both national and international) and partly due to the decline in job opportunities. In India, the role of a pharmacist is restricted to the pharmacy stores or medical shops unlike in the west where the pharmacist has a major role to play between the physician and the patient.

Chapter IV of the Act deals entirely with experimentation on animals and specifies the legal constraints on the sale and acquisition of animals for research purposes.[4] Section 14 of the Act permits experiments (including experiments involving surgery) on animals for the purpose of advancing new discovery of physiological knowledge, which will be useful for saving or for prolonging life, alleviating suffering or combating disease, whether of human beings, animals, or plants. The Act (Chapter IV, Section 15) also instigated the constitution of a committee, called the CPCSEA. From 1960 until 1998, the CPCSEA was chaired by the Director General of Indian Council of Medical Research (ICMR) and remained dormant. It was revived in 1998 and the new rules, "Breeding of and Experiments on Animals Rules 1998," were promulgated by the Indian Parliament.[5] These rules were again amended in 2001.[6] Under these rules, registration of animal establishments for breeding or for experimentation was made mandatory. Registration of the facility necessitates the submission of design and floor plan of the animal facility and the requirements of the institution. Approval of a facility is subject to an inspection of the facility by a person nominated by the Member Secretary of the CPCSEA and establishment of the IAEC, the equivalent to the Institutional Animal Care and Use Committee (IACUC) in the United States.

The CPCSEA constituted under the PCA Act 1960, Chapter IV Section 15, regulates all animal experiments in the country. The act permits all diploma and degree holders in biological sciences to conduct experiment on animals. However, the named or attending veterinarian is vested with the legal authority to supervise the conduct of animal experiments. The presence of the named veterinarian in the IAEC has therefore been mandatory. In an institution, it is mandatory to hold an IAEC meeting every 6 months. The IAEC has been empowered to approve project proposals on rodents and rabbits. The project proposals using larger animals such as cats, dogs, primates, sheep, and other larger animals need to be submitted with the recommendations of the IAEC to the CPCSEA National Committee for approval.

The IAEC's oversight emphasizes mostly the ethical aspects of animal usage and the animal welfare measures adopted. The Three Rs (3Rs) principles (replacement, reduction, and refinement) are normally the basis of most of the discussions. The CPCSEA rules insist that researchers submit their requisition for animal experimentation in the prescribed "Form B" to the IAEC.[7] Form B contains a clause instructing the researcher/experimenter to review the literature and declare whether the experiment has already been done in vitro/in vivo elsewhere. It requires the researcher to explain why the proposed experiment cannot be performed in vitro. In addition, the CPCSEA has introduced the *fourth R* that denotes *rehabilitation* of the animals after use in experiments. The CPCSEA insists that the fourth R be applied to all large animals. Implementation of the fourth R principle in case of rodents and rabbits, which are the most commonly used animals, could adversely impact future research, and experience has already shown that application of the fourth R in case of primates in couple of instances did not give the desired results. The present approach may be untenable and deserves further discussion and strategic planning.

Besides the aforementioned Act, there are other Acts of the Indian Parliament that govern animal experimentation and research in India. The Indian Veterinary Council (IVC) Act 1984 (Act 52 of 1984)[8] read along with its Amendment of 1992 [GSR 395(E) Section 2(f)][9] includes laboratory animal medicine, animal experimentation, and vaccine production within the definition of the veterinary profession. Section 30 of the IVC Act 1984 legally authorizes only veterinarians registered with the IVC to practice laboratory animal medicine and animal experimentation. It is mandatory for all graduates in veterinary sciences in India to register with the respective State Veterinary Councils to practice veterinary medicine and surgery and the rights of the registered practitioner are protected by the Council. This is irrespective of whether the individual is working in the government or in a private institution.

The animal facilities of small educational institutions typically consist of 2–4 rooms (mice, rat, guinea pig, and rabbit rooms) with a separate area for washing cages and, if available, a separate room for storing feed and cleaned cages. Such facilities have a working window/split air conditioner and the facilities are used only once or twice a year for short durations for experiments ranging between 14 and 90 days. The provisions for animal housing are the bare minimum and the personnel working in the facility are mostly the research scholars. A majority of the animal facilities do not have personnel trained in laboratory animal science. The veterinarians in such facilities generally work as consultants on an ad hoc basis. There is an urgent need for training personnel from such institutions.

In the government sector, especially India's premier national research institutions have good animal facilities and sound programs of animal care and use; however, challenges remain to improve the quality of animals from such institutions. A majority of the animal facilities in such institutions are headed by veterinarians who have had their training in laboratory animal science/medicine either in Europe or the United States (US), but the animal technicians have not had the benefit of formal training. On the research front, scientists and researchers working in these government institutions are not required to have formal training in laboratory animal sciences or the application of in vivo animal research methodologies and this may impact the quality of animal experimentation and research.

More than three dozen pharmaceutical companies in India have attained Good Laboratory Practices (GLP) certification and over 20 institutions have also achieved AAALAC International accreditation reflecting the strength of their programs of animal care and the quality of their facilities. Many of the institutions in the pharmaceutical industry as well as in the government sector have now switched over to individually ventilated cage systems in an effort to upgrade laboratory animal management practices. However, the quality of the animals (both microbiologically and genetically defined) needs improvement. Similarly, the personnel working in these institutions need training in hygiene, sanitation, sterilization techniques.

The pharmacy, medical, veterinary, and basic science colleges, which are the largest number of registered institutions with the CPCSEA, use the commonly available random bred laboratory rodents and rabbits for experimentation. In the government sector, several institutions have been using outbred and inbred mice, rats, Syrian golden hamsters, and New Zealand white rabbits. Almost all the facilities with a couple of exceptions predominantly use random bred animals; however, a few government institutions have imported unique or specialized stocks, inbred strains, and other transgenic and knockout mice for their particular research purposes. The initial stocks of the aforementioned strains of animals were procured from the National Institutes of Health, Bethesda, United States, and high-quality commercial vendors in the United States, Europe, or Asia. However, no database is available about the different strains and numbers of animals imported from various sources. Similarly, information about the availability, or the practices used for the maintenance of these strains in various institutions in India is not available.

In the Pharmaceutical industry, the spectrum of research animals used is more or less similar to that of government sector. However, few industries where research and development in drug discovery is in progress, do import specialized strains, transgenic and knockout animals from global suppliers. Multinational laboratory animal breeders and suppliers have recently started breeding and supply of some strains of rats and mice in India. These animals are considered to be Specific Pathogen Free; however, third party testing could be of immense help to establish the claims.

Organizations that require importing animals should submit the proposal to the IAEC of the respective institution. Once the IAEC approves the proposal, a No Objection Certificate (NOC) issued by the CPCSEA must be obtained. The proposal with the NOC goes to the Director General of Foreign Trade (DGFT). However, import of larger animals such as Beagle dogs and others requires permission from the CPCSEA National Committee. Until recently, Beagles were imported from a commercial vendor with breeding facilities in Europe or China. Recently, a European supplier started a Beagle dog breeding facility in India and the animals are now available in India.

Looking at the distribution of animal facilities in the Indian Subcontinent, the majority of the registered animal facilities are present in the states of Maharashtra (272), Andhra Pradesh (230), Karnataka (190), Tamil Nadu (184), and Gujarat (138). The presence of animal facilities in different states of India also reflects the developments in the pharmaceutical, biotechnology, and contract research organizations.

LEGAL AND REGULATORY SCOPE/APPLICABILITY

The CPCSEA utilizes the very inclusive definition of "animal" provided in the Indian Penal Code[2] and the PCA Act[3] as noted above, which means "any living creature other than a human being." Chapter I of the PCA Act also provides definitions for "captive animal" and "domestic animal," but these subcategories have minimal representation in or bearing on the Indian regulations pertaining to experimentation on animals. The privilege to conduct animal experimentation is protected in Chapter IV of the PCA Act, which states: "Nothing contained in this Act shall render unlawful the performance of experiments (including experiments involving operations) on animals for the purpose of the advancement by new discovery of physiological knowledge or knowledge which will be useful for saving or prolonging life or for combating disease, whether of human beings, animals or plants."

About 2000 animal facilities have registered with the CPCSEA. A majority of the animal facilities are attached to the pharmacy, medical, veterinary, and basic science colleges (1150), followed by government research, training, and testing institutions (410) and private pharmaceutical and biotech industry (450).

In all the educational institutions, the use of animals for education and training has been banned and replaced with other alternate methods such as audiovisual aids.

REGULATORY AUTHORIZATION AND ENFORCEMENT/INSPECTION MECHANISMS

All activities related to animal welfare and cruelty in India are governed and regulated by the AWBI under the Ministry of Environment, Forests and Climate Change, Government of India. The AWBI derives its powers from the PCA Act 1960.[3] However, the power to authorize, regulate, control, and supervise experiments on animals lies with the CPCSEA. The CPCSEA has recently been reconstituted on February 17, 2016 (Official Gazette Notification No. S.O. 503 (E)). It has 23 members, 10 of whom are government officials, 11 researchers from life sciences, 1 veterinarian, and 1 animal activist. Stake holders from professional laboratory animal science associations, representatives of pharmaceutical industry, contract research organizations, the biotechnology industry, and the breeders and suppliers of laboratory animal in India somehow could not find a place in the committee. Inclusions of some of these stake holders would have been valuable.

Registration of animal facilities entails submission of the Form "A" with details of the establishment, objectives of the organization, type of animal facility, purpose of registration (research for education purpose/research for commercial purpose/research/breeding for in-house use/breeding for trade/production of hyperimmune plasma or serum), details of the animals to be housed, location of the facility, training details of the staff, availability of postexperimental rehabilitation (in case of larger animals), constitution of the IAEC, curriculum vitae of the IAEC members along with their consent letter accepting the responsibility as a member of the IAEC, and minutes of the IAEC meeting in which the proposal for registration with the CPCSEA has been approved with the signatures of all the IAEC members. Once the application along with the registration fee cheque, Indian Rupees 10,000 for registration of facilities for the categories of research for education, research for commercial, research, and nonresearch commercial—production of hyperimmune plasma/serum; Indian Rupees 15,000 for registration for breeding for in-house use; and Indian Rupees 25,000 for registration for the purpose of trade need to be paid. The registration would be valid for a period of 5 years. The renewal fee after the expiry of 5 years would be Indian Rupees 2500 for all categories. Revision in the IAEC for individual members would be charged Indian Rupees 1000 for every revision.

Once the facility is ready, the CPCSEA authorizes one of its nominees (most nominees are personnel already working in animal facilities, have basic qualifications and background in laboratory animal science, and are appointed on honorary basis) to inspect the facility and submit a report to the CPCSEA. There were instances where the CPCSEA had nominated personnel who do not have a basic degree or diploma in laboratory animal sciences to inspect a facility or as a main nominee to the IAEC. As a consequence of the nominee's lack of knowledge and preparation, facility registrations were delayed due to incomplete or inconclusive inspection reports. However, the CPCSEA has now defined the qualifications required for appointment of nominees.[10] The minimum requirements prescribed are: a degree in veterinary science or a postgraduate degree in zoology, animal sciences, animal biotechnology, life sciences; postgraduate degree in pharmaceutical sciences; or medicine with specialization in medical microbiology/pharmacology with experience in animal handling and animal research. This has been a welcome change and would definitely go a long way in improving the animal facilities as well as the discourse content in the IAECs.

The inspection is done by the CPCSEA nominee individually or is accompanied by another person nominated by the CPCSEA. The inspection team is provided with a prepared list that identifies items to be inspected and recorded during the inspection. If the inspection report submitted is satisfactory, the CPCSEA allots a registration number to the institution and appoints four members of the IAEC and approves the other members proposed by the institution. The four members appointed by the CPCSEA include the nominee, the link nominee, the scientist from outside the institution, and the nonscientific socially aware member. The minimum qualification for the nonscientific socially aware member has been prescribed. The person needs to have a graduate degree in any of the subjects' preferably biological sciences.

The IAEC is expected to make site visits to the animal facilities before every meeting. The frequency of meeting depends on the number of experiments conducted in an institution. However, a minimum of two meetings annually is mandatory and the Chairman of the IAEC and the nominee are duty bound to submit a copy of the minutes to the CPCSEA for record. Failure to submit the IAEC minutes for 3–4 consecutive meetings would attract disqualification and derecognition requiring the institution to reapply for registration.

Besides the nominee, the CPCSEA may authorize any of its representatives to inspect a facility at any given time. A person so authorized would have the powers to visit the facility at any time convenient to him and after inspection would have to submit a report to the CPCSEA. However, the authorized person has no power to seize the animals. Based on the inspection report submitted by the person so authorized, the CPCSEA national committee has the power to cancel registration or suspend registration of an organization permanently or for a specified time after giving a chance to hear from the organization. Normally, the head of the institution is held responsible and would be the person subject to penalty.

The CPCSEA had intended to revise the PCA Act 1960. A draft proposal calling it the PCA Act 2011 was circulated for public opinion.[11] In this particular revised Act, the penalties were revised and were considered to be very harsh. The scientific community had objected and had demanded a revision. In the revised Act; Section V, page 17, Section 28: Penalties, it states that any person contravening the provisions under the Act would be liable for a fine of Rs. 50,000 or imprisonment not less than 1 year extended up to 3 years. On second or subsequent offense, a fine of Rs. 75,000–100,000 with imprisonment not less than 2 years and extended up to 5 years was proposed. As per the Act, any scientist who contravenes the Act would be liable for prosecution. This would be true even in instances where the person conducting an experiment would be a public servant performing his duty in good faith. The Act has not yet been taken up for consideration and hope it remains so.

DOG STUDIES: GUIDELINES AND CONSIDERATIONS

The first condition for conducting large animal experiments is the availability of a CPCSEA approved facility to house and experiment on animals of that particular species. The same would be true for dogs. The only dog breed approved for use in scientific experiments in dogs is the "Beagle." Earlier dogs were being imported from Europe and it involved a heavy expenditure for purchase, freight cost, quarantine, and transport from either the Mumbai or Chennai ports where the quarantine facilities are available. However, since 2009 a joint venture between an Indian and European company started a Beagle dog breeding and testing facility. Beagles are now available in India at very reasonable cost and effort.

All dog experiments need to be presented in the prescribed format to the IAEC. The IAEC is expected to review the projects related to its feasibility at the facility, availability of space, housing conditions, trained man power, infrastructure to conduct toxicity, pharmacokinetic and pharmacodynamic studies at the Institution. With the recommendations of the IAEC, the protocol in the prescribed format is forwarded with the comments of the IAEC to the CPCSEA for consideration. The project is reviewed and discussed at the CPCSEA at New Delhi, which meets once in alternate months. The newly formed CPCSEA has given due consideration to the fact that experiments on animals (more so in large animals) is a necessity, and it has been judiciously approving experiments on dogs. The CPCSEA has also issued Guidelines for the Reuse and Rehabilitation of Dogs after experiments. It defines "Rehabilitation" as "the aftercare rendered to animals that have been (i) bred for the purpose of experimentation (ii) subject to any form of experimentation (iii) retained in laboratory animal houses or breeding houses for the purpose of experimentation, both for education and research, with the sole intention of alleviating the pain/distress or suffering due to the physical, physiological and psychological trauma that the animals have been exposed to and to provide the animal a life distinctly different from laboratory housing and care, until the point of natural death."[12]

The CPCSEA also permits reuse of dogs in experiments subject to the following conditions:

1. The animal should have suffered no significant adverse effects as a consequence of the first use, and the animals have not been subjected to any intervention, which compromises its welfare or suitability in scientific terms, to be used as a subject for the second or subsequent use.
2. The reuse of animals in an approved study may be considered for second/repeated use when it may serve as a way to reduce the number of animals used, without causing any incremental pain/distress to the animal, which results from second/repeat use.
3. During subsequent use the physical and psychological health and well-being of the animal must be considered.
4. Before seeking permission from the CPCSEA, the health of the animal and the opinion of the veterinarian and consent of the IAEC must be in order.
5. Health certificate for sound health and fitness of animals intended for reuse must be obtained from a qualified veterinarian and should include a complete clinical examination, including vital signs; skin condition; behavior of animals; complete blood count; temperature, pulse, and respiration (TPR); kidney function test (KFT), and liver function test (LFT). The veterinarian should also certify that there has been no adverse effects including psychosomatic disorders, by way of the first experiment/caging due to laboratory housing/procedures. Animals showing stereotypic behavior, fear, freezing on human touch; genetic or physical defects; permanent implants; etc., should be declared unfit for reuse and should be recommended for proper rehabilitation.
6. Laboratory must maintain records of reuse with detailed documentation.
7. Reusing animals as a reduction strategy can be promoted/considered by the IAEC only. The IAEC should closely monitor the end points and determine the suffering of animals.
8. The reuse of dogs in toxicity studies would be for a maximum period of 3 years for pharmacokinetic studies. If the dog shows any impairment in the LFT or KFT within this period then it cannot be reused. A 3-month wash out period between studies has been recommended.
9. Dogs used in breeding may be limited to five whelping cycles and must be rehabilitated on completion.
10. In case of telemetry studies, dogs can be used for a maximum period of 3 years subject to the condition that they show normal physiological functions or until the device is no more functional. Dogs from which devices have been explanted should not be used to implant another second device.

11. The animals should be assigned unique numbers by way of microchips.
12. Dogs after the completion of 3 years experimental period or five whelping cycles should be rehabilitated by the institute with information to the CPCSEA. They may be given for adoption by pet lovers. However, spaying/castration by the institution and adoption facilitated through trustworthy Animal Welfare Organizations after due approval from CPCSEA can be done.

NONHUMAN PRIMATES

Earlier, institutions whenever required would obtain a license from the respective State Wildlife Warden to capture wild rhesus or bonnet monkeys for research. The private vendors with license to supply wild animals were supplying the institutions with the required number of animals. In fact, due to massive deforestation, wild animals, especially monkeys in search of food, have been moving over to the urban areas where they have become a nuisance. The population of these animals both in the wild and urban areas has grown exponentially as all wildlife is protected under the law. The Wildlife authorities have now stopped issuing licenses to trap and use monkeys for research, insisting that scientists procure the animals from breeding facilities.

There are seven institutions in India in the government sector that have experimental primate facilities. This includes National Centre for Laboratory Animal Sciences, National Institute of Nutrition (ICMR) at Hyderabad; Indian Institute of Science, Bangalore; All India Institute of Medical Sciences, New Delhi; National Institute of Immunology, New Delhi; Central Drug Research Institute (Council for Scientific and Industrial Research, CSIR), Lucknow; National Institute of Virology (ICMR), Pune; and National Institute for Research in Reproductive Health (ICMR), Mumbai. The Zydus Cadila in Ahmedabad, Gujarat, is the only organization in the private sector having a nonhuman primate facility for rhesus monkeys. They also breed rhesus monkeys to a limited extent. In the government sector, the majority of the facilities use rhesus monkeys and the National Institute of Research in Reproductive Health, Mumbai, has marmoset and bonnet monkey facilities where animals are also bred in limited numbers, whereas Indian Institute of Science is maintaining the bonnet monkey colonies. Breeding of nonhuman primates especially *Macaca mulatta* (rhesus monkey) has not been taken up by any of these institutions. The National Institute for Research in Reproductive Health, Mumbai, had commenced the work to build an exclusive primate breeding facility: National Centre for Primate Breeding and Research near Mumbai.

The ICMR approved the establishment of National Animal Resource Facility for Biomedical Research (NARF-BR), the 33rd institution of ICMR on December 3, 2015 at the Genome Valley, Biotechnology Park, Shameerpet near Hyderabad. The Telangana state government has sanctioned 100 acres

(40 ha) of land and the Government of India has sanctioned an amount of 3.4 billion rupees for the establishment of the facility. The NARF-BR has plans to establish breeding facilities for several large animals used in research.

USE OF EQUINES IN PRODUCTION OF HYPERIMMUNE PLASMA/SERUM: GUIDELINES AND CONSIDERATIONS

More than a dozen organizations in India are in the business of antiserum production against snakes, scorpions, spider venoms, rabies, tetanus, and diphtheria antigens. Some of these organizations have been raising anti-snake venom antiserum against several snakes from the African continent, the Middle Eastern, and West Asian countries and have been exporting the antiserum. Most of these organizations are registered with the CPCSEA and have had fully functional IAEC. Almost all these organizations also have small laboratory rodent facilities for in-house regulatory testing of their products for potency. The CPCSEA constituted a subcommittee to formulate the guidelines for the "Care and Management of Equines used in the production of Biologicals" in 2001.[13] These guidelines were amended in April.[14] As per the CPCSEA regulations, the IAEC has been given permission to approve animal experimentation on small animals such as mice, rats, guinea pigs, hamsters, gerbils, and rabbits. Experimentation on animals larger than rabbits, such as cats, dogs, sheep, goat, and other higher animals need the permission of the CPCSEA. Over the years, the exception has been the equines, where organizations registered with the CPCSEA were following the Protocol II "Care and Management of Equines used in the production of Biologicals," guidelines 2001. The CPCSEA has now taken this issue into cognizance and has advised all organizations in the production of antisera in equines to reregister with the CPCSEA. The IAECs are also impressing upon the organizations to submit all protocols in the prescribed proforma's with its recommendations for onward transmission to the CPCSEA for approval.

In all the organizations, the common equines used for hyperimmune plasma production are ponies, mules, and rarely horses. The CPCSEA, in April 2005, laid out some changes to the Protocol II, 2001. The lower and upper age limit for equines to be used in sera production has been fixed. The bleeding of horses (phlebotomy) less than 5 years age is not permitted though the animals can be inducted into the hyperimmune plasma production program. The rules also insist that the animals are subjected to strict health, blood and biochemical analysis to ensure that the animals are physically and clinically sound and immunologically responsive. The minimum recommended body weight for animals under the hyperimmune plasma production program for ponies is 150 kg; for mules, minimum is 250 kg, and for horses, minimum is 300 kg. The upper age limit is fixed at 18 years for ponies, 22 years for horses, and 25 years for mules. Earlier the Protocol II, 2001 had fixed the period of hyperimmune plasma production program for 3 years, which was relaxed up to a maximum age limit of 21 years.

The recommended bleeding schedule and quantum of blood collection recommended are 5% of circulatory blood volume or 0.5% of total body weight repeated every 2 weeks with plasmapheresis and transfusion or 15% of circulatory blood volume or 1.5% of total body weight repeated every 4 weeks with plasmapheresis and transfusion. The guidelines insist that the responsibility of rehabilitation of equines after their use in hyperimmune sera production will remain with the respective organizations that are using them. The CPCSEA will identify and notify the nongovernmental organizations that are willing to take up the rehabilitation of these equines.

GENETICALLY ALTERED ANIMALS: SPECIAL CONSIDERATIONS

As such there are no laws restricting the development of genetically altered animals in India. The CPCSEA rules permit the researcher to import transgenic and knockout mice. However, all genetically modified animals are on the restrictive import list and need to go through an elaborate import procedure.

An end user can only import an animal species/strain that is not available in India. The user needs to submit a research proposal to the IAEC. The application form along with all the necessary documents (viz., NOC from IAEC, registration certificate, invoice of animals/free gift letter, gazette notification, and application fee) should be submitted to the Joint Director General of Foreign Trade of the respective states of India. One set of application should be sent to the DGFT at New Delhi. The DGFT office forwards the application to the Animal Husbandry Commissioner, Ministry of Agriculture, Government of India, New Delhi for their NOC. Subsequently, the DGFT either approves or rejects the importation of animals and directs the respective state Joint DGFT to issue the license. The animals can be imported through any of the airports in India where an approved quarantine facility exists. Presently, the quarantine facilities exist at Mumbai, Chennai, New Delhi, Kolkata, Hyderabad, and Bangalore airports.

The Biological Diversity Act, 2002 (Act No. 18 of 2003) governs the use and/or trade of any biological material in the country.[15] Under the provisions of the Act biological samples cannot be imported or exported out of the country without proper approvals from the competent authority.

INSTITUTIONAL AND DESIGNATED PERSONNEL

The CPCSEA holds the Head of the Institution/Organization responsible for the maintenance, upkeep, and smooth functioning of the animal facility. The Head of the Institution is also the individual who would be held responsible for any eventuality or accident in the animal facility. The overall responsibility of care, breeding, management, and experimentation in an organization lies directly with the veterinarian and staff working in the animal facility. However, the IAEC has the overall responsibility to oversee the functioning of the facility and the conduct of animal experiments.

From the institutional point of view, the three most important persons representing the institution in an IAEC are the Director/Officer in charge of the Institute (equivalent to the Institutional Official as designated by the *Guide for the Care and Use of Laboratory Animals*[16] and AAALAC International) who typically also is the Chairman of the IAEC according to the CPCSEA rules, the Member Secretary who arranges all the IAEC meetings and ensures that all animal experimental protocols that need to be discussed and approved in the IAEC are in the prescribed format and the third most important person is the veterinarian (the attending veterinarian) who has the responsibility to procure the animals and ensure that the experiments are conducted as per the prescribed laws and also ensure animal welfare. The attending veterinarian is also authorized to suspend any animal experiment, if he/she finds that the animals are under extreme stress or animal welfare laws are being flouted.

From the view point of the CPCSEA, the nominees appointed by them are the most important persons in implementing the laws and ensuring that the animals are not being misused or subjected to unnecessary stress. The presence of the main nominee, the scientist from outside the institute, and the nonscientific socially aware member are necessary for the conduct of the IAEC meeting. The minimum quorum required for a successful IAEC is six persons among whom the presence of main nominee and the nonscientific socially aware member are mandatory. The link nominee in most instances attend the first IAEC meeting for the purpose of introduction and later would be called only in case, the main nominee expresses his/her inability to attend and designates the work to the link nominee.

OVERSIGHT AND ETHICAL REVIEW

The purpose of the IAEC is to control and supervise the experiments on animals performed in an establishment, which is constituted and operated in accordance with procedures specified for that purpose by the CPCSEA.[17] The IAEC will review and approve all types of research proposals involving small animal experimentation (such as mice, rats, guinea pigs, hamsters, and rabbits) before the start of the study. For experimentation on large animals, the research protocol should be forwarded to the CPCSEA in prescribed manner with recommendations from the IAEC. The IAEC is required to monitor the animal experimentation throughout the study and after completion of study through periodic reports and visit to animal house and laboratory where the experiments are conducted. The committee has to ensure the compliance with all regulatory requirements, applicable rules, guidelines, and laws.

The members are appointed to the IAEC for a period of 5 years (revised rules effective from September 1, 2015) coincident with the CPCSEA registration of the IAEC. When the IAEC is reconstituted, at least half of the members must be replaced. Members are expected to observe absolute confidentiality and declare conflicts of interest pertaining to IAEC discussions and actions. The IAEC also is expected to formulate standard operating procedures (SOP) for its working requirements and follow it in all the meetings.

The IAEC (a mandatory provision in the law) consists of eight members. Five members of the committee are from the institution (a biological scientist; the scientist in charge of the animal facility; the named veterinarian; two scientists from a different biological discipline) and other three members (the main nominee; scientist from outside the institution; and a nonscientific socially aware member) are appointed by the CPCSEA. The CPCSEA also nominates a link nominee who may serve in the capacity of the main nominee if the main nominee conveys his/her unavailability for the meeting in writing to the IAEC Chairman. A minimum of six members (out of the eight members) are required to compose a quorum and all decisions should be taken in meetings and not by circulation of project proposals. The presence of CPCSEA nominee is required unless excused by written notification of unavailability to the Chairman IAEC, in such cases the link nominee may substitute. The socially aware member's presence is compulsory in large animal protocol review and also at least in one meeting in a calendar year. Minutes of IAEC meetings must be prepared by the Chairman or his/her alternate and submitted to the CPCSEA Member Secretary within 15 days for the meeting to be valid.

Meetings of the IAEC are held on scheduled intervals as prescribed in the IAEC's SOP and additional meetings may be held if there are reasons to do expedited review. The proposals are sent to members at least 15 days prior to the meeting. Decisions are derived through consensus after discussions, and dissenting view points are recorded in the minutes. If consensus is not reached, the case should be referred to CPCSEA. Researchers will be invited to offer clarifications if necessary, and independent consultants/experts may be invited to offer their opinion on specific research proposals if needed. All decisions are recorded in the minutes of IAEC meeting and the Chairperson's approval is confirmed in writing with signature of all the IAEC members present.

The registered animal facility must maintain appropriate records such as curriculum vitae of all members of IAEC including training programs in animal ethics/welfare if attended; copy of all study protocols with enclosed documents, progress reports; minutes of all meetings duly signed by the Chairperson and the members; copy of all existing relevant national and international guidelines on animal ethics and laws along with amendments; copy of all correspondence with members, researchers, and other regulatory bodies; final report of the approved projects; record of breeding of animals, supply, etc., if breeding of animals is undertaken; record of import of animals with species, source, quantity, usage, etc.; record of all contract research, if conducted at the institute; health records of animals and care takers; stock register; water and feed analysis reports; and record of rehabilitation of large animals if done. All documents should be archived for period as prescribed by the IAECs; however, this should not be less than 1 year.

IAEC Members should be made aware of all relevant new guidelines and amendments to the rules and PCA Act. Also, members should be encouraged to attend national and international training programs/workshops/conferences in research ethics for maintaining quality in ethical review and be aware of the latest developments in this area.

In addition to the submission of minutes of IAEC meetings to CPCSEA within 15 days, the IAEC must also submit the inspection report of the animal house with photographs once in a calendar year to the CPCSEA. If action is required, the facility must provide the action taken report within 30 days of inspection.

Associations such as the Laboratory Animal Scientists' Association of India have been striving to update the professionals by conducting regular conferences, workshops, and training programs. Collaborations and global partnering with international organizations and associations have given a tremendous exposure to the professionals. The Federation of European Laboratory Animal Science Associations (FELASA) "C" training program in reality has commenced with more than 100 participants being trained. The FELASA accreditation has given a boost to this program conducted at the Tamil Nadu Veterinary and Animal Sciences University in Chennai.

OCCUPATIONAL HEALTH AND SAFETY

The Article 21 of the Indian Constitution guarantees the protection of life and personal liberty of a person. The right to health, medical aid to protect the health, and rigor to a worker while in service or postretirement is a fundamental right under Article 21 of the Indian Constitution read with Article 39(e), 41, 43, 48 A. The Factories Act 1948, the Mines Act 1952, the Dock workers (Safety, Health and Welfare) Act 1986 are some of the laws that regulate health of workers in an establishment.[18] The Employees' State Insurance Act 1948 and the Workers' Compensation Act 1923 are compensatory in nature. However, the implementation of the Occupational Health Laws in India reportedly is inconsistent.

In majority of animal facilities in India, neither the technical nor the scientific staff is aware of the diseases of laboratory animals. In the laboratory animal facilities, most of the acts mentioned previously are irrelevant; hence the application of these acts is limited. The CPCSEA rules are silent on the issue of occupational health of people working in animal facilities. In several animal facilities both in the government and private sector, the animals have a wide variety of infections and infestations[19] have been recorded. The exposure of the personnel working in these animal facilities and their immediate families is not known.

CONDUCT OF EXPERIMENTAL PROCEDURES

According to the IVC Act 1984, Section 30 and its amendment of 1992 G.S.R 395 (E), laboratory animal medicine, animal experimentation, and vaccine production comes under the veterinary profession.[8] As per the Act,

Registered Veterinary Practitioners alone are permitted as per the law to practice laboratory animal medicine and perform animal experiments. The Ministry of Environment & Forests, Government of India, passed the S.O. 134 (E) [15/2/2001]—The Breeding of and Experiments on Animals (Control and Supervision) Amendment Rules 2001.[6] According to the regulations, the IAEC is expected to ensure that experiments are conducted with care and humanity, and the IAEC may impose conditions as it may deem fit to ensure that animals are not subjected to unnecessary pain or suffering before, during, or after the performance of experiments on them.[5] While constituting the IAEC, it has made the named/attending veterinarian a mandatory member of the committee. Further, the rules have given permission to degree holders in medicine or veterinary sciences, postgraduate and above in life sciences/pharmaceutical sciences or any other natural sciences, degree or diploma holders in pharmacy, diploma or certificate in laboratory animal techniques sciences from a recognized institution as identified by the CPCSEA, the liberty to perform animal experiments. All the professional degree courses mentioned previously, with the exception of veterinary sciences where laboratory animal sciences are included in the syllabus, do not have laboratory animal sciences or laboratory animal medicine in their curriculum. The permission to conduct animal experiments without significant education and training in laboratory animal sciences has led to a situation where some research has progressed unfettered under conditions that may compromise the quality and conduct of the research.

The ethical concern and laboratory animal welfare is gaining more importance in India because of the strict guidelines stipulated by the regulatory authority, the CPCSEA.[20] The regulations and Form B do not specifically draw attention to the application of the 3Rs and offer no schema for severity classification of experiments. Only some researchers in some of the recognized government institutions and multinational companies in the private sector are aware of the 3Rs principles. In the case of regulatory data generation, most institutions follow the Organization for Economic Cooperation and Development (OECD) guidelines or the Drug Controller General of India (DCGI) rules, which are almost similar to the OECD guidelines. The LD50 test has been replaced with the fixed dose method or the up and down method. The Draize test has been replaced by the vaginal mucous membrane test in rabbits and the pyrogen test has been replaced by the limulus amebocyte lysate (LAL) test using the blood from horse shoe crab. However, the Indian regulatory authorities still continue to accept the in vivo pyrogen test results from rabbits.

The CPCSEA rules insist that the researcher use animals that are lower on the phylogenetic scale. However, the use of earth worms and frogs in education and teaching at high school (10+2) level and at undergraduate level has been banned since several species of frogs are now listed on the endangered list.

EUTHANASIA

The IVC Act 1984 (Act 52 of 1984) authorizes only the registered veterinary practitioners to practice veterinary medicine and surgery.[8] As per the law, only the registered veterinarians are supposed to anesthetize or euthanize animals, conduct necropsies, and issue health or death certificates in animals. In addition, the law authorizes only the registered veterinarians to give evidence at any inquest or in any court of law as an expert on any matter relating to veterinary medicine. As per the CPCSEA rules, the attending veterinarian has been entrusted with the authority as well as duty to administer anesthesia or euthanasia to animals.[5] Other professionals have also been permitted to use anesthetics and euthanasia agents, but under the supervision or advice of the attending veterinarian. The guidelines also emphasize that the animal should remain under veterinary care till it completely recovers from anesthesia and postoperative stress.

The CPCSEA Rules (SO 42 (E) 14/10/2001) Breeding of and Experiments on Animals (Control and Supervision) Amendment Rules 2005 have laid down the parameters for application of euthanasia in experimental animals.[21] An animal is to be euthanized if it is paralyzed and is not able to perform its natural functions; if it becomes incapable of independent locomotion; if it can no longer perceive the environment in an intelligible manner; if it has been left with a recurring pain during the course of an experimental procedure wherein the animal exhibits obvious signs of pain or suffering; or if the continuation of life of the experimental animal will be life threatening to human beings or otherwise.

The CPCSEA Guidelines provide information on the recommended methods to be followed for euthanasia. Under physical methods, the guidelines recommend exsanguination in rats, hamsters, guinea pigs, rabbits, and cats but not in mice, dogs, and monkeys. Decapitation and cervical dislocation are recommended in mice, rats, and hamsters. Inhalation euthanasia agents that have been approved are: carbon monoxide in all animals, carbon dioxide in all animals except dogs and monkeys; a combination of carbon dioxide and chloroform in all animals except dogs and monkeys and halothane in all animals; overdose of barbiturates in all animals; chloral hydrate only in rabbits, cats, dogs, and monkeys and not in mice, rats, hamsters, and guinea pigs. Ketamine has been approved for use in all animals; similarly, Sodium pentathol (Sodium pentobarbitone) has been approved for use in all animals. The table of acceptable euthanasia methods from the CPCSEA Guidelines is reproduced below in Table 8.1.[22]

The methods that have been disapproved are electrocution, decompressio,n and stunning under physical methods. Nitrogen flushing and Argon flushing under the inhalation drugs category are not recommended. Drugs such as curare; nicotine sulfate, magnesium sulfate, potassium chloride, strychnine, paraquat (N,N′-dimethyl-4,4′-bipyridinium dichloride), dichlorvos (2,2-dichlorovinyl dimethyl phosphate), and air embolism are not acceptable.

TABLE 8.1 Methods of Euthanasia in Laboratory Animals as per CPCSEA Guidelines

	Mouse	Rat	Hamster	Guinea Pig	Rabbit	Cat	Dog	Primate
a) Physical Methods								
Electrocution	NR	NR	NR	NR	NR	NR	NR	NR
Exsanguination	NR	A	A	A	A	A	NR	NR
Decapitation (for analysis of stress)	A	A	A	NR	NR	NR	NR	NR
Cervical dislocation	A	A	A	NR	NR	NR	NR	NR
b) Inhalation of Gases								
Carbon monoxide	A	A	A	A	A	A	A	A
Carbon dioxide	A	A	A	A	A	A	NR	NR
Carbon dioxide plus chloroform	A	A	A	A	A	A	NR	NR
Halothane	A	A	A	A	A	A	A	A
c) Drug Administration								
Barbiturate overdose (route)	A (IP)	A (IP)	A (IP)	A (IP)	A (IV, IP)	A (IV, IP)	A (IV, IP)	A (IV, IP)
Chloral hydrate overdose (route)	NR	NR	NR	NR	A (IV)	A (IV)	A (IV)	A (IV)

Continued

TABLE 8.1 Methods of Euthanasia in Laboratory Animals as per CPCSEA Guidelines—cont'd

	Mouse	Rat	Hamster	Guinea Pig	Rabbit	Cat	Dog	Primate
Ketamine overdose (route)	A (IM/IP)	A (IM/IP)	A (IM/IP)	A (IM/IP)	A (IM/IV)	A (IM/IV)	A (IM/IV)	A (IM/IV)
Sodium pentathol overdose (route)	IP	IP	IP	IP	IV	IV	IV	IV

A, methods acceptable; *CPCSEA*, Committee for the Purpose of Control and Supervision of Experiments on Animals; *IM*, intramuscular; *IP*, intraperitoneal; *IV*, intravenous; *NR*, not recommended.

Methods not acceptable for any species of animals.

a) Physical Methods
(i) Decompression, (ii) Stunning

b) Inhalation of Gases
(i) Nitrogen flushing, (ii) Argon flushing

c) Drug Administration
(i) Curariform drugs, (ii) Nicotine sulfate, (iii) Magnesium sulfate, (iv) Potassium chloride, (v) Strychnine, (vi) Paraquat, (vii) Dichlorvos, (viii) Air embolism

EDUCATION, TRAINING, AND COMPETENCE OF PERSONNEL

Institutions recognized by the Indian government as eligible to conduct scientific experimentation in animal subjects have the requirement to ensure the education, training, and competence of personnel in several areas according to the CPCSEA Guidelines.[23] The guidelines specify education and training expectations at every level of involvement in the program of care and use of research animals including the IAEC, the supervisory veterinarian or scientific personnel, investigators and technical staff involved in animal experimentation, and personnel involved in the basic care of animals. Training of all CPCSEA nominees has been made mandatory by the CPCSEA. The training programs (3 and 5 days) are being conducted at the National Institute of Animal Welfare regularly.

The task to train laboratory animal science personnel was taken up in a big way by the National Centre for Laboratory Animal Sciences, National Institute of Nutrition (ICMR), Hyderabad. Recently, the 36th Supervisors Training program (12-week hands-on training) and 48th Technicians Training program (6-week hands-on training) completed successfully. Over 1000 personnel have been trained over the last several years. A new 5-day program on Orientation in Laboratory Animal Sciences called National Certification Program in Laboratory Animal Sciences for Research Scholars is under consideration. The Tamil Nadu Veterinary and Animal Sciences University, Chennai, along with the CPCSEA, Laboratory Animal Science Association (LASA), and the FELASA commenced the FELASA level "C" training in 2013. Four batches have already completed the training. Other institutions such as the Indian Institute of Science, Bangalore; Central Drug Research Institute, Lucknow; Central Food Technological Research Institute, Mysore; and Advanced Centre for Training Research Education in Cancer, Navi Mumbai, have been contributing by conducting training programs in laboratory animal sciences periodically.

VETERINARY CARE

The veterinary care in the CPCSEA guidelines is closely aligned with and appears to be derived from the recommendations of the Guide for the Care and Use of Laboratory Animals, eighth edition.[16] As mentioned previously, adequate veterinary care must be provided although the direct involvement of a veterinarian is not required. Daily observation of animals can be accomplished by someone other than a veterinarian provided that a mechanism of direct and frequent communication allows timely and accurate information on problems in animal health, behavior, and well-being is conveyed to the attending veterinarian. Other areas of potential veterinary contribution are also recognized including the establishment of appropriate policies and procedures for ancillary aspects of veterinary care, such as reviewing protocols and proposals, animal husbandry and animal welfare; monitoring occupational health hazards containment, and zoonosis control programs; and supervising animal nutrition and sanitation. Institutional requirements will determine the need for full-time or part-time or consultative veterinary services. The CPCSEA guidelines highlight other interrelated and integrated programmatic considerations typically involving veterinary direction and oversight. These include animal procurement, source assessment, and receiving practices; quarantine, stabilization, and separation practices; and disease surveillance, diagnosis, treatment, and control provisions.

HOUSING AND ENRICHMENT

The CPCSEA guidelines recognize the importance of the cage environment and the quality of the animals. The guidelines also suggest that cages should be constructed of sturdy, durable materials, and designed to minimize cross-infection between adjoining units, and additional information is provided on the type of materials that are suitable for cage construction for small and large laboratory animal species. The space recommendation for laboratory animal housing in CPCSEA guidelines closely matches with the recommendations of the Guide for the Care and Use of Laboratory Animals, eighth edition.[16]

The cage sizes recommended in the table provided in Annexure 3 of the CPCSEA guidelines[24] are expected to accommodate the normal behavioral activities of laboratory animals adequately. The importance of monitoring animals for behavioral aberrations is mentioned in the CPCSEA guidelines in connection with veterinary care, species separation in housing to avoid interspecies conflicts, the acclimatization to restraint devices, and population density. The PCA Act[3] and CPCSEA guidelines do address the importance of the social environment to the expression of the basic natural behavior of the species in question and the ability of animals, especially those housed for long duration to exhibit natural locomotor patterns. Environmental enrichment has not been discussed as behavioral management tool or as a means to expand the repertoire of the natural behaviors that might be expressed by laboratory animals.

HUSBANDRY AND ENVIRONMENT

Information on all dimensions of the husbandry of laboratory animals is provided in the CPCSEA guidelines in a narrative, closely aligned, albeit much more general and abbreviated, with other national and international guidance documents in this area, particularly the Guide for the Care and Use of Laboratory Animals, eighth edition.[16] Topics summarized encompass housing systems; sheltered or outdoor housing: social environment: activity; food; bedding; water; sanitation practices; assessing the effectiveness of sanitation; waste control; vermin control; emergency, weekend, and holiday care; record-keeping and guidelines for laboratory animal management SOP.

Facility systems important to the control of environmental factors also receive mention, citing widely accepted general standards for temperature and humidity control, ventilation, power, lighting, and noise control. The parameters given in these aspects of environmental control closely mimic those provided in other national and international guidance documents.

TRANSPORTATION

Regulations and guidance for the transportation of laboratory animals is contained in the PCA Act,[3] the Transport of Animals Rules,[25] and the Transport of Animals (Amended) Rules,[26] as well as in the CPCSEA Guidelines and Annexure 4.[27] The CPCSEA guidelines note the importance of careful attention to transportation of laboratory animals citing the mode of transport, container type, population density in cages, food and water during transit, protection from transit infections, injuries, and stress as relevant considerations. Animals can be transported by many acceptable methods provided that transport stress is avoided and the containers used are of an appropriate size so as to enable these animals to have a comfortable, free movement and protection from possible injuries. The food and water should be provided in suitable containers or in suitable form to ensure adequate provisioning during transit. The transport containers (cages or crates) should be of appropriate size and each container should be populated to avoid overcrowding and infighting. Annexure 4 of the CPCSEA regulations provides specifications for containers and population densities for animals during transport.

ACCREDITATION OF ANIMAL FACILITIES

The registration of the animal facility and a fully functional IAEC is designated as an accredited facility by the CPCSEA. However, some organizations, especially Contract Research Organizations carrying out preclinical toxicology studies for the pharmaceutical industry felt the need to obtain international accreditation to attract business from other countries. Responding to the need of the hour, the Indian LASA had taken the initiative in 2009 to invite AAALAC International during its national convention. Since then, more than 20 organizations in the private sector have acquired AAALAC International accreditation. There have been instances where adherence to the AAALAC requirements would go contrary to the laws of the land. These issues need to be addressed.

REFERENCES

1. The Constitution of India, part IV, Directive Principles of State Policy, 48. Organization of Agriculture and Animal husbandry. Available from: http://lawmin.nic.in/olwing/coi/coi-english/Const.Pock%202Pg.Rom8Fsss(7).pdf.
2. Indian Penal Code 1860. Act No. 45 of 1860. Available from: http://ncw.nic.in/acts/THEINDIANPENALCODE1860.pdf.
3. The Prevention of Cruelty to Animals Act, 1960. No.59 of 1960. Available from: http://www.awbi.org/awbi-pdf/Act%20&%20Rules%20-%20English.pdf.

4. *S.O.732 (E), [26/8/1998] – The Experiment on Animals (Control and Supervision) (Amendment) Rules*. 1998. Available from: http://envfor.nic.in/legis/awbi/awbi04.pdf.
5. *S.O.1074, [15/12/1998] – The Breeding of and Experiments on Animals (Control and Supervision) Rules*. 1998. Available from: http://envfor.nic.in/legis/awbi/awbi10.pdf.
6. *S.O.134 (E), [15/2/2001] – The Breeding of and Experiments on animals (Control and Supervision) Amendment Rules*. 2001. Available from: http://envfor.nic.in/legis/awbi/awbi11.pdf.
7. *Standard operating procedure for institutional animal ethics committee, guidelines on the regulations on scientific experiments on animals, form B*. Ministry of Forestry and Environment, Animal Welfare Division; 2010. p. 106–10. Available from: http://cpcsea.nic.in/WriteReadData/userfiles/file/SOP_CPCSEA_inner_page.pdf.
8. The Indian Veterinary Council Act 1984 (Act 52 of 1984). Available from: http://www.vci.nic.in/writereaddata/IVC%20Act%201984.pdf.
9. The Veterinary Council of India [Standards of Professional Conduct, Etiquette and Code of Ethics for Veterinary Practitioners] Regulations 1992, GSR 395. Available from: http://www.vci.nic.in/writereaddata/code-cond-regulation-1992.pdf.
10. CPCSEA communication No. F. No. 25/197/2013-CPCSEA (Vol-III) dated 19th August 2016. Available from: http://cpcsea.nic.in/WriteReadData/LnPdf/Advertisement%20for%20Nominees%20of%20CPCSEA-1.pdf.
11. The Animal Welfare Act, 2011 (Draft). Available from: http://www.awbi.org/awbi-pdf/draftawact2011.pdf.
12. *CPCSEA report of the consultative group on review of the norms and practices for regulation of animal experimentation*. 2004. Available from: http://www.moef.nic.in/sites/default/files/CPCSEA_Mtg_Agenda_%20201204.pdf.
13. *CPCSEA care and management of equines used in the production of biologicals*. 2001. Available from: http://www.moef.nic.in/sites/default/files/equines.pdf.
14. CPCSEA amendment in the protocol for use of Equines for production of hyperimmune-sera (ASVS). Available from: http://cpcsea.nic.in/WriteReadData/userfiles/file/Amenment%20in%20Equine%20guidelines.pdf.
15. Biological Diversity Act, 2002 (18 of 2003). Available from: http://lawmin.nic.in/ld/P-ACT/2003/The%20Biological%20Diversity%20Act%202002.pdf.
16. NAS 2011 National Academy of Sciences. *Guide for the care and use of laboratory animals*. Washington, D.C.: National Academy Press; 2011.
17. *Standard operating procedure for institutional animal ethics committee, guidelines on the regulations on scientific experiments on animals*. Ministry of Forestry and Environment, Animal Welfare Division; 2010. p. 1–6. Available from: http://cpcsea.nic.in/WriteReadData/userfiles/file/SOP_CPCSEA_inner_page.pdf.
18. Pingle S. Occupational safety and health in India: now and the future. *Ind Health* 2012;**50**:167–71.
19. Manjunath S, Kulkarni PG, Nagavelu K, Samuel RJ, Srinivasan S, Ramasamy N, Hegde NR, Gudde RS. Sero-prevalence of rodent pathogens in India. *PLoS One* July 9, 2015. http://dx.doi.org/10.1371/journal.pone.0131706.
20. Bayne K, Ramachandra GS, Rivera EA, Wang J. The evolution of animal welfare and the 3Rs in Brazil, China, and India. *J Am Assoc Lab Anim Sci* 2015;**54**(2):181–91.
21. Ministry of Environment and Forests. *Breeding of and experiments on animals (control and supervision) amendment rules*. 2005. Available from: http://envfor.nic.in/legis/awbi/awbi42(e).pdf.

22. *Standard operating procedure for institutional animal ethics committee, guidelines on the regulations on scientific experiments on animals.* Ministry of Forestry and Environment, Animal Welfare Division; 2010. p. 53. [Annexure 6] Available from: http://cpcsea.nic.in/WriteReadData/userfiles/file/SOP_CPCSEA_inner_page.pdf.
23. *Standard operating procedure for institutional animal ethics committee, guidelines on the regulations on scientific experiments on animals.* Ministry of Forestry and Environment, Animal Welfare Division; 2010. p. 55. [Annexure 7] Available from: http://cpcsea.nic.in/WriteReadData/userfiles/file/SOP_CPCSEA_inner_page.pdf.
24. *Standard operating procedure for Institutional Animal Ethics Committee, guidelines on the regulations on scientific experiments on animals.* Ministry of Forestry and Environment, Animal Welfare Division; 2010. p. 43–50. [Annexure 3A–3G] Available from: http://www.aaalac.org/resources/SOP_CPCSEA_inner_page.pdf.
25. *No. 18-6/70-LDI, [23/3/1978] – The Transport of Animals Rules.* 1978. Available from: http://envfor.nic.in/legis/awbi/awbi07.pdf.
26. *S.O.269(E), [26/3/2001] – The Transport of Animals (Amendment) Rules.* 2001. Available from: http://envfor.nic.in/legis/awbi/awbi16.pdf.
27. *Standard operating procedure for Institutional Animal Ethics Committee, guidelines on the regulations on scientific experiments on animals.* Ministry of Forestry and Environment, Animal Welfare Division; 2010. p. 51. [Annexure 4] Available from: http://cpcsea.nic.in/WriteReadData/userfiles/file/SOP_CPCSEA_inner_page.pdf.

Chapter 9

Oversight of Animal Research in China

Kathryn Bayne[1], Jianfei Wang[2], Wanyong Pang[3]
[1]AAALAC International, Frederick, MD, United States; [2]GlaxoSmithKline R&D China, Pudong, P.R. China; [3]Sanofi R & D, Sanofi China, Beijing, P.R. China

INTRODUCTION

The most recent statistics from the World Health Organization (WHO) note that the global population growth and a concomitant increase in the longevity of the human population are leading to a rapid increase in the total number of middle-aged and older adults, with a corresponding increase in the number of deaths caused by noncommunicable diseases.[1] The WHO projects that the annual number of deaths due to cardiovascular disease will increase from 17 million in 2008 to 25 million in 2030, with annual cancer deaths increasing from 7.6 to 13 million. These health statistics also predict that the total number of annual deaths from noncommunicable disease will reach 55 million by 2030, while deaths from infectious disease are projected to decrease over the same period. Countries with rapidly expanding populations and resources to invest in health research, such as China, will clearly be among the leaders in advances in health research. Because health-based research generally involves animals at some stage of the research, animal-based research will continue to have a pivotal role in the health research portfolio of China for the foreseeable future. The Pan American Health Organization report on "Health in the Americas"[2] states: "The world's increasing connectivity, integration, and interdependence in the economic, social, technological, cultural, political, and ecological spheres…is one of the greatest challenges confronting the health sector." Thus, the linkage between China's approach to the conduct and oversight of animal research has significant implications for the rest of the world.

As evidence of this linkage, in recent years China has positioned itself to be an attractive provider of outsourcing capabilities in preclinical safety assessment studies, recognizing that biomedical research conducted in accordance with global standards can be a strong economic engine. China has also dramatically ramped up its animal research program to benefit the health of the

Laboratory Animals. http://dx.doi.org/10.1016/B978-0-12-849880-4.00009-X
Copyright © 2018 Elsevier Inc. All rights reserved.

domestic Chinese population. In addition, the cosmetic industry has recently been increasing its presence in China, both in terms of sales to Chinese women and for safety testing purposes. These initiatives have prompted increased attention to global standards of laboratory animal care and use resulting in the construction of new research facilities, additional government-approved producers of laboratory animals, increased government support for biotech and pharmaceutical companies, training opportunities for researchers and research support personnel, and the adoption of regulations that incorporate key globally accepted principles in laboratory animal care and use.

China's emergence onto the global stage of biomedical research and testing has not gone unnoticed. China is not immune to attention from animal rights/protection organizations. Animal activists have criticized China for requiring animal testing of numerous chemicals and products. For example, People for the Ethnical Treatment of Animals (PETA) Asia–Pacific advances ceasing animal experimentation on its website. The Chinese Animal Protection Network addresses many animal issues including what they refer to as "academic research of animal ethics" and an "awareness campaign of lab animal protection" and in 2008 organized the first World Lab Animal Day in China. Multinational corporations and academic institutions see opportunities for collaboration and expansion of scientific enquiry. The general public views this development with caution due to the number of recalls of products manufactured in China, ranging from pet food to toys, clothing, tainted candy sold in Asia, and other goods. Yet, China's rapid pace in maturing its biomedical research enterprise is a strong indicator that the culture readily embraces new knowledge and positive change.

The National Science Foundation (NSF) reports[3] that the pace of growth over the past 10 years in China's overall research and development (R&D) sector is approximately 20% annually, placing it as the second largest R&D country in the world. Indeed, since 1999, the percentage of gross domestic product (GDP) invested in R&D has almost tripled to 1.7%. Multinational companies are continuing to invest in Asia, and especially China. As the NSF reports, "the share of R&D performed by Asia-located affiliates (other than in Japan) increased from 5.3% to 14.4% from 1997 to 2008. In particular, the share of U.S.-owned affiliates R&D performed in China, South Korea, Singapore, and India rose from a half percentage point or less in 1997 to 4% in China…." To support this burgeoning field of research in China is a similar increase in the number of scientists, with the number tripling between 1998 and 2008, in sharp contrast to the reduced proportion of researchers in the United States and Europe that comprise the global researcher pool over the same period. From 1995 to 2002, the United States and Europe experienced an annual growth rate in number of researchers of 3%–4%, while the annual growth rate for China was 12%. If one views the number of research publications as an indicator of the contribution a country makes to the body of scientific knowledge, then the trend exemplified by China is noteworthy. The NSF report notes that the combined share of published articles by Europe and the United States declined from 69% to 58% in the period from 1995 to 2009, while Asia's share increased from 14% to 24%, largely due to China's 16% average annual growth in publications, with China displacing Japan as having the second highest publication rate in science and engineering. A *Newsweek Magazine* article entitled "It's China's World; we're just living in it"[4] highlighted the pivotal role China has assumed in major global issues, including trade, climate change, currency, and technology. As the NSF data demonstrate, China has assumed a similarly prominent role in life sciences research.

Regardless of the countries involved, when considering outsourcing R&D or engaging in an international collaboration that involves the use of animals, any potential differences in key animal care and use program elements should be identified and the impact of those differences on the animals' welfare and the quality of the science assessed. Critical areas that could directly impact laboratory animal welfare include animal procurement, transportation of the animal from the vendor to the place of study, the provision of adequate veterinary care, the training and competency of the personnel associated with the animal program, the animal's environment (both in the primary enclosure and in procedure areas), and the method of review and approval of the proposed work (i.e., ethical review).[5] Each of these program elements may be influenced by the economic, religious, and cultural experience of the personnel at the institution, as well as the degree of their training and level of authority within the institution. Exposure to animal facilities outside of one's own country is extremely informative; however, the cost and logistics of setting up such travel can be a factor that impedes this form of training. But the value of such interactions cannot be overstated. Indeed, China's "sea turtles" (individuals who train outside of China and then return to the country with specific expertise) return with an expanded concept of animal use and often retain contact with their overseas colleagues, thereby keeping open channels of discourse, learning and awareness of changing global societal mores.[6,7] Also, increased international collaborations and participation in international conferences further inform Chinese scientists and regulators regarding global standards in the care and use of laboratory animals.[7]

OVERSIGHT OF RESEARCH ANIMALS

The use of animals for research in China dates back to 1918, but the first formal national laboratory animal science meeting, organized by the Ministry of Science and Technology (MOST), was held in 1982.[8] In China, MOST is the responsible government agency for establishing regulations pertaining to the conduct of research. In October 1988 the State Council approved the "Statute on the Administration of Laboratory Animal Use," which was promulgated by Decree No. 2 of the State Council in November 1988. Thereafter, the MOST provided oversight of laboratory animal science development in China. Implementation of the Statute was achieved thorough the issuing of provincial laws, beginning with Beijing in 1996.[8] The regulations address quarantine, import/export, animal quality, personnel training, licensing for production and use of laboratory animals, biosafety, required documentation, and other relevant subjects. Specific topics within these regulations were subsequently expanded and updated, and new overarching regulations were published (most recently in 2006, "Guidelines on the Humane Treatment of Laboratory Animals").[9] The Statute has been under revision, and it is expected that laboratory animal welfare and the Three Rs (3Rs) principles will be incorporated into the new regulation (personal communication with the revision committee). Over the years, the State Regulations and standards regarding the care and use of laboratory animals have been supplemented with guidelines and policies developed in the provinces, autonomous regions, and municipalities. A Provincial Department of Science and Technology (PDST) oversees laboratory animal use at the provincial government level, with a local Administration Office of Laboratory Animal Use (e.g., the Beijing Administration Office of Laboratory Animal, BAOLA) being established to enforce regulations, issue use licenses, as well as conduct inspections and related management activities.[10] Both individual and institutional licenses are issued. In addition, the voluntary accrediting organization, AAALAC International provides a third party evaluation of animal research, testing, and teaching programs around the world and accredits approximately 60 institutions in China at the time of this writing. Fig. 9.1 shows the oversight framework for laboratory animals in China.

Certification and accreditation in China are governed by the Certification and Accreditation Regulation of China (State Council, 2003; http://www.cnca.gov.cn/rjwzcfl/flfg/fl/46822.shtml).[11] The national standard, "Laboratory animal institutions—General requirements for quality and competence," proposed and administered by Technical Committee 261 of the Standardization Administration of China (SAC/TC261) became effective in October 2014.[12] By referring to internationally recognized management tools and scientific achievements, this Standard promotes the humane treatment of research animals and the reduction in or avoidance of the use of animals in research through the full cycle of

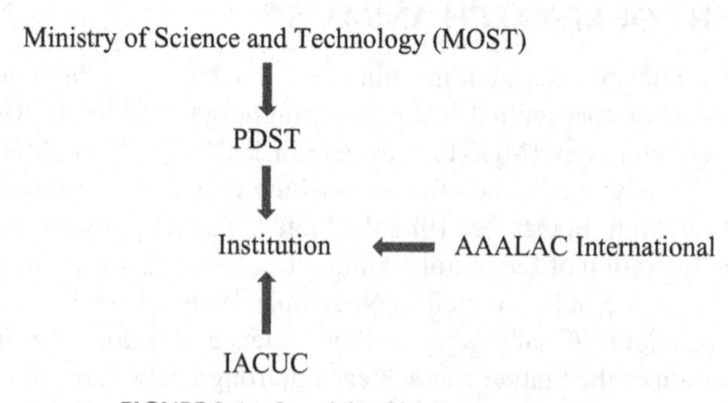

FIGURE 9.1 Oversight of laboratory animal use.

animal breeding and use. In addition, the Standard ensures the quality of laboratory animals and animal experiments, enhances the occupational health and safety of personnel, ensures protection of the environment, and promotes the scientific endeavors. The China National Accreditation Service for Conformity Assessment has initiated a trial accreditation program, using this standard as the reference, though such accreditation is not mandatory.

ANIMAL WELFARE PRINCIPLES

Kong and Qin have summarized the progression of inclusion of animal welfare in Chinese regulations.[13] Reference to research animal welfare dates back to 1988 in the Statute on the Administration of Laboratory Animals, which essentially addressed animal welfare through the provision of nutritious food, potable water, and qualified personnel working with the animals. Several years later, in 1997, mention is made of the 3Rs in MOST documents. Since 2005, the guidelines published by MOST have increasingly strengthened recommendations related directly to laboratory animal welfare, to include a requirement for provincial ethical review of proposed work with animals and regulations addressing husbandry, use, and transportation issues. Indeed, MOST has funded projects to study and translate animal welfare laws, regulations, and policies from other countries, such as the United States and Europe. The 2006 regulation (Guidelines on the Humane Treatment of Laboratory Animals), which is applicable to laboratory animal research institutions and scientists in China, cites noncompliance with laboratory animal policies as one of six "dishonorable behaviors," which can result in disciplinary action.[9] The MOST guidelines promote the 3Rs principles, and require each institution to establish a local committee to oversee all aspects of the care and use of laboratory animals. Five aspects of animal welfare are described, which include general requirements, husbandry, experimentation, transport, and implementation measures. The guidelines address space allowance (to allow natural behaviors) and environmental enrichment; providing an exercise opportunity for large animals such as nonhuman primates and canines; promoting the development and implementation of alternative methods to the use of animals; ensuing the minimization of animal fear and pain; restraining and restraint devices; using nonhuman primates only if required and ensuring "retirement" care for nonhuman primates. Implementation measures include an administrative licensure system, Institutional Animal Care and Use Committee (IACUC) approval of the protocol before initiation of the study, and personnel training. In addition, the guidelines encourage consideration of alternative methods to the use of animals, in accordance with scientific principles. They specifically note that humane endpoints should be used whenever possible and that euthanasia should be done in a manner that avoids pain and not in the presence of other animals. The MOST guidelines are extended by the Beijing Regulation for Administration of Laboratory Animals, which instructs investigators to design animal experiments to conform with the 3Rs principles.[8]

The Guideline of Beijing Municipality for Review of Welfare and Ethics of Laboratory Animal states: "The animal experiment procedure should be refined for the purpose of laboratory animal protection; reducing unnecessary number of animal used; the alternative methods should be used for replacement of animal experiment after evaluation."[10]

In addition to the MOST guidelines, the Chinese Food and Drug Administration (CFDA) regulations include requirements pertaining to animal care and use, for GLP studies. The Chinese Veterinary Medicine GLP regulations, which took effect on December 9, 2015, mandate the establishment of an institutional animal welfare review body and basic animal husbandry requirements in nonclinical research of veterinary medicines [Government Bulletin No. 2336, Ministry of Agriculture (MOA)].[14]

Of particular note, the final draft of Laboratory Animals—Codes of Welfare and Ethics, the first National Standard on Laboratory Animal Welfare in China, has been submitted to the National Technical Committee on Laboratory Animal Science, SAC for review. It is expected that the standard will be approved for implementation in late 2016. The National Standard will regulate the welfare, ethical review, and administration of the production, transportation, and utilization of the laboratory animals, including the technical requirements for the ethics review body, the principles of ethical review as well as the recommended ethical review processes and criteria, personnel qualifications, facilities, the responsibilities of the veterinarians, resources, technical procedures, animal care and use, occupational health and safety, transportation, record keeping, and documentation. The codes are applicable to the review, as well as the quality management of laboratory animal welfare and ethics (unpublished, personal communications).

The MOA has assigned the Chinese Veterinary Medical Association (CVMA) to draft Welfare Requirements of Laboratory Animals, a standard which is applicable to animals used in the agricultural research. The main references for this standard are the World Organization for Animal Health (OIE) Terrestrial Animal Health Code (2015)[15] and European Directive 2010/63/EU.[16] The standard is under final review at the MOA (unpublished, personal communications). This standard, as a minimum recommendation on laboratory animal welfare, could be applicable to all veterinarians, no matter what industry sector they are working in, since veterinarians in China are regulated by the MOA.

Examples of steps taken to implement the 3Rs include the acceptance of the Swordtail fish (*Xiphophorus helleri*) as a laboratory animal by the State Evaluation Committee of Fisheries Stock, for use in quality monitoring of water, testing of chemical toxicity, and as an animal disease model. In addition, the chicken egg has been established as an alternative to the Draize test and for evaluation of cosmetics ingredients. Also, the limulus amebocyte lysate (LAL) test has been embraced as an alternative to rabbit pyrogen testing. Notably, the proposed Animal Protection Law of the People's Republic of China, Article 89,[17] affirms "The State encourages domestic and international sharing of experimental data and information to reduce the number of experimental animals; promote alternative experimental methods, to reduce unnecessary animal experiments; optimization of the experimental method, technology, content and procedures, and avoid unnecessary suffering to the animals and injury." Reinforcing this approach, Article 95 of the proposed Animal Protection Law[17] states that procedures should be selected that use the least number of animals and that produce the minimum pain or distress that can produce satisfactory results. The draft was then renamed as Prevention Cruelty to Animals Act; however, at the time of this writing it has not yet been approved by the People's Congress.

AUTHORIZATION FOR THE USE OF LABORATORY ANIMALS

National Level

A Regulation on the Management of Laboratory Animal License System was issued by MOST in 2002,[18] which created a system of licensing of institutions that breed or use animals for research, is administered at the level of the province, municipality, or autonomous region. The Regulation details requirements and processes for license application, review, and approval as well as facility inspection and supervision. Kong and Qin report that 29 of 31 provinces have established administrative offices of laboratory animals (AOLAs).[13] In Beijing, for example, animals cannot be bred for research or used in research without the appropriate license. Institutions using animals for research are obligated to use "qualified" animals, in other words, animals produced at licensed breeding facilities, which meet specific genetic and microbiological criteria. The institution applying for a license must have standard operating procedures in place; have animal welfare and biosafety management systems; have provisions in place for training of staff and for the health protection of staff; meet the national standards for facility environment; meet the national standards for quality of feed and water, bedding, cages, and support equipment; as well as employ staff who must receive the training organized by local administrative office for laboratory animals, have knowledge of laboratory animals, and have passed the qualifying examinations. There are additional requirements for animal research institutions that conduct studies using infectious agents, radiation, or chemicals (i.e., where public safety may be compromised) or that conduct research on wildlife. The application submitted to BAOLA must be accompanied by several supporting documents, such as testing reports for the facility environment, proof of medical examination of the personnel, a description of the ethical review system, animal facility plans, etc. Site evaluation by an expert panel, as arranged by BAOLA, is usually conducted using a check list, with reference to relevant national standards. Facility and personal licenses are valid for a period of 5 years. Upon expiration, a reinspection process is initiated for facility license renewal. An individual has to attend the training organized by the local AOLAs again if he or she wishes to maintain the license. Continuing education for the license holder is encouraged. Should the institution wish to change the scope of work conducted or make facility modifications, must apply to the Science and Technology Commission of the Municipality/Province for authorization of the change(s).

Annual inspections of animal facilities are also under the authority of the Science and Technology Commission of the Municipality/Province. The inspection process relies on a combination of random sampling and a "self-check." The inspection includes a review of the work records; records of training of personnel; records of physical examinations of personnel; sources of the animals and methods of transportation to the facility; the quality of feed, bedding, cages, and ancillary equipment; the management system and operating rules; the operating conditions of the animal facility; implementation of biosafety practices; animal welfare and ethical review systems; and whether the number and types of animals used conforms with the license. Violations or forging of the license is punishable by law. Work conducted outside the scope of the license renders the animal experiments invalid. The results of the inspection are submitted to the State Science and Technology Commission of the People's Republic of China. Institutions that fail the annual inspection have a 3 month period to effect corrections and are then subject to reinspection. Should the institution again fail the inspection, the license is revoked. However, the institution can reapply for a license once all corrections have been made and the institution complies with the standards.

Local Level

A local authorization system serves as an adjunct to the national licensing system. The Regulations of Beijing Municipality for the Administration of Affairs Concerning Laboratory Animals[19] defines a laboratory animal as "those animals that are artificially fed and bred, the microorganisms and parasites on or in whose bodies are kept under control, whose genetic backgrounds are definite or whose sources are clear and that are to be used in scientific research, teaching, production, examination and verification and other scientific experiments. In accordance with the controls on microorganisms and parasites, laboratory animals are classified into conventional animals, clean animals, animals carrying no specific pathogens and animals carrying no bacteria."

In Beijing, the Welfare and Ethics Committee (referred to as the Ethics Committee) is responsible for the inspection of the welfare of the animals and for implementing an ethical review system. The Committee must be comprised of at least five people. The Chairman of the Ethics Committee should be a specialist in laboratory animals, and there is a preference for a veterinary surgeon to fulfill this role. By regulation, the institution is responsible for providing training of the Ethics Committee members and to ensure Committee membership is maintained in accordance with the

3–4 year appointments. Also, in accordance with the regulations, the Ethics Committee members "shall commit to preserve the welfare and ethics of laboratory animals."

LABORATORY ANIMAL QUALITY CONTROL

To guarantee the quality of laboratory animals, various national standards (referred to as "GB") have been established in China. In 1994, the China State Bureau of Technical Supervision issued a series of national standards, 47 of which applied to laboratory animals. The 2001 revision of these standards increased the number to 83.[20] In 2005, the National Technical Committee on Laboratory Animal Science of the SAC was established and set 93 standards for laboratory animal quality control covering five areas: microbiology, genetics, feedstuff, environment, and specific pathogen-free (SPF) chickens. These national standards are mandatory requirements, and are referred to wherever appropriate, with regard to laboratory animals, such as when the CFDA is certifying GLP laboratories. These standards are engineering standards, leaving very limited flexibility for a performance-based approach. In addition, other government agencies, such as the ministries of health and agriculture, have set laboratory animal quality standards in their sectors. Also, some of the more developed provinces and cities, such as Beijing (e.g., laboratory fish standards, laboratory minipig standards, technical requirements for laboratory animal environments and facilities), Shanghai (laboratory minipig standards), Guangdong (laboratory animal microbiologic testing, laboratory animal diet standards, and macaque care standards), Jiangsu (caging standards, laboratory pig standards), and Yunnan (e.g., Tree Shrew standards) have developed and implemented their own standards.

The Laboratory Animal Development Program for the Ninth Five-Year Plan established a national lab animal quality control network, which determines requirements for laboratory animal quality monitoring centers overseen by MOST and the PDSTs. MOST subsequently issued the "Guideline and Detailed Criteria for the Review of Provincial Laboratory Animal Monitoring Centers."[21] The task of these centers is to explore research on testing techniques, train technicians, carry out site inspections and annual review of provincial laboratory animal monitoring centers, and arbitrate disputed monitoring results. The quality monitoring network comprises six national centers covering microbiology, parasitology, genetics, pathology, diets, environment, and facilities, as well as 26 provincial centers in 23 provinces.

OVERSIGHT AND ETHICAL REVIEW PROCESS

In accordance with the "Examination Guideline for the Welfare and Ethics of Laboratory Animals," Article 4, the institution must establish a Welfare and Ethics Committee comprised of staff from management, scientific and technical personnel, professional veterinary staff, and persons who are not members of the institution. The Committee is responsible for the welfare and ethical review of laboratory animals and supervision and administration of this activity. The Committee's role is increasing in importance as MOST implements reforms to discourage scientific misconduct.[22]

The Beijing Experimental Animal Welfare Ethics Review Guidelines specify several activities of the Ethics Committee.[10] These duties include review and supervision of breeding, transportation, feeding, experimental design, and procedures. The Ethics Committee is charged with ensuring implementation of animal welfare and ethical principles. The Committee is also instructed to take into account the interests of animal welfare and the research in its assessment of the proposed work. The research application (protocol) should include information regarding the significance of the project, the need for animals, the expected harm to animals, and a detailed description of the animal welfare and ethical issues involved in the project. The animal welfare issues that must be addressed include overall general requirements of the animal, husbandry and care, experimental objective, transportation, and procedural methods. Prevention and mitigation of pain and distress are a particular focus of the Guidelines, which call for adherence to the 3Rs during each stage of the animal experiment. In addition, the final disposition of the animals must be described, and the method of euthanasia must be the least painful. Any activity involved with animal husbandry and animal experimentation shall start only after approval from the committee, and is expected to receive routine supervision (i.e., postapproval monitoring). The composition, independence, roles and responsibilities of IACUC, or Ethics committee are described in this guideline, and recommendations on general principles for protocol review are also provided. Of note is that applicants can request to be present during the review, for example, to answer questions, and can request a specific member(s), who may have violated confidentiality or impartiality during the review, to recuse himself/herself. The Guidelines state that the review by the Ethics Committee should be independent, impartial, scientific, democratic, transparent, confidential, not influenced by politics, commercial interests, or self-interest. Finally, the Guidelines outline several reasons that would result in a protocol being rejected or suspended. Activities of the local Ethics Committees that are found to be noncompliant with the Guidelines are subject to a "rectification decision" by the Beijing Experimental Animal Management Office.

The Ethics Committee is responsible for establishing institutional regulations, review procedures, supervision systems, conducting regular meetings, establishing a reporting system, establishing training programs, and reporting the list of members to the overarching Welfare and Ethics Committee of Laboratory Animals of Beijing. Of note, the standard requires that the Ethics Committee work independently in its inspections and supervision of the research, breeding, feeding, production, management, and transportation of the laboratory animals and in confirming that the experiments conform with the animal welfare and ethical principles, to include a harm-benefit analysis. Specifically, the Ethics Committee is charged with taking into account the benefits of the animal research while "comprehensively evaluating the injuries that animals suffer from the necessity of using animals." The Committee must provide a report of this ethical review. To assist the Committee in this determination, the study application must describe the significance of the project, its purpose, and the expected harm to the animal.

A draft "Animal Protection Law of the People's Republic of China" was developed in 2009 by a committee chaired by Chang Jiwen, Director of Social Law Research at the Chinese Academy of Social Sciences.[17] The draft law is broadly encompassing, addressing wildlife, companion animals, food animals, laboratory animals, and other animal uses. Chapter 6 of the draft law focuses on "Legal Protection for Laboratory Animals." Of note is a recommendation that a national laboratory animal ethics committee should be established to supervise and manage the scientific use of laboratory animals, which is comprised of at least one veterinarian and one representative of an animal protection organization. There is a further recommendation that each province also establish an animal ethics committee with similar composition to the national committee. The draft law proposes that the national and provisional ethics committees publish basic principles or guidelines on animal experimentation.

REUSE

The Guideline on Humane Treatment of Laboratory Animals addresses in general terms the welfare of animals used in research, but does not specifically touch on the topic of reuse.[9] However, it should be noted that the subject is under consideration. Specifically, the draft "Animal Protection Law of the People's Republic of China" (Article 95) proposes, "The recording of the number of times laboratory animals are used and the aims of the experiments should be the responsibility of a specified person."[17] The draft law also states (Article 96) that "The same animal may not undergo the same experiment repeated in the same test cycle. If such an experiment is necessary it should be examined and approved by the laboratory animal ethics committee."

SETTING FREE/REHOMING

The draft "Animal Protection Law of the People's Republic of China" does not support the release of research animals.[17] Specifically, it states in Article 99, "Work units, individuals and organizations engaged in animal experimentation should safeguard biosecurity and public health and safety, and the abandonment or release into the wild of laboratory animals is prohibited."

OCCUPATIONAL HEALTH AND SAFETY

China has established a classification system for hazardous agents that is based on pathogenicity to people or animals, as follows:[23–25]

- "Hazard Level IV (High Individual Hazard, High Group Hazard)": pathogens that can cause serious disease; they cannot be cured in general cases; and can be transmitted between people or between animals and people, or between animals through direct contact, indirect contact, or "casual" contact.
- "Hazard Level III (High Individual Hazard, Low Group Hazard)": pathogens that can cause serious disease or cause serious economic loss, but cannot be transmitted through casual contact in general cases or be treated with antibiotics or antiparasitic drugs.
- "Hazard Level II (Medium Individual Hazard, Limited Group Hazard)": pathogens that can cause people or animals to become infectious, but do not pose a serious hazard to healthy operators, groups, domestic animals, or the environment in general cases. Laboratory infections do not cause serious disease; there are effective therapeutic and preventive measures and a limited transmission risk.
- "Hazard level I (Low Individual Hazard, Low Group Hazard)": bacteria, epiphytes, viruses, parasites, and other biological agents that do not harm healthy operators or make animals infectious.

Microorganisms in Hazard Levels III and IV are collectively referred to as "highly infectious" agents and can only be studied in laboratories that are authorized by the government as Class III or IV. Protective measures for the use and transportation of these agents depend on the biosafety level of the agent. Both engineered safety features in the facility (e.g., airlocks, high efficiency particulate air (HEPA) filtration, directional airflow, biological safety cabinets, and safety hoods), and use of protective equipment at the individual level are required to achieve personnel protection (e.g., goggles, safety shield, mask or respirator, gloves, shoes). In addition, procedures such as hand washing, minimizing aerosols in the workplace, safe use and disposal of sharps, etc. are described for each biosafety level. Biohazard signage is required for level 2 and higher areas and must be prominently displayed. Biosafety equipment must be inspected annually to ensure appropriate function and personnel protection.[24]

Chemical and radiological safety standards address storage, handling, use, and disposal practices. Documentation of control measures is required and a safety supervisor is designated at the institution.[26]

A plan for emergency evacuation of personnel must be in place at the institution.[23] Personnel must participate in at least one fire drill each year and must be familiar with the evacuation plan.

In addition to the national standards addressing occupational health and safety, Beijing Municipality regulations (Article 12)[10] require that "Institutions dealing with laboratory animals shall take protective measures to ensure the health and safety of employed persons….[and] employed persons shall be physically examined each year. Persons who are unsuitable for undertaking affairs relating to laboratory animals due to health conditions shall change their positions."

All laboratory animals are subject to isolation and quarantine for infectious disease control before their entry into destination buildings or research laboratories. The relevant vaccination programs and infectious disease control policies must be followed, according to the Law on Animal Disease Prevention and Control. Biosafety regulations for laboratories using microbiologic pathogens must be followed, as appropriate. The regulations on medical waste disposal (State Council, 2003; http://www.lascn.com/Item/12024.aspx)[27] is the only regulation relatively applicable to the management of waste generated in laboratory animal production and research.

ALTERNATIVES

Consideration of the use of alternatives in animal testing is gaining significant momentum in China. The MOST Guidelines[10] encourage consideration of the 3Rs when designing an animal study, and laboratories in government agencies have been funded to evaluate Replacement strategies (e.g., swordtail fish, silkworm, horseshoe crab, earthworm), especially in toxicology studies. Factors contributing to this include increasing public attention within China to animal welfare issues, such as harvesting bear bile and the consumption of shark fin soup, as well as China's entry into the World Trade Organization. China could encounter trade restrictions if animal testing procedures are not conducted at an international level or, in some cases, if nonanimal alternatives are not used. Much of this pressure has been felt in the cosmetics industry. A 2003 amendment to the EU's Cosmetics Directive called for a timetable to implement a progressive ban on the use of laboratory animals in testing cosmetic products. Since 2004, animal testing of final cosmetics products has been banned and, as of 2013, cosmetics tested on animals cannot be marketed in the European Union. A result of this action is the pressure on countries, such as China, that manufacture cosmetics ingredients.

The First International Symposium on Cosmetics—Alternatives to Animal Experimentation for Cosmetics was held in Beijing in 2011 and participants called for more focused action to reduce the number of animal toxicity tests and to increase the use of alternatives, with a specific request for assistance from the West in developing nonanimal alternatives.[28] The CFDA requires animal toxicology test reports prior to licensing a new cosmetic ingredient; however, this requirement is undergoing review due to concern expressed by countries where nonanimal testing methods have been demonstrated to be valid and due to the economic pressure on Chinese companies that conduct the nonanimal tests for the export of ingredients, but still must conduct animal tests for the

CFDA and domestic Chinese market. In February 2012, the CFDA proposed the first in vitro method for cosmetic phototoxicity.[29] The test uses a BALB/c 3T3 fibroblast cell line to test acute phototoxic effects of cosmetic ingredients on the skin. The test conforms with "OECD guidelines for the testing of chemicals 3T3 NRU phototoxicity test (no. 432, 2004)."

Despite this significant step toward increasing reliance on nonanimal testing methods, recent actions taken by the CFDA in recategorizing some cosmetics will likely increase the use of animal testing.[30,31] In China, cosmetics are categorized as either ordinary use (e.g., hair care, nail care, skin care, perfumes, etc.) or special use (e.g., hair growth, hair color, hair removal, spot removal, sun block, etc.).[32] Ordinary cosmetics are reviewed at the provincial level and often are sold without animal testing. Special use products require a State Registration, which necessitates animal testing for eye and skin irritation. In recent years there have been reports of safety issues associated with nonspecial use cosmetics. The proposed recategorization is intended to strengthen consumer safety by moving some cosmetics into the special use category, which requires more intense testing, including animal testing to produce safety assessment data. Nevertheless, in a long run, the cosmetics industry may become the first sector to give up animal testing as a result of legislation. But, for the pharmaceutical, agrochemical and the chemical sectors in general, there is no clear timetable for replacing animal testing.

EDUCATION AND TRAINING

The training provided to personnel working in the animal research environment is generally determined by the provincial government.[33] For example, in Beijing, the BAOLA is responsible for the professional training of supervisory staff and assessing their training. The supervisory staff at an institution are provided with a training syllabus for the personnel in their workplace. Staff are required to take an examination established by BAOLA, using a randomized set of questions based on the syllabus, and administered at the institutional level. In Beijing, for personnel to work with animals, they must obtain a "Position Qualification Certificate of the Laboratory Animal Practitioners in Beijing." Continuing education and a training plan for personnel must be in place. Records of training must be maintained. Supervisory staff must be university graduates with a degree in a relevant field (e.g., medical science, biology, zootechnical science, veterinary medicine).

The Chinese Association for Laboratory Animal Science (CALAS) was established in 1987, and although it is independent from MOST, it has a nationwide significant role in the self-regulation of laboratory animal science through supervising provincial associations and in providing continuing education and training. Several colleges of medicine, veterinary medicine, pharmacy, or biotechnology throughout the country offer undergraduate and graduate programs in laboratory animal science.

There is a professional society standard entitled "Laboratory Animals—Requirements for Laboratory Animal Practitioner" TB/CALAS 00001-2016, which was issued in December 2015 and became effective on January 1, 2016.[34] According to this standard, laboratory animal practitioners are grouped into six categories: laboratory animal technicians (assistant laboratory animal technician [ALAT], laboratory animal technician [LAT], and laboratory animal technologist [LATG]), managers of laboratory animal resource (manager of laboratory animal resource [MAR], senior MAR [SMAR]), laboratory animal veterinarians (assistant laboratory animal veterinarian [ALAV], laboratory animal veterinarian [LAV], senior LAV [SLAV]), laboratory animal researcher (junior laboratory animal researcher [JLAR], laboratory animal researcher [LAR], and senior LAR [SLAR]), laboratory animal associate support practitioner (LASP), and laboratory animal phased practitioner (LAPP). The corresponding qualifications, training, certification, competency requirements are described in the standard.

China has a long-established open attitude to inviting nonnationals to provide training. As early as 1992 China initiated training of laboratory animal scientists and technicians through collaboration with the Japanese International Cooperation Agency (JICA). The formal collaboration continued through 1997. The goal was to organize a system of study and to elevate training to an international level, with an emphasis on breeding, managing, and using laboratory animals. A significant output of this initiative was the development of knowledgeable professionals who could then assume the role of trainers, thereby leading to sustainability of the project. This pattern of inviting subject experts into China to provide topic-specific training has continued at a steady pace. Symposia and workshops are routinely organized to disseminate training in IACUC function, veterinary care roles, environmental enrichment, and other related subjects throughout their research community. Examples include the annual China Pharmaceutical R&D Summits; the "Shanghai Laboratory Animal Welfare Sharing Conference" cohosted by the Office of Shanghai Administrative Committee for Laboratory Animals, Global Research Education & Training (GR8), and AAALAC International; Peking University; and others.[35]

IMPORT, EXPORT, AND TRANSPORTATION OF LABORATORY ANIMALS

Institutions that plan to import breeding animals from overseas as animal "seeds" must register in the designated laboratory animal seed centers administered by MOST. Institutions that plan to export laboratory animals must apply to the relevant PDST, before exportation.

Quarantine of imported and exported laboratory animals is covered by the Law of the People's Republic of China on the Entry and Exit Animal and Plant Quarantine, 1992[36] (Adopted at the 22nd Meeting of the Standing Committee of the Seventh National People's Congress and promulgated by Order No.

53 of the President of the People's Republic of China on October 30, 1991, http://www.npc.gov.cn/englishnpc/Law/2007-12/12/content_1383874.htm) and associated regulations, Law of the People's Republic of China on the Protection of Wildlife (1989),[37] Convention on International Trade in Endangered Species of Wild Fauna and Flora[38] (CITES, the CITES Office is affiliated with General Administration of Forestry in China), and Quarantine and Health Requirements for SPF Animal to be Imported into China (AQSIQ, trial, personal communication), and Quarantine and Health Requirements for SPF Animal Genetic Material to be Imported into China (AQSIQ, trial, personal communication).

Transportation of animals should comply with domestic regulations and guidelines as well as relevant IATA requirements. The MOST guidelines[9] stipulate that the cages used to transport animals must provide adequate ventilation, they must be clean, and the animals must be secure. Mixing of animal species or strains/lines in the same cage is prohibited. Food and water, as well as veterinary care, must be available if the duration of transport is long. In addition, Beijing Municipality Regulations[10] stipulate that the transfer facilities and transportation equipment must be in line with the pathogen status of the animal. An animal health certificate/quarantine certificate is needed from a local competent veterinary authority for cross-province animal transport in China and for international transport. It is important to note that for cross-province purchase and transport of nonhuman primates, an application must be submitted to and an approval be granted from the local competent Forestry authority.

Animal rights activists have successfully pressured Chinese airlines to cease shipments of nonhuman primates overseas. In February 2012, Hainan Airlines stopped a specific shipment of monkeys from China to Canada, and announced it would cease "its engagement in such shipments." In April 2012, the US Department of Agriculture (USDA) cited Air China for improper transport procedures after a monkey was injured in transit from Beijing to New York. In May 2012, the USDA again cited Air China for two violations of the US Animal Welfare Act based on the escape of a monkey on a flight from Beijing to New York City and for handling the monkey in a way that posed harm to the animals. A subsequent e-mail and telephone campaign by animal rightists directed at Air China resulted in the company also ceasing shipment of nonhuman primates internationally.

HOUSING AND ENRICHMENT

Cage size standards for common species of laboratory animals used in China are contained in the standard GB 14925-2010.[39] These are summarized in Table 9.1 (small laboratory animals) and Table 9.2 (large laboratory animals and chickens). Standards for other species are being developed. For example, a standard for zebrafish established by the Hunter Biotechnology, Inc. has been adopted as the local provincial standard in Zhejiang, and is being discussed for adoption as the national standard. For institutions accredited by AAALAC International, cage size recommendations in the *Guide* apply.[40]

TABLE 9.1 Cage Sizes Required in GB 14295-2010 for Small Laboratory Animals

	Mouse (g)		Rat (g)		Guinea Pig (g)		Hamster (g)		Rabbit (kg)	
	<20	>20	<150	>150	<350	>350	<100	>100	<2.5	>2.5
Individual (m^2)	0.0067	0.0092	0.04	0.06	0.03	0.065	0.01	0.012	0.18	0.2
Mother with litter or in group (m^2)	0.042		0.09		0.76 (height 0.21)		0.08		0.42 (height 0.4)	
Minimum height (m)	0.13	0.13	0.18	0.18	0.18	0.21	0.18	0.18	0.35	0.4

TABLE 9.2 Cage Sizes Required in GB 14295-2001 for Large Laboratory Animals and Chickens

	Cat (kg)		Dog (kg)			Nonhuman Primate (kg)			Minipig (kg)		Chicken (kg)	
	<2.5	>2.5	<10	10–20	>20	<4	4–8	>8	<20	>20	<2	>2
Individual (m^2)	0.28	0.37	0.60	1.0	1.5	0.5	0.6	0.9	0.96	1.2	0.12	0.15
Mother with litter (m^2)	–		–			–			–		–	
Minimum height (m)	0.76	0.76	0.8	0.9	1.1	0.8	0.85	1.1	0.6	0.8	0.4	0.6

Concern for animal welfare in Chinese society is increasing across multiple sectors of animal "use." As an example, Davey and colleagues have documented this both in student attitudes toward animal welfare[41] in general and in zoos.[42] In the first instance, he polled students about their level of concern for animal welfare across a variety of issues. His results affirmed that Chinese society has generally positive attitudes toward animal welfare initiatives. In the latter study he and his colleagues evaluated the impact on the Chinese public of zoo exhibits that included environmental enrichment. The enriched exhibit promoted greater zoo visitor interest, measured by the duration of the visit to the exhibit and other parameters. Specific to the laboratory animal environment, the *Guidelines on the Humane Treatment of Laboratory Animals* state that the facility and environment must provide for the animals' behavioral and physiological needs, including the provision of environmental enrichment.[9] Also, the topic of the animal's housing environment, including enrichment, is being discussed at conferences more often (such as http://www.ibclifesciences.com/china/overview.xml). Davey and Wu[42] assessed attitudes of Chinese university students about animal research. Of the university students surveyed, concern was strong regarding the use of animals for safety testing of cosmetics and household products and there was a tendency for students to disagree with the statement that "humans have the right to use animals as we see fit." As the authors noted, these students will enter the workforce and some may become directly involved in animal use. Trends regarding the treatment of laboratory animals in China will likely be shaped by their views, and thus topics such as the provision of environmental enrichment will be of increasing importance.

HUSBANDRY AND ENVIRONMENT

The National Standard of Laboratory Animals,[20] first published in 1991 and revised in 2001, 2006, 2008, and 2010, defines the quality of the animals and the conditions for their housing and care. The National Technical Committee 281 on Laboratory Animal Science of the SAC (TC281) was created in 2005 and expanded the process of implementing standards. The standards address the pathogen status and genetic monitoring of the animals, environment and housing facilities, as well as feeding and nutrition. The standards apply to both breeding and research institutions and require that the facility accommodate the animals' behavioral and physiological needs, to include the provision of environmental enrichment. The current number of national standards is 93; however, MOST is evaluating the implementation of additional standards, such as regarding the transportation of animals. The Beijing Municipality Regulations[10] require that "Production environments and facilities of laboratory animals shall comply with standard requirements of laboratory animals of different classes. According to the corresponding standards, laboratory animals of different classes and breeds shall be managed in different environments and facilities. Up-to-standard supplies such as feedstuffs, cages and bedding materials etc.

shall be used." The regulations (Section IV, Article 24) also prohibit conducting experiments in the same room on animals that may interfere with each other (behaviorally or due to their pathogen status).

Cages should not contain toxic materials and should resist corrosion and high temperatures; they should be impact resistant; and they should be easy to clean, disinfect, and/or sterilize. Bedding material should provide good absorption, should have little dust, no "foreign smells," should be nontoxic, and contain no grease or impurities. Feed should only be purchased from licensed manufacturers. The feed should be accompanied by a certificate of conformity. Drinking water provided to the animals should conform with the standards for domestic potable water in the city, or in the case of barrier housed animals, water must be sterilized.

VETERINARY CARE

Veterinary medical training in China has traditionally been associated with agriculture and the production of healthy animals to feed the country's growing population, and had a more technical emphasis.[43] The veterinary medical education program is typically 4–5 years postsecondary education and addresses a wide range of relevant basic science and veterinary medical topics. In addition, the cultural roots of Traditional Chinese Veterinary Medicine remain a key component of contemporary veterinary medical education in China. However, it has been suggested that students do not learn to synthesize the individual courses into a cohesive approach to evaluate, diagnose, and treat the patient due to a gap in clinical training.[44] In March 2012 the CVMA and the International Veterinary Collaboration for China (IVCC), a consortium of veterinary schools in the United States, the United Kingdom, and Zoetis, formalized a memorandum of understanding to enhance veterinary education in China by supporting Chinese students (who have already completed a 4- to 5-year program in a veterinary school in China) in pre–veterinary courses for 1 year (10–15 candidates per year), followed by an opportunity to apply to veterinary school at one of six participating overseas universities. Funding for the students is made available by the China Scholarship Council and covers all of their expenses, including travel back to China after they complete their Doctor of Veterinary Medicine education. This represents a significant step in advancing the quality of veterinary medical practice and teaching that will be made available in China, and is in accord with a recommendation made for input from international educators several years ago.[43]

In addition, the Chinese Ninth Five-Year Plan called for government actions to strengthen institutions of higher education, specifically with the goal of assisting universities "reach the advanced international standards for the overall quality of teaching, scientific research, and the training of professional manpower, so as to establish their international prestige and position among universities in the world," referred to as Project 211.[33] The NSF has reported that research and development (R&D) expenditures in China increased by a record 28% in 2008–09—a time when R&D growth stopped and decreased in the US and Europe—and the 2010 data show an additional 22% increase.[3] Therefore, China's investment in intellectual development is well placed and strategically sounds.

The CVMA, established in 2009, is the first professional organization representing veterinarians in China, and it is taking a leadership role in improving the quality of education provided to veterinary students. As China's population grows, there is increasing pressure for well-trained veterinarians to support greater numbers of food and companion animals, and these initiatives represent a central recognition of the demand for a better educated professional to meet these socioeconomic demands. Nation-wide veterinary licensing in China also began in 2009. Eligible candidates must pass the Chinese veterinary licensing examination to be licensed to practice veterinary medicine in China (Regulation on Administration of Licensed Veterinarian, MOA, 2008, amended in 2013, http://www.gov.cn/gzdt/2008-12/09/content_1172844.htm).[45]

Concomitantly, there is an amplified need for veterinarians who are skilled and knowledgeable in the specialty of laboratory animal medicine (LAM). The inadequacy of specialty training in the Chinese veterinary medical education program is recognized within and outside of China,[35] though there is a branch established under the CVMA for veterinarians working with laboratory animals, and numerous training sessions have been held for Chinese veterinarians to begin to close the gap. Bayne et al.[6] have proposed a laddered system of training options (ranging from a simple mentor system to board certification) to address inadequacies in veterinary training and qualifications around the world, and follow-up actions are taking hold.

For example, CALAS provides educational opportunities, and certificates of education are offered by several universities that provide training in laboratory animal science topics, to include medicine.[8] In 2016, CALAS set up a LAM working group. The working group has been assigned to set up a LAM curriculum, training plan, and certification program that are suitable for China, referencing the aforementioned TB/CALAS 00001-2016, Laboratory Animals—Requirements for Laboratory Animal Practitioner,[34] and taking into account the status of Chinese veterinary medical training the regulatory environment and the like.

Despite the changes occurring in the education system for veterinarians, other obstacles must still be overcome. Prominent among these is the limited number of pharmaceutical-grade anesthetics and analgesics available in China.[44] Agents that may result in human addiction are not routinely accessible, thus the choice for optimal drugs is severely restricted. Permits for the importation of certain drugs may be applied for, but the process can be quite time consuming and challenging.

The perception of the veterinarian as a professional is increasing, due in part to demands from Western clients working with contract research organizations or pharmaceutical companies based in China. Western clients express clear expectations that the veterinarian will be a partner in the research enterprise, will be trained and competent in the management and care of the species used, and will have a credible role on the IACUC. In addition, institutions that seek accreditation of their animal care and use programs by AAALAC International must conform with the standards for veterinary care in the *Guide*[40] and with AAALAC's Position Statement on the roles and responsibilities of the veterinarian (http://www.aaalac.org/accreditation/positionstatements.cfm#vetcare).

CONDUCT OF EXPERIMENTAL PROCEDURES

Researchers using animals must obtain a personal license issued by MOST, through the PDSTs, that is based on attending a training course and examination, as well as supervision. The license is valid for 5 years. In addition, an IACUC must review the proposed work and must approve it before the work can begin. The institution where the research is conducted must also be licensed for animal use (and animal production facilities undergo a separate licensing process), and similarly, this license is valid for 5 years, though annual reviews and site visits are performed. It is illegal for procedures to be conducted without a license. The "Regulations of Beijing Municipality for the Administration of Affairs Concerning Laboratory Animals," Section IV, Article 26,[10] states that "Individuals dealing with animal experiments shall design experiments based on the principles of substitution, reduction and optimization." In this manner, it is stipulated that the 3Rs must be addressed in the proposed research. As a corollary to the consideration of the 3Rs, in accordance with MOST guidance,[9] humane endpoints should be used whenever possible for reducing or eliminating unnecessary pain or distress.

The draft "Animal Protection Law of the People's Republic of China"[17] recommends that "The State encourages the sharing of experimental data and material domestically and internationally, in order to reduce the numbers of laboratory animals used; the State also promotes alternative [replacement] experimental methods, in order to reduce the number of unnecessary animal experiments; and refinement of experimental methodology, technology, content and procedures in order to avoid causing animals unnecessary suffering and harm." The draft law further suggests (Article 97), "If intense pain is going to be caused to laboratory animals, anesthesia, painless methods or other methods which avoid pain, suffering, distress or lasting harm, should be used. In cases where owing to the particular aims of the experiment, anesthesia cannot be used or there is no practicable way of using anesthesia in the experiment, the requirements of the basic guidelines on animal experimentation should be followed."[17]

EUTHANASIA

In accordance with MOST guidance,[9] euthanasia should be carried out in a manner that avoids pain and other animals should not be present when euthanasia is performed. The draft "Animal Protection Law of the People's Republic of China" states in Article 98, "On conclusion of the experiment, the health of the laboratory animals should be examined immediately.[17] If as a result of the experiment an animal has lost part of a limb or organ or is injured, and in continuing to live it will have to endure lasting pain, suffering, distress or harm, in compliance with the prerequisites of the basic principles or guidelines on animal experimentation, the animal may be promptly destroyed using a humane method."

EQUIPMENT AND FACILITIES

Project 985 (named for the year (1998) and month (May) of its implementation) was launched to promote the development of China's higher education system, to include the construction of new research centers and the improvement of facilities. To that end, China has invested in the construction of animal facilities that provide physical environments for the conduct of sophisticated animal research, though older facilities also remain in use.[46] In 2010, Kong and Qin[8] stated that there were 320 production facilities and more than 1530 facilities licensed for laboratory animal use. Approximately 100 standards have been issued by the Chinese government to promote quality control, and animal housing and use facilities are addressed in these standards. However, the Laboratory Animal—Requirements of Environment and Housing Facilities (GB 14925-2010)[39] and Architectural and Technical Code for Laboratory Animal Facility (GB 50447-2008)[46] are the main national standards with respect to lab animal facilities.

In addition, the "Regulations of Beijing Municipality for the Administration of Affairs Concerning Laboratory Animals," Section IV, Article 25 (The Utilization of Laboratory Animals)[10] states, "The use of up-to-standard laboratory animals and environments and facilities at the corresponding levels shall be taken as basic requirements in the research projects submitted for approval and in assessing the results of such projects, performances of verifications and inspections as well as in producing products with laboratory animals as productive materials. The results of animal experiments obtained by using non-conforming laboratory animals or in non-conforming laboratory environments are invalid. Moreover, products produced thereof shall not be sold."

In Beijing, construction requirements stipulate that interior walls should be smooth and even so that they can be cleaned and disinfected easily. Materials that are water resistant, shock resistant, and "hard to peel off" are required for interior walls of the facility. The floor must not be slippery and should resist abrasion and corrosion. Its surface should be smooth and should not have cracks. Additional standards are imposed for barrier facilities that address the environment outside the facility as well as inside construction requirements. For example, the regulations state: "The outdoor environment shall be in good order and hygienic. No puddles, weeds, garbage, diluvial soil or breeding ground for mosquitos and flies shall exist."[19] Insect and rodent pest control must be implemented outside the facility. The interior elements of the barrier are also described, to include walls, floors, ceilings, corridors, doors, power supply, activity areas for dogs and monkeys, air supply and exhaust, water supply, temperature, humidity, etc.

OVERSIGHT OF NONHUMAN PRIMATES IN RESEARCH

China is one of the richest nonhuman primate habitats in the world. Of the more than 200 living species of primates in the world, 4 families, 7 genus, 23 species, and 39 subspecies are found in China. Six subspecies of *Macaca mulatta* are distributed in more than 20 provinces in China. For example, in the Yunnan province, 15 species of nonhuman primates are found, making it the richest primate resource in China, which in turn facilitated it becoming the earliest province to undertake primate research.

During the mid-20th century, Chinese scientists began investigating the diversity of nonhuman primate species, their geographic distribution, habitat, ecology, morphology, etc. In the 1960s–1970s, scientists started to breed nonhuman primates for research in physiology as well as disease prevention and treatment. In the early 1980s, laboratory nonhuman primates, such as rhesus monkeys, were bred at a large scale at the Shanghai Laboratory Animal Center, Chinese Academy of Sciences. In 1982, MOST organized the first meeting on laboratory animals and as a result four National Laboratory Animal Centers were set up; one of the four centers was a primate breeding center. Since the late 1980s, more than 40 monkey farms or nonhuman primate resource centers that breed rhesus and cynomolgus monkeys have been established. Those breeding farms are located mostly in the Guangdong, Guangxi, and Yunnan provinces. In 2006, there were approximately 108,940 monkeys in China, of which 86,200 were cynomolgus and 22,740 were rhesus monkeys. However, the number of nonhuman primates has increased over time. To date, there are more than 200,000 cynomolgus and 30,000 rhesus monkeys in China. All animals are in a closed breeding environment; the animals used in biomedical research and drug discovery are F2 animals with a well-defined microbiological status (Table 9.3). As of 2011, it was estimated that 25,000 monkeys were used domestically and about 25,000 were exported to overseas such as the United States and European countries.[47]

Primate facilities must strictly follow the national standard (GB 50447-2008) for architectural and technical requirements.[46] Monkey breeding and transportation, export of animals, as well as tissue and blood are conducted under a permit from the Ministry of Forest in China and are also subject to MOST regulation. Each monkey farm is allowed to provide a specific, limited number of animals for export and domestic use yearly. Violation of this quota is subject to severe punishment, including a significant fine and even imprisonment.

TABLE 9.3 National Microbiological Standards for Nonhuman Primates

	Bacteria	Viruses	Parasites
Conventional	*Salmonella* spp. Pathogenic dermal fungi *Shigella* spp. *M. tuberculosis*	Cercopithecine herpesvirus type 1 (BV)	Ectoparasites *Toxoplasma gondii*
Specific pathogen free (Conventional plus...)	*Yersinia enterocolitica* *Campylobacter jejuni*	Simian retrovirus D (SRV) Simian immunodeficient virus (SIV) Simian T lymphotropic virus type 1 (STLV-1) Simian pox virus (SPV)	All Helminths *Entamoeba* spp. *Plasmodium* spp. Flagellates

OTHER

China has established a network of national laboratory animal production centers to supply national needs. The goal is to "scientifically protect and manage the resources of laboratory animals…so as to achieve an assurance of the quality of breeds."[48] To participate in the National Resource Centre for Laboratory Animals, the institution must go through a formal application process and meet certain standards for quality control.

In addition, attention is paid to genetically modified animals through the Statute on the Administration of Transgenic Organism Biosafety in the Agriculture System issued by the MOA.[49] This statute requires the establishment of a cross-ministry collaborative network to address the biosafety of transgenic organisms. MOA has also issued related regulations covering safety evaluation, export and import supervision, labeling systems, product review, and advertising. Further oversight is achieved by the Regulation on Sanitation Oversight of Transgenic Food[50] and the Regulation on the Import and Export of Transgenic Products[51] issued by the Ministry of Health (MOH) and the General Administration of Quality Supervision, Inspection and Quarantine of China (AQSIQ), respectively. These regulations aim to prevent genetic contamination of humans, the environment, and the ecosystem.

CONCLUSIONS

China has established a complex matrix of regulations at the state, provincial, and local level to address laboratory animal welfare. It is clear that China intends to establish an environment for the conduct of biomedical research that is on a similar level with the international community. Yet, like the oversight framework in any country, establishment of regulations is just a first step. Implementation of the regulations and guidelines must be enforced with tangible consequences (e.g., suspension of funding) should noncompliance be detected. In many countries this is achieved by inspections from government representatives and the submission of annual reports that summarize the institution's use of animals (e.g., types of research, species and numbers of animals used, etc.). While the regulatory environment is important, regardless of how the standards are developed and implemented by oversight bodies, daily compliance by personnel at the institution is key. The institutional "culture" has a profound impact on the degree of compliance with requirements as well as attention to quality science and animal welfare. Klein and Bayne[52] propose that the establishment of a culture of care, conscience, and responsibility is as or more important than the regulatory framework itself because it promotes an environment of workplace integrity, ethics-based decision making, good communication of institutional expectations, clear lines of authority, and a system for continuous development and improvement of the animal care and use program. When this institutional culture is established, regardless of the details of the regulatory framework, institutional personnel will ground their decisions and actions in an ethical context that will promote quality science and animal welfare.

China is a very diverse region overflowing with energy, talent, and potential. There is no doubt that China is increasingly becoming a global focus for the conduct of animal-based biomedical research. Kong and Qin[8] estimate that 19 million laboratory animals were produced and 16 million animals were used in China in 2010. It has been estimated that over 100,000 people work in the field of laboratory animal science. Likely, those numbers have increased in the intervening time and will continue to grow. It is important that collaborations with Chinese colleagues continue with the goal of promoting training and competency in LAM and science for IACUC members, veterinarians, researchers, and other staff involved in the production and use of laboratory animals as one way to harmonize standards in laboratory animal welfare and quality science.

REFERENCES

1. World Health Organization. *World Health Statistics 2012*. Available from: http://www.who.int/gho/publications/world_health_statistics/2012/en/index.html.
2. *Health in the Americas*. 2007 ed. vols. 1 and 2. ISBN 9275116229. Pan American Health Organization/World Health Organization. Available from: http://www.paho.org.
3. National Science Foundation. *Science and engineering indicators 2012*. Available from: http://www.nsf.gov/statistics/seind12/c4/c4h.htm.
4. Foroohar R, Liu M. It's China's World; we're just living in it. *Newsweek Mag* March 22, 2010:36–9.
5. Bayne K, Bayvel ACD, Williams V. Laboratory animal welfare: international issues. In: Bayne K, Turner PV, editors. *Laboratory animal welfare*. New York: Elsevier; 2014.
6. Bayne K, Bayvel D, MacArthur Clark J, Demers G, Joubert C, Kurosawa T, et al. Harmonizing veterinary training and qualifications in laboratory animal medicine: a global perspective. *ILAR J* 2011;**52**(3):393–403.

7. National Research Council. *International animal research regulations: impact on neuroscience research: workshop summary*. Washington, DC: The National Academies Press; 2012.
8. Kong Q, Qin C. Laboratory animal science in China: current status and potential for the adoption of Three R alternatives. *ATLA* 2010;**38**:53–69.
9. Ministry of Science and Technology (MOST). *Guideline of humane treatment of laboratory animals*. 2006. Available from: http://www.most.gov.cn/fggw/zfwj/zfwj2006/200609/t20060930_54389.htm.
10. MOST. *Guideline of the Beijing municipality on the review of welfare and ethics of laboratory animals*. Beijing, People's Republic of China: Beijing Municipality Administration Office of Laboratory Animals (BAOLA); 2005. Available from: http://bjxkz.lascn.com/system_manager/news_manager/UploadFile/250/250_3.doc.
11. State Council. *Certification and accreditation regulation of China*. 2003. Available from: http://www.cnca.gov.cn/rjwzcfl/flfg/fl/46822.shtml.
12. Technical Committee 261 of the Standardization Administration of China (SAC/TC261). *Laboratory animal institutions – general requirements for quality and competence. GB/T 27416-2014*. 2014. Available from: http://www.chinesestandard.net/PDF-English-Translation/GBT27416-2014.html.
13. Kong Q, Qin C. Analysis of current laboratory animal science policies and administration in China. *ILAR J* 2010;**51**:E1–10. Available from: http://dels-old.nas.edu/ilar_n/ilarjournal/51_1/PDFs/v51(e1)Kong.pdf.
14. Ministry of Agriculture. *Government Bulletin No. 2336. People's Republic of China*. 2015.
15. World Organization for Animal Health (OIE). *Terrestrial animal health code*. 2016. Available from: http://www.oie.int/international-standard-setting/terrestrial-code/access-online/.
16. The European Parliament and the Council of theEuropean Union. Directive 2010/63/EU of the European Parliament and of the Council of 22 September 2010 on the protection of animals used for scientific purposes. *Off J Eur Union* 2010. L 276/33–79.
17. *Animal Protection Law of the People's Republic of China*. Available from: http://www.actasia.org/uploads/file/Feedback%20on%20draft%20APlegislation.pdf.
18. MOST. *A regulation on the management of laboratory animal license system*. 2002. Available from: http://www.lascn.net/policy/law/nationlaw/200805/17.html.
19. Beijing Municipal Government, Committee of Science, & Technology. *Regulations of Beijing municipality for the administration of affairs concerning laboratory animals*. 2005. Available from: http://www.baola.org/ContentDetail/zhengceFG.aspx?CategoryId=48&contentId=7.
20. National Bureau of Quality Monitoring, Inspection & Quarantine. *National standards for laboratory animals*; 2001, 2006. Available from: http://www.baola.org/ShowContentList2.aspx?CategoryId=24.
21. *Guideline and detailed criteria for the review of provincial laboratory animal monitoring centers*. Available from: http://www.lascn.net/policy/ShowClass.asp?ClassID=4.
22. Hennig W. Bioethics in China. *EMBO Rep* 2006;**7**(9).
23. *Law of the People's Republic of China on prevention and control of infectious diseases*. Available from: http://www.chinacdc.cn/flfg/wsfl/200507/t20050725_41129.htm.
24. *Laboratory bio-safety management regulation for microbiological pathogens*. Available from: http://www.moa.gov.cn/fwllm/zxbs/xzxk/bszl/201104/t20110416_1970325.htm.
25. *Classification list of animal pathogens*. Available from: http://www.moa.gov.cn/fwllm/zxbs/xzxk/bszl/201104/t20110416_1970333.htm.
26. *Fire control law of the People's Republic of China*, Sixth, Article 14, [Chapter 2]. Available from: http://www.mps.gov.cn/n16/n1282/n3493/n3763/n4198/434859.html.
27. State Council. *Medical waste management regulations*. People's Republic of China State Council; 2003. Order No. 380. Available from: http://www.lascn.com/Item/12024.aspx.

28. *First international Symposium on cosmetics – alternatives to animal experimentation for cosmetics*. Beijing, China: China Cosmetics Research Center (CCRC) of the Beijing Technology and Business University with major support from the Food Licensing Department of the SFDA; 2011.
29. China FDA. *Comment letter on alternative method of phototoxicity study through 3T3 neutral red*. Letter no. 45. 2012.
30. China FDA. *Comment letter on the management of ordinary cosmetics*. Notice no. 57. 2012.
31. China FDA. *Comment letter on amended management of ordinary cosmetics*. Notice no. 263. 2012.
32. China Ministry of Health. *Hygienic standard for cosmetics*. 2007.
33. China Education and Research Network. *Project 211: A brief introduction*. Available from: http://www.edu.cn/20010101/21852.shtml.
34. Chinese Society of Laboratory Animal Standards. *Laboratory animals- requirements for laboratory animal practitioner. TB/CALAS 00001-2016*. 2016. Available from: https://translate.google.com/translate?hl=en&sl=zh-CN&u=http://www.calas.org.cn/upimg/2016/0829/201608292.pdf&prev=search.
35. Bayne K. Animal welfare in Asia: the AAALAC international experience. *ALN World* 2011;**4**(1):11–9.
36. Law of the People's Republic of China on the Entry and Exit Animal and Plant Quarantine. *Order of the president of the People's Republic of China No. 53*. 1992. Available from: http://www.asianlii.org/cn/legis/cen/laws/lotprocoteaeaapq722/.
37. Law of the People's Republic of China on the Protection of Wildlife. *Order of the president of the People's Republic of China No. 9*. 1989. Available from: http://www.china.org.cn/english/environment/34349.htm.
38. *Convention on International Trade in Endangered Species of Wild Fauna and Flora (1973 et seq.)*. Available from: https://www.cites.org/eng/disc/text.php.
39. General Administration of Quality Supervision, Inspection and Quarantine of China. *Laboratory animal – requirements of environment and housing facilities (GB 14925–2010)*. Standardization Administration of the People's Republic of China.
40. National Research Council. *Guide for the care and use of laboratory animals*. Washington, DC: The National Academies Press; 2011.
41. Davey G. Chinese university students' attitudes toward the ethical treatment and welfare of animals. *J Appl Anim Welf Sci* 2006;**9**:289–97.
42. Davey G, Wu Z. Attitudes in China toward the use of animals in laboratory research. *ATLA* 2007;**35**:313–6.
43. Yin JC, Li GX, Ren XF. An overview of veterinary medical education in China: current status, deficiencies, and strategy for improvement. *J Vet Med Educ* 2006;**33**(2):238–43.
44. James A, Kelly H. East meets west: the reformation of veterinary education in China. *ALN Mag* May 7, 2012. Available from: http://www.alnmag.com/article/east-meets-west-reformation-veterinary-education-china?page=0,4.
45. Ministry of Agriculture. *Decree No. 18. Measures for the administration of licensed veterinarian*. 2013. Available from: http://www.gov.cn/gzdt/2008-12/09/content_1172844.htm.
46. General Administration of Quality Supervision, Inspection and Quarantine of China. *Architectural and technical code for laboratory animal facility (GB 50447-2008)*. Standardization Administration of the People's Republic of China. Available from: http://www.lascn.net/policy/ShowClass.asp?ClassID=4.
47. China Laboratory Primate Breeding and Development Association. Available from: http://clpa.org.cn/organization.asp; 2012.

48. MOST. *Guideline for national resource Centre for laboratory animals*. 1998. Available from: http://www.la-res.cn/la-res/website/publicinfo/showArticleInfo.jsp?articleId=7021b060-c7de-4a31-8daf-3c7074532678.
49. Ministry of Agriculture (MOA). *Statute on administration of transgenic organism bio-safety in agriculture system*. 2001. Available from: http://www.agri.gov.cn/xzsp_web/bszn/t20031105_133969.htm.
50. Ministry of Health (MOH). *Regulation on sanitation administration of transgenic food*. 2002. Available from: www.fsi.gov.cn/news.view.jsp?id=17596.
51. General Administration of Quality Supervision, Inspection and Quarantine of China (AQSIQ). *Regulation on the entry and exit transgenic product*. 2004. Available from: http://www.aqsiq.gov.cn/xxgkml/jlgg/zjl/20032004/200610/t20061027_12236.htm.
52. Klein HJ, Bayne KA. Establishing a culture of care, conscience, and responsibility: addressing the improvement of scientific discovery and animal welfare through science-based performance standards. *ILAR J* 2007;**48**:3–11.

Chapter 10

Laws, Regulations, Guidelines, and Principles Pertaining to Laboratory Animals in Far East Asia

Tsutomu M. Kurosawa[1], Jae-Hak Park[2], Chou-Chu Hong[3]
[1]*Kagoshima University, Kagoshima, Japan;* [2]*Seoul National University, Seoul, Korea;* [3]*Level Biotechnology, Inc., New Taipei City, Taiwan*

JAPAN

Background and Current Situation

The importance of laboratory animal science was becoming more recognized in the early 1950s in Japan. The oldest national association in Asia is the Japanese Association of Laboratory Animal Science (JALAS) founded in 1951. Biomedical scientists have contributed to the establishment of the national legislation system in their own countries by the reflection of public concern over animal welfare. The Japanese government does not formally publish its legislative documents in English and therefore there is a paucity of Japanese laboratory animal welfare information in other countries. However, the advancement of translation technology through the Internet may help us to understand the Japanese legislative framework in this issue.

The Act on Welfare and Management of Animals (Law No. 105, 1973) (Act) was published in 1973.[1] The Act was influenced by Western countries, in particular the United Kingdom. The Act was revised several times and the most recent revision was made in 2012. The Act included for the first time the Three Rs tenets (Replacement, Reduction, and Refinement) at the revision in 2005. Before the revision in 2005, Refinement had been the only one stated, and Reduction and Replacement were then added to the Act concerning the use of laboratory animals. Under the Act, the Standards Relating to the Care and Management of Experimental Animals (Notice No. 6 of the Prime Minister's Office 1980) was published.[2] The Standards were revised several times and the new revision was made in 2013. At the revision in 2006, the title of the

Laboratory Animals. http://dx.doi.org/10.1016/B978-0-12-849880-4.00010-6
Copyright © 2018 Elsevier Inc. All rights reserved.

Standards was changed and the current title is the Standards Relating to the Care, Management and Relief of Pain of Experimental Animals. Euthanasia of laboratory animals is regulated by the Standards Relating to the Care and Management of Experimental Animals (Notice No. 6 of the Prime Minister's Office 1980).[2] The usage of genetically modified animals is restricted by the Law Concerning the Conservation and Sustainable Use of Biological Diversity through Regulations on the Use of Living Modified Organisms, which is in accordance with the Cartagena treaty.[3] This law may be the first law to restrict the usage of animals for research in Japan. Many violations of this law have been reported because of the paucity of practical information for compliance in research settings. The importation of nonhuman primates (NHPs) is restricted by the Invasive Alien Species Act[4] and the Act on Prevention of Infectious Diseases and Medical Care for Patients of Infectious Diseases.[5] The institutions using NHPs should be registered and the importers of NHPs need a license for importation and quarantine. The Japanese government restricts countries from which NHPs can be imported to Japan.

With these laws and standards, the Ministry of Education, Culture, Sports, Science and Technology and the Ministry of Health, Labor and Welfare compiled fundamental guidelines for the proper conduct of animal experimentation and related activities in academic research institutions under the jurisdiction of the Ministry of Education, Culture, Sports, Science and Technology, and basic policies for the conduct of animal experimentation in the Ministry of Health, Labor and Welfare. The Ministry of Agriculture and Fisheries also published a guideline for the proper conduct of animal experimentation.[6]

The Science Council of Japan prepared the Guidelines for Proper Conduct of Animal Experiments (the Guide) to serve as a reference material or a model when research institutions compile their own regulations for animal experimentation in accordance with the foregoing fundamental guidelines and basic policies in 2006.

These guidelines indicate the responsibility of the president of each research institution to carry out animal experimentation. The president of each research institution was asked to make an institutional regulation to control animal experiments and to establish an Institutional Animal Care and Use Committee (IACUC). However, the duty of the IACUC is thought to be limited to a review the protocol but not to oversight of the animal care program. The involvement of a veterinarian as a member of the IACUC is not stated. There are no descriptions of veterinary care and laboratory animal veterinarians' responsibilities to look after laboratory animal wellbeing in these guidelines. Generally speaking, so-called self-regulation is emphasized and the legal restriction of scientific activities is thought to trounce the research, therefore the legal system for laboratory animal welfare is not stringently enforced in Japan.

Scope

The Act on Welfare and Management of Animals[1] covers all aspects of the use and management of animals in Japan. Under the Act, the Standards Relating to the Care, Management and Relief of Pain of Experimental Animals regulates usage of animals for scientific purposes including research, education, testing, and manufacture of biological products. In the Act and Standards, animals are defined as mammals, birds, and reptiles. Neither fish nor cephalopods are included as laboratory animals.

The Principles

The Act on Welfare and Management of Animals[1] focuses on the use and management of animals but does not emphasize laboratory animal welfare. The relief of pain of laboratory animals is mandatory as Refinement, but Replacement and Reduction are both just suggestions. Laboratory animal welfare and the legislation of research institutions using animals are not prescribed.

Ethical Review

There is no indication for IACUCs in the Act,[1] which means they are not mandatory by law. However, the IACUC should be the body for ethical review of the protocol for animal experimentation as indicated by the various guidelines.[6] The president of the research institution is responsible for overseeing animal experimentation. The IACUC members are nominated by the president including researchers conducting animal experiments, laboratory animal specialists, and other persons with knowledge and experience.

Institutional and Designated Personnel Responsibilities and Qualifications

The Guide by the Science Council of Japan[6] describes the following positions. The president of the institution bears the final responsibility for all experiments conducted in the institution, prepares the facilities considered necessary for the proper care and management of the laboratory animals and the proper and safe conduct of the animal experiments, and appoints a manager of the facilities and a laboratory animal manager who has knowledge and experience related to laboratory animals. The president of the institution also provides education for related persons including researchers and animal technicians with the cooperation of the facility manager and the laboratory animal manager to inform them of related laws and policies.

In each institution, in-house regulations including the authority and responsibilities of the president of the institution, standard operating procedures for the conduct of animal experiments, proper care and management of laboratory animals, and methods of maintenance and management of facilities should be established based on the policies.

According to the guidelines[6] an IACUC should be established in each institution. The president of the institution should request the IACUC to review the animal experiment protocols submitted by principal investigators based on scientific rationale and in consideration of animal welfare. The president of the institution then approves or does not approve the protocol based on the report of the IACUC. After completion of the animal experiment, the president of the institution examines the results obtained and instructs the principal investigator and manager to make improvements based on advice of the IACUC.

However, the competencies and qualifications of these personnel are not stated.

Personnel Training, Education, and Competency

The Guide of the Science Council of Japan[6] states the following.

The president of the institution should endeavor to provide separate education and training of the laboratory animal manager, researcher(s), and animal technicians as required. Education and training should be provided prior to engaging in animal experiments and should also be provided later as required.

Education and training should be conducted in accordance with in-house regulations, and the dates of instruction, educational content, and the names of the instructor and those receiving instruction should be recorded and retained.

Education and training content should be specified in the in-house regulations taking into account activities undertaken in the institution. From the viewpoint of proper conduct of animal experiments, the topics included in education and training should be related to:

- pertinent laws and ordinances, bylaws, guidelines, in-house regulations;
- the conducting of animal experiments, etc., and the handling of laboratory animals;
- care and management of laboratory animals;
- safety assurance;
- use of facilities.

Animal Care and Use Areas Addressed by the Regulations

The Act on Welfare and Management of Animals[1] does not address the different areas of a laboratory animal care and use program. The Standards[2] describe several conditions for the proper facilities to hold laboratory animals.

The Guide of the Science Council of Japan[6] defines the proper conditions for laboratory animal facilities.

Animal Transportation

Laboratory animal transportation is not regulated by the Act on Welfare and Management of Animals.[1] The transportation of genetically modified animals is regulated by the Law Concerning the Conservation and Sustainable Use of Biological Diversity through Regulations on the Use of Living Modified Organisms.[3] The prevention of escape of animals is a main concern. Animal welfare during transportation of laboratory animals is briefly addressed in the Standards.[2] Under Article 6 of the Standards, when laboratory animals are transported, the following items have to be considered to ensure the health and safety of laboratory animals and also to prevent a hazard to the public:

a. The method of transportation should be selected with the shortest time to minimize exhaustion, pain, and distress.
b. Proper feeding and watering as necessary and ventilation of the vehicle to maintain a proper temperature should be carried out.
c. With consideration of the physiology, ecology, and behaviorology of laboratory animals, vehicle sand cages should of suitable size and design to ensure the health and safety of laboratory animals and to prevent their escape.
d. Necessary measures should be taken to prevent environmental contamination by microorganisms held by laboratory animals.

Physical Plant and Environmental Conditions

According to the Guide of the Science Council Japan,[6] the following conditions are defined.

Respecting the opinions of the laboratory animal manager, the facility manager should provide the conditions necessary to meet the requirements for conduct of the research, for animal physiology, ecology, and behavior, and sanitary management while establishing and administrating the facilities. Cleaning and disinfecting floors, inside walls, ceilings, and auxiliary equipment in the facilities should be easy, and the structure should facilitate maintenance of sanitary conditions to eliminate the possibility of laboratory animals being injured by projections, holes, depressions, inclined planes, etc. Inspections and maintenance of cages should be conducted to prevent laboratory animals escaping, to keep laboratory animals physically comfortable, and to make hygiene management and daily operations easy.

Housing or Enclosures

The following items should be considered when setting up facilities according to the Guide of the Science Council of Japan[6]:

- Equipment for the care and management of laboratory animals, sanitation equipment for cleaning and disinfecting materials and devices, and experimental equipment.
- Assurance of structures and strength to prevent entry of wild animals.
- Assurance of structures and strength to prevent laboratory animals escaping.
- In facilities where animal experiments on infection with pathogens or using radioactive materials are conducted, effective equipment to prevent escape of infected animals, laboratory animals administered chemicals, and laboratory animals treated with radioactive materials.

- Facilities with the necessary structure to reduce odor and noise, and storage facilities required for waste material.
- Within the limits that do not interfere with achieving the objective of the animal experiment, proper space, temperature, humidity, ventilation, lighting, etc., to ensure laboratory animals are not subjected to excessive stress.
- Depending on the animal species and the objective of the experiment, air conditioning equipment necessary to maintain a constant environment in the facilities.
- In facilities for rearing laboratory animals extremely susceptible to infectious diseases, sanitation and air conditioning equipment necessary to control microorganisms.
- Assurance of structures that enable researcher(s) and animal technicians to work without risk.
- Facilities and equipment such as safety cabinets, draft chambers, and localized air exhaust equipment to prevent work-related accidents as required. Also education and training of researcher(s) and animal technicians to familiarize them with safe operating procedures.
- When conducting gas sterilization of rearing equipment such as vinyl isolators, the wearing of gas masks as required.
- For autoclaves and ethylene oxide gas sterilizers, periodic inspections as indicated by laws and ordinances, as well as daily checks before starting work.

Behavioral Management and Environmental Enrichment

In neither the Act,[1] Standards,[2] nor Guidelines[6] are behavioral and environmental management and enrichment clearly stated for laboratory animals in Japan.

Occupational Health and Safety

In Japan there is a stringent occupational health and safety legal framework for most industrial institutions including biological hazards. However, specific hazards in laboratory animal settings are not clearly defined.

Veterinary Care

There are no requirements for veterinary care in laboratory animal regulations in Japan.

Euthanasia

The euthanasia of laboratory animals is regulated by the Standards Relating to the Care and Management of Experimental Animals.[2] An overdose of anesthetics and carbon dioxide inhalation are recommended for most laboratory animal species. For rodents, decapitation and cervical dislocation by experienced personnel are also recommended.

Special Considerations

Nonhuman Primates

The importation of NHPs is restricted by the Invasive Alien Species Act[4] and the Act on Prevention of Infectious Diseases and Medical Care for Patients of Infectious Diseases.[5] The institutions using NHPs should be registered and the importers of NHPs need a license for importation and quarantine. The Japanese government restricts the countries (USA, China, Indonesia, Philippines, Vietnam, Cambodia, Guyana, and Suriname after the official inspection of the breeders) from which NHPs can be imported to Japan.

However, there are no regulations for ethical treatment and care of NHPs as laboratory animals.

Genetically Altered Animals

The usage of gene-modified laboratory animals is restricted by the Act on the Conservation and Sustainable Use of Biological Diversity through Regulations on the Use of Living Modified Organisms, which is in accordance with the Cartagena treaty.[3]

Reuse of Animals

There are no restrictions for the reuse of laboratory animals in Japan. These should be reviewed by the IACUC during the protocol review process.

Other Areas

Annual Reports

Annual reports of animal experimentation are recommended by various guidelines[6] and standards[2] including the number of protocols, number of laboratory animals used, and the results of animal experimentations. However, they are only suggested examples.

Care and Use of Fish

Because fish are not defined as laboratory animals, there are no regulations for the care and use of fish for research in Japan.

SOUTH KOREA

In South Korea there are two kinds of Acts regarding animal experimentation. One is the Animal Protection Act (APA),[7] which is managed by the Ministry of Agriculture, Food and Rural Affairs (MAFRA); and the other is the Laboratory Animal Act (LAA),[8] which is managed by the Ministry of Food and Drug Safety (MFDS). The Acts mandate that animal facilities must establish an IACUC.

Registration in government is the duty of both the experimental and production animal facilities. Workers must be certified for training from the government. In addition, the animal research protocols must be approved by the IACUC. The LAA started to designate excellent animal testing facilities with the introduction of certification schemes to entice ethical and scientific animal experiments. Each experimental animal facility creates its own guidelines, based on the APA and LAA, to perform the ethical operation and maintenance of the animal facility. In this chapter, the ethical use of experimental animals, based on the APA and LAA, is described.

General Framework

The Korean Association for Laboratory Animal Science (KALAS) was established in 1985 to promote science and technology in laboratory animal sciences. From the mid-1970s, experimental animal sciences were introduced in Korea. Accordingly, researchers and institutions have felt the need for exchange and cooperation. This led to the establishment of KALAS on May 1, 1985. Currently, KALAS has about 4800 individual members, 55 affiliates, and 105 organization members (www.kalas.or.kr). The inaugural meeting of the Korean Society for Alternative to Animal Experiments was held on February 23, 2007 with the aim of developing new methods that can replace the use of laboratory animals and ultimately minimize the number of experimental animals and animal suffering (www.ksaae.org). The Korean College of Laboratory Animal Medicine (KCLAM) was established in October 25, 2006 for the purpose of (1) enhancing a humane technique for animal experiments, (2) developing knowledge in the fields of disease of experimental animals, operation, anesthesia, pain relief, animal welfare, and animal protection, (3) training for the development of laboratory animal veterinarians, and (4) establishing technical standards. Certification of laboratory animal veterinarians consists of a qualification review and written examination. A candidate must achieve specified standards to be certified (www.kclam.org). By the LAA, the Korean Association for Laboratory Animals (KAFLA) was established in 2009. The main roles of KAFLA are education for administrators of the animal experimental facilities, laboratory animal suppliers, and laboratory animal handlers, the standardization of experimental animal products, and consultancy on management (www.kafla.kr).

The purpose of the APA[7] is to provide adequate protection and management of animals including experimental animals by preventing their mistreatment, and to guide citizens in the care, safety, and respect of animals. To respond to the consultation of the animal welfare policy of MAFRA, an animal welfare committee must be established in MAFRA. This committee is different from the IACUC existing in the animal facilities. The animal welfare committee will carry out the follow works: (1) matters relating to the establishment and enforcement of a comprehensive plan regarding animal welfare; (2) matters concerning guidance and supervision of animal experiments such as the configuration of the IACUC; (3) policy relating to husbandry and certification of animal welfare; and (4) matters relating to animal welfare and protection such as rescue and prevention of abuse of animals. The animal welfare committee shall be constituted by 10 people including one chairman. The minister of MAFRA can impose the role from persons who fall under each of the following categories: (1) a veterinarian who has plenty of experience and knowledge regarding animal welfare and animal protection; (2) a person who has received a recommendation from nongovernmental organizations with relevant knowledge and experience related to animal welfare issues; (3) a person with expertise in animal welfare matters; and (4) other people who meet the eligibility criteria specified by the ordinance of MAFRA. In the same Act, to prohibit animal abuse, no one shall commit the following acts toward animals: (1) killing by brutal methods, such as hanging; (2) killing animals in public or in other animals' presence; and (3) any other act of killing without a valid reason set by a decree from MAFRA, such as veterinary care reasons or reasons related to emergency situations affecting human life, health, and finance. Also no one shall commit the following abusive acts toward animals: (1) an act of harming an animal with instruments or substances. However, this is permitted in certain circumstances approved by the IACUC as prescribed by the decree of MAFRA, such as for preventing or curing disease; (2) an act of harming an animal's body, extracting bodily fluids, or installing devices to extract bodily fluids while it is alive. However, this is permitted in certain circumstances approved by the decree of MAFRA, such as for preventing or curing disease; (3) an act of harming an animal for the purpose of gambling, advertisement, entertainment, or amusement. However, there are exceptions, such as for traditional games, by the decree of MAFRA; and (4) any other act of killing without valid reason set by the decree from MAFRA, such as reasons or reasons related to emergency situations affecting human life, health, and finance. No one shall capture and sell or kill any abandoned animal. Also the owner or caretaker of an animal shall not abandon his or her animal. If so, the owner of the abandoned animals shall be imposed with a fine of 1 million KRW or less. MFDS manages the LAA whose purpose is to contribute to the development of life sciences and the improvement in national health by enhancing the ethics and reliability on animal testing through appropriate administration of laboratory animals and animal testing. The duties of MFDS for the LAA are: (1) formulation and promotion of policies concerning the use and administration of laboratory animals; (2) support for the establishment and operation of animal testing facilities; (3) support for the maintenance, conservation, and development of laboratory animals in animal testing facilities; (4) research support for the improvement in quality of laboratory animals; (5) support for the collection and management of information, and education in connection with laboratory animals; and (6) formulation and promotion of policies concerning the development and approval of methods that can substitute animal testing. The Korean Center for the Validation of Alternative Methods (KoCVAM; http://www.nifds.go.kr/en/inter/kocvam.jsp) was established based on the LAA in November 2009 within the National Institute of Food and Drug Safety Evaluation. KoCVAM seeks to find alternative methods through various activities to build cooperative relationships with both domestic and foreign organizations and to review and validate proposed alternatives. KoCVAM also intends to keenly respond to these international trends by globally promoting alternative test methods developed by Korean organizations.

The Principles

In the APA,[7] the principles of animal protection are based on the Five Freedoms: (1) animals must be maintained with the original prototype of habit and body and normal lifestyle; (2) freedom from hunger and thirst by ready access to fresh water and a nutritional diet; (3) freedom to express normal behavior and not experience inconvenience; (4) freedom from pain, injury, and disease; and (5) freedom from fear and distress. In the same Act, the principles of animal experiments are different from the general principles of animal protection, and are described as follows with consideration to the Three Rs: (1) animal experiments must be performed after considering the promotion of human welfare and the dignity of animal lives; (2) when considering the use of animals in experiments, a replacement must be considered as preferential; (3) experiments with animals must be performed by someone who possesses knowledge and experience in the ethical handling and scientific usage of the experimental animals, and only the minimum of the required number of animals should be used; (4) animals with low sensitivity should be used for experiments necessarily involving pain, and appropriate measures must be taken to lessen the pain, using veterinary methods that employ analgesics, sedatives, and anesthesia; and (5) a person who has conducted experiments with animals shall inspect the subject animals immediately after the conclusion of experiments and if the animals will suffer lasting pain or injury, then such animals shall be put to euthanasia as quickly as possible in the most painless way.

Scope/Applicability

The term "animal testing" used in the APA[7] and LAA[8] means testing conducted on laboratory animals for the scientific procedure for scientific purposes, such as education, testing, research, and production of biological medicines.

In the APA, animal testing facilities are defined as facilities set in the national institutions; local governments; government-funded research institutes; institutions for safety and efficacy testing of drugs, cosmetics, and medical instruments; universities; agricultural and fishery cooperatives; hazardous chemicals and pesticides testing and research institutions; and food manufacturing institutions.

While the APA covers all the animal testing facilities, the LAA applies to the animal testing facilities related to development and safety/effective tests of foods, functional health foods, medical and pharmaceutical products, nonmedical and pharmaceutical products, biomedicines, medical appliances, cosmetics, and narcotics. Therefore the animal facilities defined by the LAA are those that are related to the manufacturing, importing, or selling of food, health supplements, drugs or quasi-drugs, cosmetics, narcotics, medical devices, or which are delegated from MFDS to do the same jobs. The management systems of the animal facilities have been defined by the LAA. The responsible person at the institution shall establish guidelines for the scientific use and administration of laboratory animals, education of those who carry out and engage in animal testing, preferential consideration of methods that can substitute animal testing, formulation of plans concerning the appropriate disposal of waste matter as a result of animal testing, and the safety of workers. As part of this process, a laboratory animal management committee should be set within the animal facilities. Its members shall be appointed by a responsible member of the animal facilities, while the IACUC members based on the APA shall be appointed by the CEO, not by an operator or institutional official (IO) of an institution.

The term "laboratory animal" was defined as any vertebrate used or raised for the purpose of animal testing in the LAA. The APA defines animals as cattle, horse, swine, dog, cat, rabbit, chicken, duck, goat, sheep, deer, fox, mink, and other species as designated by the presidential decree of the APA such as mammalians, birds, fish, and reptiles. Therefore all kinds of vertebrates are regarded as laboratory animals. However, the APA prohibits the use of abandoned animals and animals that have served for humans, such as guide dogs, rescue dogs, police dogs, military dogs, and drug and explosive detection dogs, except for cases when it is necessary to conduct research on the species' health or disease. A person who has performed experimentation with abandoned or serving animals shall be punished with a fine of not more than 500,000 KRW. Exceptional cases are the diagnosis and treatment or research on zoonotic diseases or pest control from these animals. These investigations should be done after specific IACUC approval.

Authorization of User-Breeding Institutions/Inspection/Penalties by Competent Authorities and Institutional and Designated Personnel Responsibilities

Any person who intends to establish animal testing facilities by the LAA[8] shall register them with the minister of MFDS. The manager should be a qualified person who has graduated from college/university and who has worked in animal facilities for over 3 years. The minister of MFDS needs documents to prove the qualifications of the manager, the status of the structure, and the layout of the facility and the area. Animal rooms must be separated by type of animals and from laboratory or waste storage rooms; floor and wall finishes must be constructed with materials convenient to clean and disinfect; temperature and humidity must be adjustable; and laboratory, surgery, or autopsy room, waste storage room, and animal carcasses room should be properly equipped. Also a standard operating procedure for disinfection and animal husbandry shall be prepared.

When a researcher uses laboratory animals in the animal testing facilities, the LAA asks the researcher preferentially to use laboratory animals produced in animal testing facilities or excellent laboratory animal production facilities. When a researcher intends to use laboratory animals imported from abroad, they shall use laboratory animals of which qualification has been certified by a government or institution.

The minister of MFDS may, for the appropriate use and administration of laboratory animals, designate high-quality animal testing facilities (in terms of personnel, resources, and management) as "excellent animal testing facilities." Any person who intends his/her animal testing facilities to be designated as "excellent animal testing facilities" shall apply for designation.

Any person who has registered facilities as animal testing facilities or excellent animal testing facilities shall be guided and supervised by the minister of MFDS.

Any person who intends to engage in the business of production, importation, or sale of laboratory animals shall register with the minister of MFDS. However, this shall not apply to cases where he/she supplies laboratory animals produced in the process of maintenance or research in animal testing facilities. A supplier of laboratory animals shall administer laboratory animal production facilities and laboratory animals so that no harm may be caused to health and hygiene, and safety may be secured; when he/she transports laboratory animals, he/she shall do so by methods suitable to the ecology of such laboratory animals.

As with testing facilities, the minister of MFDS may designate "excellent laboratory animal production facilities." The requirements for being excellent laboratory animal production facilities are almost same as those of the criteria for excellent animal testing facilities. Any person who has registered as a supplier of laboratory animals or has been designated as excellent laboratory animal production facilities shall be guided and supervised by the minister of MFDS.

A person who has failed to register animal testing facilities or a person who has illegally attached a mark of excellent laboratory animal production facilities or a similar mark or has publicized it shall be punished by a fine for negligence not exceeding 1 million KRW.

Nonhuman Primates

In Korea, there are several NHP research facilities. There was no national guideline for the care and use of laboratory NHPs in South Korea, so the Seoul national university hospital has prepared its own guideline for laboratory NHPs.[9]

The guideline includes: purposes and definition, laboratory primate husbandry, veterinary care, behavioral and psychological management of primates, animal care and health, and planning and execution of animal experimentation. Also the environmental enrichment program including social behavior and positive reinforcement training was considered in the facility.[9]

Genetically Altered Animals: Special Considerations

A person who wants to produce or import genetically modified organisms must report to or obtain permission from the head of the relevant central administrative organization according to the Genetically Modified Organisms Act to move between countries.

Oversight and Ethical Review Process

By the APA,[7] an IACUC must be established in all animal facilities to oversee the protection and ethical treatment of experimental animals. Heads of institutions that have animal facilities must submit research proposals to the IACUC. If the institutions that have animal facilities carry out animal experiments without going through the deliberations of the IACUC, they will be imposed with a fine of 1 million KRW or less.

The IACUC is to perform the following functions: (1) deliberations shall be made on the proposals regarding animal experiments; (2) the committee will direct animal experimentation so that it will be carried out in accordance with the principles of animal experiments; (3) the committee can request that the head of the institutions that have animal facilities takes necessary measures for the protection and ethical treatment of experimental animals.

A committee member who is taking part in the experiments may not take part in that review. A member of the committee must not disclose or fraudulently use the information learned during his duty. Members of the IACUC who steal or leak the secrets of proposals shall be punished by a fine of not more than 5 million KRW.

The committee shall consist of 3–15 members including one chairman. Members of the committee have to include, among others: (1) a veterinarian who specializes in experimental animals (KCLAM) or who has worked for at least 1 year in laboratory animal facilities or who has completed training animal welfare in accordance with Animal and Plant Quarantine Agency; (2) a person who is very learned and experienced in animal protection, recommended by a nongovernment organization, selected by presidential decree, and who meets the criteria as follows.

At least one-third of the members must not be employed by the research institution or be affiliated with it. The term of a member of the committee is 2 years. Heads of any institution not complying with the IACUC requirements shall be imposed with a fine of 1 million KRW or less.

The LAA requires the animal facilities to set up a laboratory animal management committee, which shall be established and managed within animal facilities to review the proposals and secure the ethics, safety, and reliability of animal testing. The member organization of the laboratory animal management committee is almost the same as the IACUC defined in the APA.

Reuse

The decision to reuse or not may be different depending on the IACUC. There is no provision for reuse in the legislation. Reuse is rare because most of the experiments will finish with the pathological examination of animals after euthanasia.

Setting Free/Rehoming

Rehoming of companion animals that were used in experiments is encouraged if pain is not present at the end of the experiment and there is no risk of pathogen transmission to humans or animals. Recently, parliamentarians are making legislation to adopt dogs and cats that were used in the experiment as companion animals.

Occupational Health and Safety

The responsible person in charge of animal testing facilities conducting animal testing using substances or pathogens that may cause problems must take the necessary measures for preventing harm to humans and animals. When a disaster caused by animal testing facilities and laboratory animal production facilities is deemed detrimental to the public, the administrator shall take the necessary measures immediately, such as closure and disinfection, and then report the results to the government.

When an operator of animal testing facilities intends to use any biologically harmful substances prescribed by ordinance of the Ministry of Health and Welfare for animal testing, he/she shall report it in advance to the government. Waste matter, such as the carcasses of laboratory animals from animal testing facilities and laboratory animal production facilities, shall be disposed of pursuant to the Wastes Control Act.

Education, Training, and Competence of Personnel

The government has carried out regular training for IACUC members. IACUC members recommended from nonprofit organizations should participate in training offered by MAFRA and obtain certification. In addition, the following persons should participate in educational programs offered by MFDS; (1) a person who has established animal testing facilities; (2) an administrator; (3) a supplier of laboratory animals; and (4) a person who conducts animal testing. Persons willing to conduct research in each animal experimental facility must attend training for animal management and the ethical handling of animals offered by each institution before they perform the experiments.

Transport

According to the APA,[7] a person using a vehicle to transport any kind of animal must comply with the followings (1) feed and water suitable for animals must be supplied, and shock and injury caused by abrupt departure or braking must be avoided; (2) animal transport vehicles must be designed as a structure that will minimize the pain caused by sudden temperature changes and dyspnea; (3) young or pregnant or sick animals or animals with a sucker should be shipped with the installation of a partition to escape from contact with other animals; (4) animals shall be carefully treated s to avoid impact and injury when they are shipped or unloaded; and (5) electric tools shall not be used for transportation.

On the other hand, the LAA[8] indicates that suppliers of laboratory animals must use transportation vehicles that maintain adequate environmental conditions such as temperature, humidity, ventilation, and space dependent on the species, and the health and safety of laboratory animals from cross-infection shall be maintained.

Veterinary Care

Veterinary medical care is an essential part of an animal care program to provide animals with preventive medical care and treatment, humane and ethical management, and to perform effective animal experiments. It is highly desirable to have appropriate veterinary management under the supervision of an attending veterinarian with wide experience in laboratory animal science. In addition, one must keep in mind that medical examination and treatment on dogs, cats, and other animals comply with the Veterinary License Act.

1. Animal Procurement/Transportation and Preventive Medicine
   a. All animals must be acquired lawfully.
   b. An investigator and facility manager must confirm the means of animal transportation. It is important for them to observe and record the condition of animals' health upon their arrival.
   c. Quarantine must be enforced to avoid providing unhealthy animals for experiments. Those newly received animals that will be used in experiments must be separated from the existing group of animals until they are proven to be healthy and acceptable for the experiments. The veterinary medical staff should evaluate the health, the pathogen status of newly received animals, and, if necessary, treat any disease. For those animals from highly reliable suppliers, the foregoing procedures can be substituted with their own pathogen monitoring test records. This quarantine period can be a refinement period for these animals to adjust to a new environment; such a period helps them show stable physiological conditions and behaviors.
   d. Vaccination during the quarantine period—preventive veterinary medical treatment including extermination of entozoa and epizoa must be carried out. Dogs must be registered with each region's animal registration section according to the rabies prevention regulation, which stipulates that dogs must receive a vaccine injection once a year.

2. Surveillance, diagnosis, treatment, and control of disease

All animals should be observed for signs of illness, injury, or abnormal behavior by investigators, facility managers, and technicians in cooperation with each other. In addition, it is imperative to observe conditions of food, water, and cage utensils. As a rule, this should occur daily, but more frequent observations might be warranted, such as during postoperative recovery or when animals are ill or have a physical deficit. Unexpected death and signs of illness, distress, or other deviations from normal in animals should be reported promptly to the attending veterinarian or investigator to ensure appropriate and timely delivery of veterinary medical care. Animals that show signs of a contagious disease or injury should be treated immediately. Methods of disease prevention, diagnosis, and therapy should be those currently accepted in veterinary practice. Diagnostic laboratory services facilitate veterinary medical care and can include gross and microscopic pathology, clinical pathology, hematology, microbiology, clinical chemistry, and serology. Regular microbiological monitoring shall be performed four times a year. The choice of medication or therapy should be made by the veterinarian in consultation with the investigator. Examples of infectious agents that can be subclinical but induce profound immunological changes or alter physiological, pharmacological, or toxicological responses are Sendai virus, Kilham rat virus, mouse hepatitis virus, lymphocytic choriomeningitis virus, and *Mycoplasma pulmonis*. The principal method for detecting viral infections is serological testing. Other methods of detecting microbial infections, such as bacterial culturing, histopathology, and DNA analysis using the polymerase chain reaction, should be used in combinations that are most suitable for specific requirements of clinical and research programs. Transplantable tumors, hybridoma, cell lines, and other biological materials can be sources of murine viruses that can contaminate rodents. The mouse–antibody–production, rat–antibody–production, and hamster–antibody–production tests are effective in monitoring for viral contamination of biological materials and should be considered. Animals that show signs of contagious disease should be isolated from healthy animals in a colony. If an entire room of animals is known or believed to be exposed to an infectious agent, the group should be kept intact during the process of diagnosis, treatment, and control.

Conduct of Experimental Procedures

Degree of Pain on Animals

It is the investigator's responsibility to recognize and minimize the degree of pain on animals in experiment management. However, it is not easy to be aware of such degrees of pain. For example, the only way to study scientific knowledge of the sense of pain on animals is through studies of comparative anatomy, physiology, or pathology as a basis. Thus investigators must understand animals' pain with knowledge not only from scientific studies, but also from humane perspectives. Furthermore, it is essential to take special considerations to reduce pain.

The degree of the pain has been classified from A to E in which animals are under severe pain without pain relief. Surely this is not a definite standard, but it is imperative to follow such an index as a form of standard to place each experiment into a certain category. Investigators must do their best to perform experiments as a lower categorized one.

Pain, Analgesia, and Anesthesia

The proper use of anesthetics and analgesics in research animals is an ethical and scientific imperative. Such use requires an effort to avoid or minimize anxiety, pain, or distress. It is imperative to use appropriate amounts of sedatives, analgesics, and anesthesias in tests that cause little ache or pain. Euthanasia might be necessary at the end of a protocol or as a means to relieve pain or distress that cannot be alleviated by analgesics, sedatives, or other treatments. The IACUC gives advice on the selection and use of necessary medications. Animal tests without using such medication must follow the IACUC's judgment.

Fundamental to the relief of pain in animals is the ability to recognize its clinical signs in specific species. Species vary in their response to pain, so criteria for assessing pain in various species differ. Some species-specific behavioral manifestations of pain or distress are used as indicators, for example, vocalization, depression, or other behavioral changes, abnormal appearance or posture, and immobility. It is therefore essential that personnel caring for and using animals are very familiar with species-specific (and individual) behavioral, physiological, and biochemical indicators of wellbeing. In general, unless the contrary is known or established, it should be assumed that procedures that cause pain in humans also cause pain in animals.

Selection of the most appropriate analgesic or anesthetic should reflect professional judgment as to which best meets clinical and humane requirements without compromising the scientific aspects of the research protocol.

Some classes of drugs—such as sedatives, anxiolytics, and neuromuscular blocking agents—are not analgesics or anesthetics and thus do not relieve pain; however, they might be used in combination with appropriate analgesics and anesthetics.

Restraint

Restraint is an act of restraining animals without causing them fear or anxiety and helps investigators perform such experiments while animals remain motionless; it is also known as stationary.

Such a restraint should minimize the pain given to animals significantly, make experiment management easier, and prevent investigators from any potential. Restraint determines the outcome of the experiment, therefore it is desirable to receive help from experienced technicians. However, a long-term restraint must be avoided; it is imperative to set the duration of such restraint. When using any restraint equipment, the investigator must be an expert in such equipment prior to the experiment.

Management of Surgery and Postoperative Care

Surgery must be performed by a well-experienced investigator or by someone under his/her supervision. In survival surgery, sufficient sterilization must be applied in the course of surgery and a properly trained person must monitor the animal with appropriate care. An appropriate amount of anesthesia should be a factor for safe surgery, therefore precautions should be taken for anesthetic management both during surgery and upon recovery from surgery.

Presurgical planning, personnel training, sterilized surgical techniques, animal care, and appropriate precautions taken throughout the course of the protocol will improve the outcome of the surgery. The influence of each factor may vary depending on the complexity of surgical procedures and/or animal species used in the surgery. A team approach toward surgical planning often brings successful outcomes with consultation from experts in different fields.

Euthanasia

Generally, every institution follows the American Veterinary Medical Association (AVMA) guidelines.[10]

TAIWAN

General Framework

The APA (Act)[11] was approved by the Legislative Yuan and promulgated by the President of the Republic of China in 1998. The Council of Agriculture (COA) of the Executive Yuan is the authority of the government that supervises the implementation of the Act. The APA consists of seven chapters and 40 articles. The Act defines the "animal" as any vertebrate that is fed or kept by people, including economic animals, laboratory animals, pet animals, and other kinds of animals. "Laboratory animal" is defined as an animal that is fed or kept for the purpose of scientific application. According to Article 16, Chapter III of the Act, institutions that perform animal experimentation shall organize an IACUC to oversee the animal care and use program for animal experiments. In 2015, there were 221 organized IACUCs in Taiwan.

The COA has conducted site visits for institutions with animals used in research, teaching, and testing since 2003. Eighty institutions are selected each year for assessment by a multidisciplinary team consisting of laboratory animal science

professionals, animal protection inspectors, and animal welfare interested groups.[12]

The COA also published the book *A Guidebook for the Care and Use of Laboratory Animals* (Guidebook)[13] as a reference for the institutions. Reports with mandates and/or suggestions for improvement will be forwarded to the institutions following the visits. Any institution violating the Act will be subjected to a fine according to the Act.

The Guidebook has 25 chapters including an introduction, animal care and use programs, animal husbandry, veterinary care, physical plant, animal breeding, occupational health and safety programs, disaster plans and emergencies, and special chapters for pigs, fish and amphibians, dogs, cats, and NHPs.

The annual report of the institutional usage of laboratory animals in the animal use and care program is to be submitted to the COA for review. The report has to describe the types and sizes of animal facilities, the numbers of animals and species used in the programs, the sources of animals, the euthanasia methods and number of euthanized animals, and the methods and quantity of carcass disposal.

In 2015, there were 221 institutions that had organized an IACUC and registered in the COA in Taiwan, of which 55% were research institutions and universities/colleges, and 45% were hospitals, contract research organizations (CROs), various types of pharmaceutical companies, and nonprofit medical research institutions.

Annually, the IACUCs held a total of 423 meetings and reviewed 5178 animal study protocols. The IACUCs approved 3994 animal study protocols without revision (77.13%) and 1137 with revision (21.96%), and rejected 47 of them (0.91%). Among the IACUC activities, the most common was the semiannual inspection of animal facilities, followed by consultation on animal breeding, experimentation, and care, while termination and/or suspension of animal experimentation was seldom executed. Further analyses of the annual reports show that the numbers of IACUC members were moderately correlated with the amount of workload of the IACUC and frequency of the IACUC meetings, but not with the numbers of amendments to animal use protocols.

The sizes of the animal facilities varied widely among institutions: 55 institutions (27.5%) had a facility size less than 20 m^2, 59 institutions (29.5%) were between 20 and 65 m^2, 25 institutions (12.5%) had 170–350 m^2, 17 institutions (8.5%) had 1700–3300 m^2, and 12 institutions (6%) were larger than 3500 m^2.

The numbers of laboratory animals used among institutions were also significantly different, ranging from fewer than 50 to more than 20,000 annually. Forty-nine institutions (34.5%) used fewer than 50 animals, 14 institutions (7%) used 51–100 animals, 43 institutions (21.5%) used 101–500 animals, 38 institutions (19%) used 1001–5000 animals, 15 institutions (7.5%) used 5001–10,000 animals, 10 institutions (5%) used 10,001–20,000 animals, and 10 institutions (5%) used more than 20,000 animals. Regarding the number of animal species used, 32 institutions (16%) did not use any animals at all in 2011, 84 institutions

(42%) used one species only, 44 institutions (22%) used two species, 17 institutions (8.5%) used three species, and 14 institutions (7%) used 5–10 species. The numbers of laboratory animals used in different types of institutions in 2011 indicated that 9.1% of animals were by CROs and pharmaceutical institutions, 42% by research institutions, 38% by universities and colleges, and fewer than 2.2% by other categories. Seventy percent of the animals used in 2011 were produced by animal breeding centers or commercial breeders domestically, 26% from institutional in-house breeding, 1.8% from local markets (mainly fish and amphibians), and fewer than 2.2% from importation or wildlife. A total of 1,526,725 animals were used for scientific purposes. Rodents accounted for 79%, fish 14%, and farm animals 7%. In terms of animal species, laboratory mice accounted for 53%, laboratory rats 25%, chicken/embryo 2.4%, and rabbits 2.0%. The fish used in the research were *Rachycentron canadum*, *Tilapia nilotica*, *Epinephelus fuscogustatus*, *Danio rerio*, *Anguilla japonica*, *Cyprinus carpio*, and others. The methods of euthanasia were included: 78% of rodents were euthanized by carbon dioxide exposure, while 22% were euthanized with a combination of carbon dioxide exposure and physical methods. Rabbits, minipigs, guinea pigs, and dogs were euthanized by injection of pentobarbital solution.

The carcasses of rodents, minipigs, dogs, amphibians, and others were mainly disposed of by incineration through licensed commercial medical waste companies, while part was disposed of by in-house incineration.

A brief retrospective examination of the previous year reveals that there was a constant decline in the number of laboratory animals used in research, teaching and testing, suggesting a substantial change of laboratory animal practice. Thus there is a better understanding of the Three Rs principles and current trend of laboratory animal use in Taiwan.

The Principles

In general provision, the APA (Act)[11] states that it is enacted to respect the lives of animals and protect them. Like most national systems that oversee the scientific use of animals, the Act incorporates the principles of the "Three Rs"[14] in the policy. The Act also states that "scientific application means the application for the purpose of teaching, experimentation, the manufacturing of biological products, laboratory merchandise, pharmaceutics, or organ transplant."

The Act indicates that the competent authority at central government level shall select experts, scholars, competent agencies, and registered civic animal protection organization representatives to periodically formulate animal protection policies and review the implementation of the Act. The number of experts, scholars, and registered civic animal protection organization representatives who do not have the capacity of governmental representatives shall not be less than two-thirds of the total number. Articles 6, 7, and 8 of the Act indicate that no one shall be allowed to harass, maltreat, or hurt animals; a responsible person shall prevent animals from the infringement of life, body, freedom, or property of others; and a competent authority at central government level shall post a public notice about those animals that are prohibited from being raised, exported, or imported.

The Act also states that responsible persons must provide the necessary medical treatment to animals that are injured or sick. The medical or surgical treatment of animals, based on their health or management need, shall be performed by veterinarians. The Act also sets requirements for meeting animals' physical and psychological requirements.

The number of live animals used in a scientific application shall be reduced to a possible minimum and only when its usage is absolutely necessary. The application shall be done in a way that inflicts the least suffering to the animals. The Animal Protection Law was amended on October 2016.[11] A person who abuses or does cruel acts to animals will be prosecuted as a criminal and sentenced for up to 5 years in prison.

Scope/Applicability

The Act[11] implements the policies for all institutions using animals for scientific purposes. This includes academic institutions, government research institutions, and private pharmaceutical and biomedical companies. The institutions that perform scientific applications for the use of animals shall organize an IACUC to supervise the scientific applications for animal experiments. By the end of each calendar year, the IACUC must submit the reports of the animal care and use programs and the usage of animals and semiannual facility inspection reports to the COA for the compliance of the Act.

As the animal study protocol review is the responsibility of the IACUC, the COA provides guidelines on the review process, including number of animals, pain and distress, anesthesia and analgesics, euthanasia, and disposal of refuse. The COA also provides two workshops annually related to the operation and management of IACUCs to new animal users, IACUC members, and/or IOs of institutions in Taiwan. The content of the course "Laboratory animal humane management" includes the following topics: (1) APA and laboratory animal-related regulations; (2) laboratory animal science application and auditing consultation; (3) IACUC roles and functions; (4) laboratory animal use—the Three Rs; (5) laboratory animal facility planning and management; (6) small laboratory animal care and transportation; (7) laboratory rodent diseases; (8) laboratory animal occupational health and safety; (9) laboratory animal pain assessment and anesthesia; and (10) IACUC practical questions and discussion.

Meanwhile, the Act declares that schools below the level of senior high school shall not teach any lessons that may cause injury or death to animals that is not included in the approved curriculum by the competent authority in charge of educational administration.

Authorization of Institutions/Inspection/Penalties by Competent Authorities

Any institution using animals for scientific research must register with the COA in Taiwan or it will be fined according to the Act.[11] An active IACUC has to be established with composition, authority, responsibilities, and functioning as defined in written terms of reference based on the supplemental document issued by the COA.[15]

The COA inspection team visits at least 80 registered institutions annually; the inspection team will review the IACUC function and the animal care and use programs to see if: (1) the overall structure of the animal care and use program in sound, effective, and well supported; (2) the IACUC remains active and functional, meeting at least twice a year, visiting the animal facilities at least once a year, and fulfilling all of the responsibilities described in the COA supplemental document; (3) the veterinary and animal care services continue to meet institution needs and Guidebook standards[13]; (4) the training and continued education program is active and provided for personnel; and (5) the animal facilities meet the institutional needs and Guidebook standards.

All the registered institutions are required to submit animal use data to the COA on an annual basis by the end of March of each year. The report is to include the types and sizes of animal facilities, the numbers of animals and species used in the programs, the sources of animals, the euthanasia methods, and the number of euthanized animals and the methods and quantities of carcass disposal. This information permits the COA to publish reasonably comprehensive data on an annual basis. Annual statistics and analysis of data are published by the COA and are available for the institution and related agencies. The annual report includes: (1) all vertebrates including fish used for research, teaching, and testing; (2) animals held, if assigned to a research protocol; (3) the euthanasia of the animals used in research; (4) the usage of animals by scientific purposes, i.e., teaching, research/development, biomedical, agricultural, pharmaceutical, and CRO services; and (5) the distribution of animal usage in the research, teaching, and testing services. Information on the statistical analysis can be found at http://www.coa.gov.tw.

Oversight and Ethical Review Process

The supplemental document issued by the COA[15] states that the structure and resources of the IACUC programs, the composition, functioning, and effectiveness of the IACUC, and the appropriateness of animal care and use practices, procedure, and facilities shall be implemented. The utmost responsibilities and authorities of the IACUC under the Act[11] include the coordination and review of the animal study protocols and animal care and use programs such as: (1) the activities and procedures related to the care of animals; (2) the standards of care and facilities for animals; (3) the training and qualification of personnel that are engaged in the care of animals; and (4) procedures for the prevention of unnecessary pain including the use of anesthetics, analgesics, and euthanasia.

Education, Training, and Competence Personnel

In Taiwan, all personnel from institutions involved with the usage of animals in scientific research should be competent and adequately trained in the principles of ethical care and use of animals. Animal users (including principal investigators, researchers, graduate students, postdoctoral fellows, research staff, study directors, etc.), animal health professionals, IACUC members, and the IO have different educational and training needs. The institution should formulate a suitable training program for them. The Chinese-Taipei Society for Laboratory Animal Sciences (CSLAS) provides three different levels of training programs for researchers, animal technologists, and animal technicians annually. The training programs are funded by the COA and executed by the CSLAS.

Responsibilities of the Laboratory Animal Veterinarian

The main responsibilities of the attending veterinarian in laboratory animal facilities is outlined by the COA,[16] which states that he/she will: (1) supervise disease detection and surveillance, prevention, diagnosis, and treatment; (2) supervise handling and restraint methods the use of anesthetics, analgesics, and tranquilizer drugs, and methods of euthanasia; (3) supervise surgical and postsurgical care; (4) supervise the promotion and monitoring of animals' physical and psychological wellbeing; (5) oversee the adequacy of the husbandry program; (6) be involved in the review and approval of all animal care and use, e.g., via a role on the IACUC; (7) train institutional staff in the care and use of laboratory animals; (8) assist in the establishment and/or monitoring of the occupational health and safety program; (9) monitor for zoonotic diseases; (10) advise on and monitor biohazard control policies and procedures relevant to the animal care and use program; and (11) monitor experimental animal care and usage through postapproval monitoring and daily runs. The duties and responsibilities of the laboratory animal veterinarian should correspond with the size and complexity of the institutional program.

Animal Husbandry, Housing, and Enrichment

In Taiwan, animal husbandry, housing, and enrichment are described in the Guidebook.[13] For general environmental parameters such as room temperature, ventilation, etc. basic recommendations are included in the Guidebook as well. More specific information relevant to individual animal species is included in the specific chapter of the Guidebook. Also the "Guide for the care and use of laboratory animals"[17] serves as a widely used reference resource.

Euthanasia

The chapter of euthanasia in the Guidebook[13] is based on the recommendation made by the ICLAS Working Group on Harmonization[18] and the 2007 version of the AVMA Guidelines on Euthanasia.[19] An overview of acceptable methods of euthanasia in the Guidebook for common species used for research, teaching, and testing is included. The methods of euthanasia were reported in the COA annual statistical analysis in 2011: 78% of rodents were euthanized by carbon dioxide exposure, while 22% were euthanized by a combination of carbon dioxide exposure and physical methods. Rabbits, minipigs, guinea pigs, and dogs were euthanized by injection of pentobarbital solution. The carcasses of rodents, minipigs, dogs, amphibians, and others were mainly disposed of by incineration through licensed commercial medical waste companies, while part was disposed of by in-house incineration.

Animal Facilities and Special Animal Biosafety Laboratories

The Act[11] requires that animal facilities for the care and use of all animals in research, teaching, and testing shall be conducive to the wellbeing and safety of the animals, provide an appropriately safe workplace for personnel, and establish a stable research environment. The Guidebook[13] provides the guidelines to promote optimal levels of animal care and facilitate good research. The guidelines are intended to be viewed as a tool for achieving acceptable standards and not as mandatory instructions. The Guidebook also encourages the adoption of the idea of a barrier system for reducing or minimizing cross-contamination.

The Taiwan Center for Disease Control also issues a specific requirement and guideline[20] to govern the animal biosafety research laboratory in the institution including ABSL 1, 2, and 3-level facilities.

REFERENCES

1. Japan. The Act on Welfare and Management of Animals (Act No. 105 of 1973).
2. Japan. The standards relating to the care, management and relief of pain of experimental animals (notice no. 6 of the Prime Minister's Office 1980; revised in 2006).
3. Japan. Act on the Conservation and Sustainable Use of Biological Diversity through Regulations on the Use of Living Modified Organisms (Act No. 97 of June 18, 2003).
4. Japan. Invasive Alien Species Act (Act No. 78 of 2004).
5. Japan. Act on Prevention of Infectious Diseases and Medical Care for Patients of Infectious Diseases (Act No. 114 of October 2, 1998).
6. Japan. Science Council. *Guidelines for proper conduct of animal experiments (the guide)*. 2006.
7. South Korea. Ministry of Agriculture, Food and Rural Affairs. Animal Protecting Act, 1991 (revised 2008).
8. South Korea. Ministry of Food and Drug Safety. Laboratory Animal Act, 2009.
9. Guide for the care and use of laboratory nonhuman primates in Seoul National University Hospital Byeong-Cheol, Kang. In: *Abstract of Winter Symposium*. KALAS; 2011.

10. American Veterinary Medical Association AVMA. *American veterinary medical association guidelines for the euthanasia of animals: 2013 edition*. 2013. Available from: https://www.avma.org/KB/Policies/Documents/euthanasia.pdf.
11. Taiwan Legislative Yuan. Animal Protection Act. Promulgated on November 4, 1999. (Amended 2016).
12. Taiwan Council of Agriculture. Enforcement Rules of Animal Protection Act, promulgated on January 19, 2000.
13. Taiwan Council of Agriculture. *Guidebook for the care and use of laboratory animals*. 3rd ed. Council of Agriculture; 2010. ISBN: 957-30437-2-6. [Chinese version].
14. Russell WMS, Burch RL. *The principles of humane experimental technique*. London, UK: Universities Federation for Animal Welfare – UFAW; 1992. p. 1959. [special edition].
15. Taiwan Council of Agriculture. *Regulations on establishment and management of laboratory animal care and use committee*. August 2013.
16. Taiwan Council of Agriculture. *Regulations on supervision and management of the institutions with scientific application of animals*. July 2013.
17. NAS, National Academy of Sciences. *Guide for the care and use of laboratory animals*. Washington, D.C.: National Academy Press; 2011.
18. Demers G, Griffin G, De Vroey G, Haywood JR, Zurlo J, Bedard M. Harmonization of animal care and use guidance. *Science* 2006;**312**(5774):700–1.
19. American Veterinary Medical Association-AVMA. *AVMA guidelines on euthanasia*. Schaumburg, IL: AVMA; 2007.
20. Taiwan. *Safety guidelines for biosafety level 1 to level 3 animal laboratory*. Center for Disease Control (CDC); November 2014.

Chapter 11

Laws, Regulations, Guidelines, and Principles Pertaining to Laboratory Animals in Southeast Asia

Montip Gettayacamin[1], Richard Grant[2], Holl Davun[3], John E. Heidrich[4], Sorn San[5], Yasmina Arditi Paramastri[6], Imelda Liunanita Winoto[7], Dondin Sajuthi[7], Goh Yong Meng[8], Abdul Rahim Mutalib[8], Ranillo Rodrigo G. Resuello[9], Maria Amelita C. Estacio[10], Joel Villa Tuplano[9], Hon Mun Wong[11], Cheryl Inguito Daludado[11], Pattamarat Kunjara[12], Jade Donavanik[13], Pradon Chatikavanij[12], Chumpol Pholpramool[14], Nguyen Bao Quoc[15], Nguyen Ba Tiep[16]

[1]AAALAC International, Samutprakarn, Thailand; [2]University of Washington, Seattle, WA, United States; [3]Royal University of Agriculture, Phnom Penh, Kingdom of Cambodia; ; [4]Central New Mexico College, Albuquerque, NM, United States; [5]Ministry of Agricultural, Phnom Penh, Kingdom of Cambodia; [6]National University of Singapore, Singapore, Singapore; [7]Bogor Agricultural University (IPB), Bogor, Indonesia; [8]University Putra Malaysia, Serdang, Malaysia; [9]Simian Conservation Breeding and Research Center, Inc. (SICONBREC), Makati City, Philippines; [10]University of the Philippines Los Banos, Laguna, Philippines; [11]Agri-Food & Veterinary Authority of Singapore, Singapore, Singapore; [12]National Research Council of Thailand (NRCT), Bangkok, Thailand; [13]Dhurakit Pundit University, Bangkok, Thailand; [14]Mahidol University, Bangkok, Thailand; [15]Nong Lam University, Ho Chi Minh City, Vietnam; [16]Vietnam National University of Agriculture, Hanoi, Vietnam

INTRODUCTION

Southeast Asia is a subregion of Asia, consisting of the following countries: Brunei or the Nation of Brunei; Cambodia or the Kingdom of Cambodia; Christmas Island (a territory of Australia); Indonesia or the Republic of Indonesia; Laos or the Lao People's Democratic Republic; Malaysia; Myanmar; Philippines or the Republic of the Philippines; Singapore or the Republic of Singapore; East Timor or Timor-Leste or the Democratic Republic of Timor-Leste; Thailand or the Kingdom of Thailand; and Vietnam or the Socialist Republic of Vietnam. All these countries are members of the World Organisation for Animal Health (OIE). The OIE *Terrestrial Animal*

Laboratory Animals. http://dx.doi.org/10.1016/B978-0-12-849880-4.00011-8
Copyright © 2018 Elsevier Inc. All rights reserved.

Health Code (the *Terrestrial Code*)[1] sets out standards for the improvement of animal health and welfare and veterinary public health worldwide, including standards for safe international trade in terrestrial animals (mammals, birds, and bees) and their products. This is achieved through the detailing of animal health measures to be used by the veterinary authorities of importing and exporting countries to avoid the transfer of agents pathogenic for animals or humans, while avoiding unjustified trade barriers. The *Terrestrial Code* is a reference document for use by veterinary authorities, import/export services, epidemiologists, and all those involved in international trade. Section 7[2] of the *Terrestrial Code* covers animal welfare. Chapter 7.1 is called "Introduction to the Recommendations for Animal Welfare," and Chapter 7.8 covers "Use of Animals in Research and Education." In addition, International Guiding Principles for Biomedical Research Involving Animals[3] provided by are well recognized and accepted in this region.

There is no specific overarching legislation applied for animal welfare relevant directly to laboratory animals for research, testing, and teaching in Southeast Asia. This chapter describes current laws, regulations, guidelines, and principles pertaining to laboratory animals used for research, teaching, and testing in some countries of Southeast Asia region including Cambodia, Indonesia, Malaysia, Philippines, Singapore, Thailand, and Vietnam. Table 11.1 summarizes legal requirements, country and institutional oversight bodies, and special requirements pertaining to laboratory animals in these countries.

CAMBODIA

Background and Current Situation

Cambodia is primarily an agricultural country with very little infrastructure for animal experimentation. More than 95% of Cambodians are Buddhists, and the humane care and treatment of animals is regarded as important in the context of those beliefs.

The following institutions are operating in Cambodia and contribute to the animal care and use landscape in this country: The Royal University of Agriculture, under the Ministry of Agriculture, Forestry and Fisheries (MAFF), graduate students with bachelor's and master's degrees of Veterinary Science and Animal Husbandry. Research being conducted at the Royal University of Agriculture, the Royal University of Phnom Penh, and the National Veterinary Research Institute (NaVRI) consists of surveys of wild animals and agricultural species for disease surveillance. NaVRI, also under the Department of Animal Health and Production (DAHP), MAFF, has as its primary responsibility, the investigation, testing, and reporting of animal health concerns throughout the country. NaVRI also has the Cambodian National Animal Health Information System, which collects, collates, and analyzes data related

TABLE 11.1 General Laws and Regulations, Legal Requirements, Legal and Institutional Oversight Bodies Pertaining to Laboratory Animals of Some Southeast Asia Countries

Countries	Laws and Regulations	Guidelines and Principles	Legal Oversight Body	Institutional Oversight Body	Remarks
Cambodia	Not applicable	Not applicable	For exporting nonhuman primates: Ministry of Agriculture, Forestry and Fisheries	Not applicable	Laboratory animals are mammals (species covered in the OIE *Terrestrial Animal Health Code*[1]).
Indonesia	Law of the Republic of Indonesia No. 18 (*Law No. 18*)[4] Law of the Republic of Indonesia No. 41 (2014)[5]	The National Guidelines on Health Research Ethics (2011, *Guidelines*)[6] Teaching Guidebook for Ethics on Health Research (2011)[7] The minimum space requirements for animal housing and enclosures in the Guidelines refer to the *Guide for the Care and Use of Laboratory Animals* (NRC, 1996)[8] The euthanasia policy in the *Guidelines* refers to the *American Veterinary Medical Association (AVMA) Guidelines on Euthanasia.* (2007)[9]	National Bioethics Committee	Not required, but some institutions maintain the Institutional Animal Care and Use Committee (IACUC) Some institutions merge the IACUC with the Health Research Ethical Committee	Conservation Law No. 5 (1990)[10] covers capturing and handling methods of animals from the natural habitat.
Malaysia	Animal Welfare Act 2015 (Act 772)[11] Animal Act (Amendment) 2013 (AA2013)[12] Animal Welfare (Animals Used in Research, Testing and Teaching) Regulations 2015 (pending approval in 2017)	The Malaysian Code of Practice for the Care and Use of Animals for Scientific Purposes (2015, pending approval)[13] The Code refers to the AVMA Guidelines for the Euthanasia of Animals: 2013 Edition[14] The Malaysian Code of Practice for the Care and Use of Animals for Teaching in Schools (2016, pending approval)	The Animal Welfare Board (AWB)—regulates the use and care of laboratory animals Department of Veterinary Services (DVS), Ministry of Agriculture and Agro-based Industries—enforcement and compliance inspection activities	Animal Ethics Committee or IACUC The IACUC is to be licensed and recognized by the AWB to carry out the necessary protocol review	Only facilities and IACUC's licensed by the DVS under the Act are authorized to use animals for scientific purposes. Research, testing, and teaching on animal species other than those covered by the Animal Welfare Act 2015 would require additional permits and approvals under the Wildlife Conservation Act 2010[15] and/or with the consent of the relevant state authorities.

Continued

TABLE 11.1 General Laws and Regulations, Legal Requirements, Legal and Institutional Oversight Bodies Pertaining to Laboratory Animals of Some Southeast Asia Countries—cont'd

Countries	Laws and Regulations	Guidelines and Principles	Legal Oversight Body	Institutional Oversight Body	Remarks
Philippines	Republic Act 8485 in 1998, known as "The Animal Welfare Act of 1998"[16] Republic Act 10631 (2013)[17]	Administrative Order No. 40 (AO40), Series of 1999 (Department of Agriculture)[18] PALAS Code for the Care and Use of Laboratory Animals in the Philippines (COP)[19] The Revised Implementing Rules and Regulations of the amended Animal Welfare Act[20]	Committee on Animal Welfare, Department of Agriculture Animal Welfare Section, Animal Health and Welfare Division (AHWD), Bureau of Animal Industry (BAI), DA	IACUC	Authorization to conduct animal research (a permit, renewed annually) is granted to institutions by the DA, BAI-AHWD based on the acceptability of the animal care and use program, including accreditation by the Philippine Association for Laboratory Animal Science (PALAS), the presence of IACUC and animal care personnel training program.
Singapore	The Animal and Birds Act (Chapter 7)[21] The Animals and Birds (care and use of animals for scientific purposes) Rules (2004)[22]	NACLAR Guidelines (National Advisory Committee for Laboratory Animal Research's Guidelines on the Care and Use of Animals for Scientific Purposes)[23] • The guiding principles for the care and use of animals for scientific purposes • IACUC Guidelines • The Training Guidelines Recommendations for floor area and height of cages are similar to the *Guide for the Care and Use of Laboratory Animals*, NRC 1996[8] There are three references on euthanasia: *Euthanasia of Animals Used for Scientific Purposes* by the Australian and New Zealand Council for the Care of Animals in Research and Teaching,[24] the Canadian Council *Guide to the Care and Use of Experimental Animals*,[25] and the AVMA *Report of the AVMA Panel on Euthanasia (2001)*[26]	AVA (Agri-Food and Veterinary Authority of Singapore)	IACUC	Any research facility, which intends to use animals for scientific purposes, must obtain a license from the AVA.

TABLE 11.1 General Laws and Regulations, Legal Requirements, Legal and Institutional Oversight Bodies Pertaining to Laboratory Animals of Some Southeast Asia Countries—cont'd

Countries	Laws and Regulations	Guidelines and Principles	Legal Oversight Body	Institutional Oversight Body	Remarks
Thailand	The Animals for Scientific Purposes Act (2014)[27]	*The Ethical Principles and Guidelines for the Use of Animals for Scientific Purposes* (1999)[28] The Standard for Institutional Animal Care and Use Committee (2012)[29] *The Ethics in Procedures on Animals for Scientific Purposes* (2015)[30] *Fundamental Principles on Designing Animal Care and Use Facilities for Laboratory Animals*[31]	Institute of Animal for Scientific Purposes Development (IAD), National Research Council of Thailand (NRCT)	IACUC	Any institution that raises or maintains, and/or uses the protected animals, must register its facilities to the NRCT. A license is required for anyone who uses or produces the protected animals. Although the animal caretakers do not need to have licenses, they have to work under a license holder. The license is valid for 4 years, and renewable. A license holder whose work involves with improvement of breeds, breeding, animal production, stem cells, genetic modification, or cloning must prepare and submit a semiannual report to the NRCT.
Vietnam	The Veterinary Law (2016)[32]	Not applicable	For exporting nonhuman primates: Ministry of Agriculture and Rural Development (MARD)	Not applicable	Major institutions and companies follow the guidelines and criteria of Ministry of Health[33] and World Health Organization Laboratory Biosafety Manual.[34] Academic institutions follow the OIE *Terrestrial Animal Health Code* (Chapter 7.1 Animal Welfare).[2]

to animal diseases. Foreign research institutions exist in Cambodia, namely, the Pasteur Institute and the US Naval Medical Research Unit (NAMRU), but those institutions report that no laboratory animal experimentation takes place. If animal experiments are performed, these foreign institutes report that they would follow their home-country requirements for animal care and use.

There are currently six nonhuman primate (NHP) breeding facilities for cynomolgus monkeys (*Macaca fascicularis*) in Cambodia; three of which are accredited by AAALAC International (AAALAC) and three that report no plans for accreditation. NHPs from these accredited facilities are primarily exported to research facilities in Korea, Japan, and the United States (US), while nonaccredited facilities export primarily to China.

Summary of Regulations or Legal Requirements

In Cambodia, guidelines for animal care are primarily focused on agricultural production and enforced by the MAFF. Governmental and academic research institutions generally follow the OIE *Terrestrial Animal Health Code*.[1] Chapter 7.1 of the *Terrestrial Code*, called *Introduction to the Recommendations for Animal Welfare,* generally recognizes the "Five Freedoms (5Fs)"[35] and the "Three Rs (3Rs)"[36] principles. There is no specific regulation or guideline covering the care and use of laboratory animals in research, teaching, and testing. No formal protocol review or prestudy approvals are required for the animal studies.

Exportation of Nonhuman Primates from Cambodia

In Cambodia, the primary policy-making and permit-issuing for exporting NHPs is the authority of MAFF involving the CITES (Convention on International Trade in Endangered Species of Wild Fauna and Flora) Office, the Department of Animal Health and Production (DAHP), the NaVRI, and the Forestry Administration (FA). There is currently a moratorium on exportation of wild-caught NHPs, enforced by the officials at FA and DAHP. The veterinary authorities of these two departments regularly inspect breeding facilities to ensure that housing and transport conditions are acceptable, and only the first-generation offspring animals (F1) or greater are exported. The following steps are the documents in chronological order required for exporting NHPs:

- CITES Permit obtained from the CITES Management Authority and MAFF
- Animal Health Certificate obtained from the NaVRI, DAHP, and MAFF
- Veterinary Certificate obtained from the DAHP and MAFF
- Transit Permit(s) as needed obtained from the transit country if animals will be transferred from one plane to another
- Cambodian Transportation Permit obtained from the DAHP and MAFF

- Cambodian Trucking Permit obtained from the FA and MAFF after FA officials perform the facility inspection
- Inspection Certificate (optional for exporting to Japan only) or Health Certificate (optional for exporting to Korea only) obtained from the DAHP and MAFF
- Veterinary Health Certificates of Cynomolgus Monkeys: Monkeys' Data and the Records of Exporting Nonhuman Primates issued by the exporter and signed by the facility's attending veterinarian (AV)

INDONESIA

Background and Current Situation

Governmental interference in animal care and health-related issues in Indonesia has been established since its very early days, as indicated by the presence of Law No. 432 in 1912,[37] which regulated transmissible diseases control, animal husbandry improvement, and veterinary health and hygiene. However, it focused mostly on farm animals.

In 1967, Law of Republic Indonesia No. 6 (*Law No. 6*), Article 22[38] gave specific attention to animal welfare by regulating the following aspects: housing, care, transportation, use, euthanasia methods, and humane treatment of animals. The Article 22 also indicated that the use of animals should be done with respect to animal welfare. Holding site and housing requirements were regulated by local government authorities at district level. Attempts should be made to limit disturbance to public facilities and compounds. The housing requirements were regulated within achievable criteria to support production growth. Efforts should be made to ensure humane treatment of animals as God's living creatures. During transportation, animals should be able to maintain normal posture, in adequate space with good ventilation. Animals should not be abused or overly used above their maximum tolerances or capabilities. Euthanasia should be done with respect to religious procedure requirements with minimal pain and distress.

In 2009, the *Law No. 6* was replaced by Law of the Republic of Indonesia No. 18 (*Law No. 18*)[4] that covers veterinary public health and animal welfare. With regard to laboratory animals, veterinary authority in the use of animals as laboratory animals and in comparative medicine is discussed in Chapter VII, Veterinary Authority, Article 74.[39] In 2014, the *Law No. 18* was amended by Law of the Republic Indonesia No. 41 (*Law No. 41*).[5] Obligation of the witnesses to report any violation of animal welfare to government authority was emphasized. The amendment includes sanction and fine for animal welfare violation, which had not written in the *Law No. 18*. Sanction and fine are applied not only for those who conduct violation to animal welfare, but also for those who witness animal welfare violation and fail to report to authority.

The laboratory animal program in Indonesia was initially established by the US Naval Medical Research Unit No. 2 (NAMRU-2) in January 1970 by the request of Indonesian Health Authorities, Ministry of Health of Republic Indonesia following their success assisting the Indonesian government in controlling the outbreak of Bubonic plague in Central Java. In 1996, the NAMRU-2 achieved the first AAALAC accreditation in Indonesia, followed by the three institutions (a university research center, a primate breeding company, and a pharmaceutical company) in 2006, 2007, and 2012, respectively. In 2010, the NAMRU-2 facility was closed and relocated to Hawaii, United States. With closing of the primate breeding facility, there are only two AAALAC accredited institutions in Indonesia. Awareness of animal welfare issues and the importance of AAALAC accreditation have been introduced continuously to laboratory animal users by the Indonesian Laboratory Animal Veterinary Medical Association (ILAVMA), a nonterritorial organization under the Indonesian Veterinary Medical Association (IVMA), and the Indonesia Association of Laboratory Animal Sciences (IALAS). Currently, laboratory animals are extensively used in teaching, testing, and research in many universities, pharmaceutical companies, vaccine companies, and research institutions. In addition, the growing number of institutions that established an Institutional Animal Care and Use Committee (IACUC) evidences the increase of awareness to promote ethical treatment of animal in animal care and use programs (ACUPs).

The National Health Research Ethics Committee (NHREC) was an independent unit established under Ministerial Decree, Ministry of Health in 2002, with the first membership appointed in 2003. This was followed by the development of National Guidelines on Health Research Ethics Committee (*Guidelines*) in 2005,[40] with the current detail on the committee function and responsibility described under Ministerial Decree 2007, Ministry of Health, Republic of Indonesia.[41] The responsibilities include establishment of ethics in health research, development of national guidelines on health research that covering both human and use of animals; and submitting annual activity reports to Ministry of Health.

The National Bioethics Committee (NBC) was established in 2014 by interdepartmental decrees of three ministries: the Ministry of Research and Technology, the Ministry of Health, and the Ministry of Agriculture.[42] The Committee membership consists of experts from various fields, including religious representatives and humanists. The main functions of the committee are as follows:

- To improve a review of the matters regarding bioethical principles
- To provide recommendations to the government on the bioethical aspects of research development and the application of biological sciences and technology
- To disseminate public understanding on bioethics

Summary of Regulations or Legal Requirements

Scope

Law No. 18 covers animal welfare in general.[4] Chapter VI describes veterinary public health and animal welfare, and Part 2 governs Animal Welfare. Articles 66 and 67 define that "animals" are any vertebrate animal and some invertebrate animals, which can feel pain such as a crab. The law covers all aspects of the efforts to support animal welfare. This includes capturing animals from the wild, handling and restraint, housing and care, transportation, euthanasia methods, and humane treatment of animals. In addition, the veterinary authority in the use of laboratory animals in comparative medicine is also covered under Chapter VII, Veterinary Authority; Article 74.[39]

The principles of ethical treatment on the use of animals in comparative medicine are described in the Regulation No. 95.[43] The detailed guidelines on the use of laboratory animals in biomedical research were developed by the NHREC, Ministry of Health, Republic of Indonesia in the National Guidelines on Health Research Ethics (*Guidelines*),[6] and the Teaching Guidebook for Ethics on Health Research (*Teaching Guidebook*).[7] However, the *Guidelines* focus primarily on the use of animals in biomedical research for human health but do not include the use of animals in research for animal health and industrial or military purposes. The Guidelines and Guidebook are mandatory for organizations under the Ministry of Health and they are recommendations for other institutions.

For the use of fetuses in health research, fetuses at postimplantation stage are treated the same as adult animals in regard to the use of appropriate anesthetics and analgesics. If a surgical procedure is conducted on any fetus and as the result, the fetus cannot grow to become a self-supported individual, euthanasia must be done after birth. For the use of endangered species, a permit must be obtained from the Forestry Department of Indonesia prior to the use of animals in health research.[44]

Applicability

There is no requirement for any facility registration, authorization, and personnel licensing for caring and using animal for research or testing in Indonesia. In addition, there is no requirement to submit the periodic report and regulatory inspection by any regulatory agency, except for the facility capturing and breeding NHPs.[10] The *Guidelines*[6] and *Teaching Guidebook*[7] are primarily applicable and enforced for the research units under Ministry of Health.

The Principles

Law No. 18 does not address the 3Rs[36] explicitly but emphasizes the importance of the 5Fs.[35] However, both the *Guidelines*[6] and the *Teaching Guidebook*[7] address the 3Rs and the 5Fs. The *Guidelines* specify the requirement of protocol to be reviewed and approved by the ethics committee and the humane treatment of animals.

Ethical Review

Article 74[39] covers the requirements of competent veterinarians to oversee the ACUP according to the ethical treatment of animals and animal welfare.

The *Guidelines*[6] cover the ethical aspects of animal use for health research, protocol review procedures by ethics committees, authority and responsibility of the ethics committees, and special considerations on study protocol reviews to include the following topics:

- Risk–benefit analysis
- Animal use justification
- Pain categories and alleviation
- Study design and statistical analysis
- Veterinary care
- Animal procurement, transportation, husbandry, sanitation, and housing
- Restraint
- Food and water restriction
- Behavioral management including physical activity and social housing of the animals
- Humane endpoints and humane treatments
- Multiple major survival surgeries
- Use of anesthetics, analgesics, and neuromuscular blocking agents
- Euthanasia methods
- Personnel qualifications and trainings
- Annual reports and final reports
- Occupational health and safety program to include annual medical checkups and vaccinations, personal protective equipment (PPE), physical plant and engineering controls, and waste management
- Use of fetuses in health research
- Use of endangered species in health research
- Animal importation and exportation procedures

Both the *Guidelines*[6] and the *Teaching Guidebook*[7] state that in order to be ethically justified, animal use for research, teaching, and testing must be covered in a protocol that is approved by the ethical committee of the institution where the activities will be done.

Many institutions have established an IACUC since this practice was initiated for the first time by the NAMRU-2 in 1991 and introduced to other institutions in 1993. This oversight body has been established in academic institutions and universities (at either the departmental level or campus-wide program), governmental institutions, research institutions, pharmaceutical companies, and contract research organizations using animals for research, teaching, and testing. Some institutions merge the oversight responsibilities of the IACUC with the Health Research Ethics Committee (HREC). For example, the Indonesian

National Institute of Health Research and Development (*Badan Litbangkes Republik Indonesia*) HREC also serves as the IACUC.

Each institution has their own animal use protocol format to fulfill their specific requirements, but basically the format covers the following items to be reviewed by the ethical committee:

- Purpose of the project (research, teaching, breeding, antibody production)
- Animal specifications (species, stock or strain, sex, age, bodyweight, number)
- Housing methods
- Length of the project
- Lay summary of the project
- Hypothesis
- Literature search to confirm no duplication of efforts and the least painful procedures are chosen
- Project description and procedures
- Ethical consideration (species justification and animal number justification)
- Pain category, the use of analgesia, justification for not using pain alleviation

Animal use procedures:

- If the project involves anesthesia: drugs, doses, routes of application, frequency, duration, methods of monitoring
- Administration of drugs, reagents, vaccines, cells, agents, substances, etc.
- Surgery (survival/non survival, number of survival surgery)
- Antibody production (the use of adjuvant)
- Behavioral testing
- The use of hazardous agents
- Blood sampling
- Food and water deprivation
- The use of paralytic agents
- Euthanasia method
- Humane endpoint
- Restraint method and duration
- Tumor induction
- Toxicity testing
- Tissue collection
- Personnel qualifications (educational degree, role in the project, training or experience)
- Genetically engineered animal use/production
- Adverse effect (unexpected outcomes)

Institutional and Designated Personnel Responsibilities and Qualifications

The *Guidelines*[6] indicate that the institution must appoint a veterinarian with expertise and experience in laboratory animal medicine and science to be in charge of and oversee the care and use of laboratory animals.

Personnel Training, Education, and Competency Requirements

The *Guidelines*[6] state that researchers and research supporting personnel must be competent and adequately trained in the care and handling of laboratory animals. Regular and periodic trainings are required for each research staff prior to the proposal submission or beginning of the study.

Animal Care and Use Areas Addressed by the Regulations

Animal Transportation

Law No. 18[4] states that transportation of animals should be done in a way that the animals are free from pain and distress. Detailed explanations of the regulation are provided in the Government Regulation No. 95 on Veterinary Public Health and Animal Welfare as follows.[43] Application of animal welfare principles on transportation must be done at minimum in the following conditions:

- Do not harm, prevent injury, and do not cause stress to the animals.
- Use clean and proper transportation means or vehicles according to the capacity.
- Provide adequate food and water to fulfill physiological need of the animals.

If animals are transported in cages, the space provided must allow the animals to move freely in the cages, as well as prevent the presence of predators and incompatible animals. Animals must be protected from direct heat and rain. Transportation must be conducted under supervision and/or recommendations of the attending/authorized veterinarian. In addition, transportation of animal by air must be done in accordance with IATA regulations.[45]

Physical Plant and Environmental Conditions

There is no specific guideline or regulation for these aspects.

Housing or Enclosures

Animal holding and housing areas must be built in a way to ensure the animals' freedom to express their natural behaviors. The regulation does not specify minimum space requirements for each animal species, but the *Guidelines* indicate that minimum space requirements stated in the NRC's *Guide for the Care and Use of Laboratory Animals*[8] should be adopted. The following guidelines are included in the Government Regulation No. 95:[43]

- Holding and housing enclosures should provide no harm and do not cause injury and/or stress to the animals.
- Facilities and equipment do not cause harm, injury, and/or stress to the animals.
- Providing separated housing of different species or allowing physical separation by species.

- Use of clean cages that provide enough space to allow free movement, provide protection from predators and incompatible animals, as well as provide protection from direct heat from sunlight and rain.

Daily Care Activities

There is no specific guideline or regulation on food and water provision, but animal care should be done in a way to ensure the animals are free from hunger and thirst and fulfill their physiologic requirements. There is no specific guideline on sanitation procedures, but animal husbandry should be done in a way to ensure the animals are free from diseases, pain, and distress.

Behavioral and Environmental Management Enrichment

The *Guidelines*[6] indicate that animals used in research must have opportunities to express their natural behaviors and display normal activities according to the biology of the species. Social species are required to have social experience. Therefore, group composition, housing structure, and placement of the animals must be considered and allow visual, auditory, and olfactory contact among their conspecifics. Territorial species could be distressed by the presence of other animals within their territorial boundary. Housing of multiple species in the same room may cause fear and distress.

Occupational Health and Safety

An Occupational Health and Safety Program as indicated in the *Guidelines*[6] consists of the following:

- Annual medical evaluation and vaccination relevant to medical risk assessment and analysis
- PPE as needed, i.e., masks, gloves, boots or shoe covers, caps, goggles, face shields, and lab coats
- Physical facilities or rooms in conformity with safety and ergonomic requirements to minimize work-related injuries and accidents
- Proper waste management to prevent environmental pollution

Veterinary Care

The care and use of animal activities relevant to comparative medicine must be done under veterinary supervision.[43] The institution must appoint a veterinarian with adequate knowledge and experience in laboratory animal medicine to be involved, responsible for, and oversee the care and use of laboratory animals.[6]

Euthanasia

As indicated in the *Guidelines*, euthanasia policy is referred to the *American Veterinary Medical Association* (*AVMA*) *Guidelines on Euthanasia*.[9] Euthanasia methods should be appropriate and without pain and distress and done by professional, qualified, or adequately trained personnel, and death following euthanasia must be confirmed. Euthanasia should be done without the presence of other animals in the same room or location.

Special Considerations

NHPs: Capturing and handling methods of animals from the original habitat must be in compliance with Conservation Law.[10]

Genetically altered animals: Indonesia signed the Cartagena Protocol in May 2000, and on October 19, 2004, the President of Republic of Indonesia signed the Law No. 21, 2004 on Ratification of Cartagena Protocol.[46] The Cartagena Protocol is the protocol for regulating safe transfer, handling and utilization of living modified organisms, the terminology used for genetically modified organisms, or genetically engineered product, which covered transgenic animals and transgenic fish.[47]

Reuse of animals: There is no specific guideline or regulation.

MALAYSIA

Background and Current Situation

The first recorded use of laboratory animals in Malaysia (or Malaya as it was called then) began in 1900 when the Institute for Medical Research was established in Kuala Lumpur by the British colonial government. The Institute's main function was study of tropical diseases, and laboratory animals were used as animal models. In 1948, the Veterinary Research Institute was established in Ipoh, about 200 km north of Kuala Lumpur, and laboratory animals were used mainly as diagnostic tools and for vaccine testing for animal diseases. To date, many universities, especially those with medical, veterinary, and life sciences schools, as well as research institutions have set up laboratory animal facilities. The regulatory framework to regulate the use and care of animals in research in Malaysia is very much shaped by the agreement, which led to the formation of the Federation of Malaysia in 1963. Being the only federated state or federation in Southeast Asia, this system of government is comprised of a union of states with clearly stipulated powers, and areas of responsibilities for both the federal and state governments, as enshrined in the Malaysian Constitution. However, the two East Malaysian states of Sabah and Sarawak retained a higher degree of local government and legislative autonomy within Malaysia, as part of the agreement to join the Federation of Malaysia in 1963. As a result, there are laws that are only applicable to the specific states within Malaysia.

The Laboratory Animal Science Association of Malaysia (LASAM) was established in 1994 and has played significant roles in the establishment of the relevant guides for the care and use of animals for scientific purposes. The LASAM also has lobbied extensively, alongside other organizations calling for the promulgation of acts and regulations that served not only to regulate, but also to promote the development of laboratory animal industry through humane use of animals in scientific activities. The LASAM has been actively involved in the education and training activities on animal use in research to the scientific community since its inception in mid-1990s. The successful organization of the sixth Asian Federation of Laboratory Animal Science Associations (AFLAS) Congress in Kuala Lumpur in 2014, among others, was an important catalyst leading to the publishing of the Animal Welfare Act 2015 (AWA2015)[11] by the Parliament of Malaysia in 2015. The promulgation of the AWA2015 paved the way for the existence of subsidiary legislations that include the Animal Welfare (Animals Use for Research, Testing and Teaching) Regulations 2015 (unavailable, pending approval, and implementation) and few other relevant regulations pertaining to the licensing, transport, and disposal of animals that are to be implemented in 2017.

There are at least 21 IACUCs in Malaysia, with more being established with the arrival of the AWA2015. Fundamentally, the IACUC in each of the respective institution adheres to what is suggested by the Malaysian Code of Practice for the Care and Use of Animals for Scientific Purposes (Code).[13] There are institutions with good organization of training, monitoring, and inspection, while some IACUCs are in the midst of establishing their workflow and refining procedures to comply with the prevailing regulations. Some organizations had multiple IACUCs within the same institution, and on the other end of the spectrum, there are ministry level organization with a centralized ministerial-level IACUC to oversee multiple institutions within that ministry itself. Therefore, it is necessary to standardize and streamline the operation of various IACUCs as per the provisions of the AWA2015 and the Code. The current document describes the necessary regulations and guiding principles that are relevant to the sustainable development of laboratory animal science in Malaysia.

Summary of Regulations or Legal Requirements

Regulations and legal requirements that regulate the use of animals in Malaysia differed between that of Peninsular Malaysia and that of the two eastern Malaysian states of Sarawak and Sabah. In general, the Malaysian AWA2015[11] and the Animal Act (Amendment) 2013 (AA2013)[12] are the principal legislations that govern the use and care of laboratory animals in Malaysia. The AA2013 traces its origin to the original legislation, the Animal Act, which came into effect in 1953. The use of nondomesticated species, which may include NHPs and other wildlife species for research, is governed by the Wildlife Conservation Act 2010 (WCA2010).[15] In addition, the protection of aquatic mammals and fisheries species is under the purview of the Malaysian Fisheries Act 1985 (FA1985).[48] The scope and geographical coverage of the laws within Malaysia was defined by the legislative framework that led to the formation of the Federation of Malaysia in 1963. States in the Malay Peninsula and the three Federal Territories (Kuala Lumpur, Putrajaya, and Labuan) enforced the

AWA2015, AA2013, WCA 2010, and FA 1985, whereas the eastern Malaysian states of Sabah and Sarawak only had legislations that focused on the extraction, conservation, and use of wildlife and fisheries in their respective states, for example, the Wildlife Conservation Enactment 1997, Sabah (WCE1997)[49]; Wildlife Protection Ordinance 1998, Sarawak (WPO1998)[50]; and the Inland Fisheries and Aquaculture Enactment 2003, Sabah (IFAE2003).[51] However, the states of Sabah and Sarawak have the autonomy and provision to adopt the federal acts where applicable, and/or when there is a necessity. The existence of the legal autonomy and overlapping has made the implementation of standardized laboratory animal welfare regulatory framework across Malaysia a challenge. Therefore a practical set of regulations for the care and use of laboratory animals in Malaysian institutions should not only be able to address animal welfare issues but also to transcend and harmonize the existing legislations that are in line with the provision of powers under the federation system of government.

The use of animals in schools other than institutions of higher learning as defined by the Malaysian Education Act 1996 (MEA1996)[52] is covered under the Section 26(5) of the AWA2015. In recognition to the immense logistical and personnel commitments required to enforce the AWA2015 in primary and secondary schools, the long-term aspiration of the Malaysian Education Ministry is to increase the use of computer-aided learning and other alternative teaching technologies in classrooms to reduce, or in some cases, to do away completely with the need to use animals. To further promote this agenda, a guideline on the use and care of animals for teaching activities in Malaysian schools has been prepared and is pending for approval in 2017.

Scope

The AWA2015[11] was issued on December 29, 2015. The definition of animal under this act covers "*...any living creature other than a human being and includes any beast, bird, aquatic animals, reptile or insect but does not include wildlife under the Wildlife Conservation Act 2010....*"[15] Transitional provisions (Section 64, Part IX) spelled out a grace period on licensing and enforcement matters, within 6 months from the date of coming into operation of the AWA2015. The AWA2015 is comprised of 9 parts and 64 sections, with evident emphasis on the 5Fs,[35] while its subsidiary regulation that governs the use of animals in research, testing, and teaching is founded on the 3Rs principles[36] as well as in line with OIE *Terrestrial Animal Health Code*.[1] Of particular interest to the Malaysian research communities will be Part II—Animal Welfare Board (AWB), Sections 3–14 that define the functions and membership of the AWB; Part III—Licensing, Sections 15–23 that deal with licensing matters for keeping and using animals; Part IV—Matters related to animal welfare, where Section 26 explicitly states that the use of animals in research, testing, and teaching is only permissible with a license from the Board, while Section 27 is on animal transportation matters; Part V—Cruelty to animals, where Section 29 lists types of cruelty offenses, and Section 30 that states the powers of Animal Ethics Committee or IACUC in the killing of research animals, apart from powers that are already given to the veterinary authority. Section 33 in Part V explicitly mentions that any person, who fails to exercise reasonable care and supervision relating to the protection of the animal from cruelty or animal welfare, commits an offense under this act. Parts VI, VII, VIII, and IX deal with enforcement powers, court action, and other general matters related to animal welfare. The AWB is an integral part of the AWA2015. It functions mainly "*…to monitor work of bodies (for example IACUC) established for the purpose of preventing trauma, pain or suffering to animals, protection of animals, to impart education in relation to the humane treatment of animals, to advise the Minister of the government on any matter relating to animal welfare…*," and most importantly to discharge all such functions authorized under the AWA2015 or any subsidiary legislations made under the AWA2015. This effectively meant that the AWB has the necessary regulatory power to regulate the use and care of laboratory animals in Peninsular Malaysia and the Federal Territories of Kuala Lumpur, Labuan, and Putrajaya. Membership of the board comprised of the Director General of the Department of Veterinary Services (DVS), who is also the chair; Director Generals of the Department of Wildlife and National Parks Peninsular Malaysia, Department of Fisheries, Department of Local Government; the Dean of a Veterinary School appointed by the Minister from a public university in Malaysia; and senior officers, one each from the Ministries of Agriculture, Education and Health. Enforcement and compliance inspection activities will be carried out by the enforcement division of the DVS, Ministry of Agriculture and Agro-Based Industries, Malaysia. Section 15 in Part III of the AWA2015 mentions explicitly that no research/activities involving animals are to be carried out without a license. For institutions, the IACUC is to be licensed and recognized by the AWB to carry out the necessary protocol review. It is also the responsibility of the IACUC to report any noncompliance practices/violations by any researcher within its institution for subsequent enforcement actions by the DVS. Most offenses spelled out by the AWA2015 carry a penalty of MYR10,000 to MYR20,000, but not more than MYR100,000, or imprisonment not exceeding 3 years or both, where applicable. Convicted individuals may also be disqualified from applying for license for up to a year (Part VIII, Section 52). However, it should be noted that for the AWA2015 to function, a series of subsidiary legislations and guidelines have to be established. These are necessary as they define the technical limits and regulatory practices that allow the AWA2015 to be enforced effectively to regulate the care and use of animals for research purposes. To date (as of November 2016), the DVS is in the process of preparing 13 sets of regulations and 12 codes of practices for various species of animals, ranging from breeding, farming, research, and animals used for pleasure, recreation, and sports activities. Of all the documents being prepared, pending approval and implementation in 2017, the Animal Welfare Regulations 2015, and the Code[13] are important documents for the research community in

Malaysia. The regulations spell out the terms of reference and membership composition of the IACUC that are in line with international practice. Apart from that, the regulation also stipulates the requirements for licensing, monitoring, required information pertaining to IACUC review and approvals, and other requirements to be included in the IACUC's annual report to the AWB. The IACUC is an important oversight instrument provided under the AWA2015 and its subsidiary regulations. The IACUC is also entrusted to monitor compliance of approved projects, ensuring all nominated personnel had the necessary training and experience to carry out the proposed procedures, and to ensure all necessary (bio)safety guidelines are adhered to. Investigation of complaints on animal use is under the purview of the AWB, through qualified enforcement officers appointed for this purpose by the DVS. In these situations, the IACUC would act mainly to assist the inquiry as it does not have enforcement powers. It should also be noted that the decisions of the IACUC are only valid when it has a quorum of at least 50% of all of the voting members. The criterion of IACUC's review is in line with the guidelines stipulated by the Code, as detailed in the next section. The Code placed heavy emphasis on individual, collective, and institutional responsibilities in ensuring justified and humane use of animals for scientific purposes. In line with this spirit, the IACUC requires the following details to be furnished for review:

i. (Lay) summary of the proposed project so that it can be understandable to anyone who comes into contact with the proposal, while allowing for justified and transparent view of the proposed project by the members of the interested public without the necessary technical background
ii. Justification and potential alternatives for the use of animals (where applicable), as well as institutional policy when approving animal use for scientific purposes. Evidence provided must be in line with the 3Rs principles advocated worldwide.
iii. Source, number, and how the animals are obtained for the proposed project
iv. Description of animal care and husbandry procedures provided, and personnel who are responsible for the care of animals
v. Description of procedures, duration, and invasiveness of the procedures
vi. Competencies of the individuals proposed to conduct the procedures on animals
vii. Defined usage endpoints and necessary actions to relieve the pain and suffering should these be inevitable
viii. Availability of emergency care for animals, euthanasia, and ways the animals are to be managed after their scientific use has ended.
ix. Details on the use and disposal of potential hazards from the experiment to the animal, environment, and the operator(s)

The descriptions above are in line with the AWA2015's aspiration to promote a sustainable self-regulation approach in dealing with animal use for scientific purposes, with periodic checks and intervention from the IACUC and

AWB (where necessary) as provided for by the Act. Apart from the use of animals for scientific purposes in higher education, research centers, and other related organizations, there is also a guideline that had been drafted to address the use of animals in Malaysian schools, as provided under section 26(5) of the AWA2015. In a glance, the guideline mirrors the recommendations and principles set forth by the Institute of Laboratory Animal Resources Commission on Life Sciences, National Research Council, United States, through their publication on the Principles and Guidelines for the Use of Animals in Precollege Education (2004).[53]

The Malaysian Code for the Care and Use of Animals for Scientific Purposes

The Code[13] was adapted and adopted with permission from the NHMRC-Australia, based on the Australian Code for the Care and Use of Animals for Scientific Purposes, 7th Edition (NMHRC, 2004).[54] However, elements from the latest version of the Australian Code for the Care and Use of Animals for Scientific Purposes, 8th Edition (NMHRC, 2013)[55] were included, for example, the specific reference to Animal Ethics Committee or IACUC membership, and aligned with the AWA2015.[11] Similar to the Australian Code that it is based on, the Code is applicable to all branches of sciences within the confines of jurisdiction (except schools) specified by the AWA2015. The Code also stresses on the personal responsibilities of investigators for animals throughout the period of experiments approved by the IACUC, in line with provisions in the AWA2015. In principle, there are five major areas addressed by the Code, namely, (1) general principles for the care and use of animals in scientific research, which require the researchers to spell out the justification for animal use, and in accordance with the principles of reduction, refinement, and replacement (the 3Rs principles)[36]; (2) responsibilities of the institution and IACUC, which describe the establishment, and terms of reference of the IACUC, workflow related to proposal review, monitoring and reporting, responsibilities of the investigators as well as matters pertaining to suitability of models, procedures, and termination and disposal of animals; (3) responsibilities of the researchers/investigators; (4) matters pertaining to the acquisition and care of animals in breeding and holding facilities, which may include transportation, quarantine, husbandry, and veterinary care; and (5) role and responsibilities of the AV.

The Code took more than 10 years from conception to where it is today, and the effort was first initiated by a group of concerned members of the scientific community who are also members of LASAM. The effort to promote the adaptation of the Code only gained traction after major institution and universities in Malaysia started to value the importance of IACUC in attracting and sustaining foreign funding and collaborations. From the commercial standpoint, having a national legislation to regulate the use and care of animals certainly is an important element of a value proposition to attract foreign investment in the area of biotechnology and pharmaceuticals. The publishing of the AWA2015 is certainly a watershed moment further underlining the importance of the Code to the Malaysian research community as the way forward.

Euthanasia and Veterinary Care

It is clear from the Code[13] and relevant regulations that euthanasia should be carried out according to the guidelines set forth by the AVMA Guidelines for the Euthanasia of Animals: 2013 Edition.[14] The Animal Welfare Regulations 2015,[11] through the Code, clearly stipulated that veterinary care, particularly the emergency veterinary care, must be made available and accessible at all times to the experimental subjects. Only veterinarians that are registered with the Malaysian Veterinary Council are allowed to provide veterinary medical care to animals of a proposed project.

PHILIPPINES

Background and Current Situation

Laboratory animal research in the Philippines started as early as the 18th century when Spanish conquistadors sent a mission to the country to study smallpox. The mission resulted in the foundation of the Board of Vaccination later renamed Bureau of Science under the American rule.[56] Since then, animals have been used primarily for the production of antivenins.

Laboratory animals are currently used by government and private academic institutions, national research institutes, local pharmaceutical companies, and clinical research companies for research, instructional, and testing purposes. A surge in usage of laboratory animals was apparent for the last 20 years because of the rise in basic and applied research such as natural products, nanotechnology, biomedical and drug development, priority research areas by the government agencies such as the Department of Science and Technology (DOST), Department of Agriculture (DA), and National Research Council of the Philippines.

Concerns regarding animal welfare and the growing awareness that policy development pertaining to laboratory animal science resulted in scientific progress led to the enactment of Republic Act 8485 in 1998, an Act to Promote Animal Welfare in the Philippines, otherwise known as "The Animal Welfare Act of 1998."[16] In 2013, Republic Act 10631[17] was enacted amending certain sections of Republic Act 8485.

The Philippine Association for Laboratory Animal Science (PALAS), a multiprofessional organization, has been advocating proper use and welfare of laboratory animals for the last 28 years through conduction of workshops and training in various institutions throughout the country to educate researchers, faculty, and students. A regular scientific conference has also been organized by PALAS yearly to bring together people in various disciplines for the purpose of scientific presentation of new research data and information dissemination on animal welfare and serve as an avenue for networking and collaboration. Recently, the Philippine College of Laboratory Animal Medicine (PCLAM), an affiliate of the Philippine Veterinary Medical Association (PVMA), was established in 2015 to further strengthen responsible animal use and care with its mission of training more veterinarians as specialist in the field of laboratory animal medicine and science to further laboratory animal research in the country.

Summary of Regulations or Legal Requirements

The amended Animal Welfare Act[17] safeguards and promotes welfare of all terrestrial, aquatic, and marine animals in the country by regulating and supervising the establishment and operation of animal facilities used for breeding, maintaining, training, diagnosing, treating, and experimentation.

The Revised Implementing Rules and Regulations (IRR) of the amended Animal Welfare Act[20] were promulgated in favor of the welfare and rights of animals and the assurance of the 5Fs.[35] The amended Act mandates that a Committee on Animal Welfare (CAW) attached to the DA shall be created. The DA-CAW composed of 14 members is tasked to formulate and issue the necessary rules and regulations to implement the law, subject to the approval of the Secretary of the Department of Agriculture.

Pursuant to Section 3 of the amended Act, the Bureau of Animal Industry (BAI) is tasked to regulate and supervise the establishment, operation, and maintenance of all animal facilities or any structure for the confinement of animals where they are bred, treated, maintained, or kept either for trade or sale or for training purposes as well as the transport of animals. Rule 3.1 of the Revised IRR mandates that the BAI shall create the Animal Welfare Section that shall implement and enforce the rules and regulations formulated by the CAW and approved by the Secretary of DA. The Animal Welfare Section is currently under the Animal Health and Welfare Division (AHWD) of the BAI.

Pursuant to the Animal Welfare Act, the Administrative Order No. 40 (AO40), Series of 1999[18] was issued promulgating the rules and regulations on the conduct of scientific procedures using animals in the country to institute the basic systems, organizations, and practices in all laboratory animal care and use establishments to safeguard the welfare of animals used in scientific procedures. The intention of the order is to provide authorization to conduct scientific procedures using animals based mainly on the existence of an IACUC and the acceptability of the ACUP.

In support of this AHWD endeavor, the Commission on Higher Education issued a circular in 2015 requiring all academic institutions at the tertiary level especially those with life science, agricultural science, medical, and veterinary programs who are engaged in animal researches to strictly adhere to the creation of their respective IACUC. Similarly, the Bureau of Secondary Education,

Department of Education, issued a memorandum to regional directors, school division superintendent, and school heads, requiring all school laboratories, facilities, and establishment to be used as experimental stations for animal research must be registered to the BAI before they are given authorization to conduct researches and procedure using animals. Further, government funding agencies such as the Philippine Council for Health Research and Development, DOST, and DA also require submission of the IACUC approval letter and PALAS accreditation prior to release of funds of approved research programs or projects.

Applicable guidelines and principles include those from the PALAS Code of Practice for the Care and Use of Laboratory Animals in the Philippines (COP),[19] the *Guide for the Care and Use of Laboratory Animals*,[57] the CIOMS-ICLAS International Guiding Principles for Biomedical Research Involving Animals,[3] and the PALAS Training Modules for Small Laboratory Animals and Non-human Primates (pending publishing at the time of writing this chapter). The two latter references will serve as the training modules for proper handling and conduction of common technical procedures by scientist, researchers, faculty, and students in the country who will conduct animal research, instructional, and testing activities.

The COP, which is currently revised by the PALAS and PCLAM officers and members to promote animal welfare and ethical use of animals for instruction, research, and testing, provides comprehensive information on the composition and responsibilities of the IACUC and categorization of scientific procedures. It also serves as an appropriate local standard guide for the development of an ACUP, including veterinary care and quality assurance procedures to ensure proper care and management of laboratory animals in the facility. This also includes the required housing, engineering standard, and environmental enrichment needed by specific laboratory animal species and further addresses the training of facility personnel, management of biohazards, proper disposal of waste and animal carcass, and appropriate plans for disaster preparedness to ensure safety of animals and personnel in the facility in the event of an emergency, disaster, or calamity.

Scope

The AO40,[18] which is currently being reviewed by the Technical Working Group (TWG) appointed by the DA-CAW, covers and regulates scientific procedures whether research, testing, or instructional activities using live vertebrate animals such as biomedical research, product testing of food, drugs, agrochemicals and cosmetics, and production of antisera or other biologicals. The abovementioned scientific procedures involving animal manipulations may cause stress, pain, injury, suffering, and even death in animals. Exempted from securing authorization requirements to conduct scientific procedures are preclinical tests on animals for evaluating veterinary products in accordance with regulatory requirements or standard procedures (TWG is recommending that this should not be exempted); recognized veterinary procedures such as therapy, prophylaxis, diagnostic, or disease surveillance procedures necessary or desirable for animal welfare; and recognized agricultural practices (e.g., castration, genetic engineering, or embryo manipulation, unless they form part of an experiment).

Registration, Authorization, and Accreditation of Institutions

All private and government institutions engaged in animal research and testing must first register their laboratory animal facility with the DA, BAI-AHWD. The registration of an entity with the BAI-AHWD is based on the acceptability of the ACUP, which adheres to the guidelines and principles of the COP and other abovementioned international guidelines, which are implemented by the AV of the Institution, existence of an IACUC, ACUP Accreditation by a duly recognized body such as the PALAS and presence of animal training programs for faculty, researchers, students, and facility personnel. The registration is valid for 2 years and renewable thereafter.

The institution must also secure from the DA, BAI-AHWD, an animal research permit or authorization to conduct scientific procedures using animals. This authorization to conduct scientific procedures in animals is granted to institutions by the BAI-AHWD based on IACUC Study Protocol Review and Approval.

The IACUC consists of at least three members, namely, (1) a chairperson who could be a licensed veterinarian with sufficient training and experience in laboratory animal science or medicine; (2) a technical member who is experienced in conducting scientific procedures in animals; and (3) a public member who is nonaffiliated with the institution and preferably with concern for animal welfare. Any additional member may be appointed provided the person is a key institutional official (IO) whose services to the institution have a direct impact on the ACUP. The IO designates all IACUC members and shall have a term of 1 year unless reappointed for the following term. A secretary is appointed to record the proceedings of the review process and prepare the meeting minutes. Whenever necessary, the chairperson may request a credible resource person to provide expert opinion during the review process.

Ethical Principles and Ethical Review

The responsibilities of the IACUC are clearly stated in the AO40. The IACUC evaluates, approves, and monitors submitted protocols for scientific procedures in animals guided by the 3Rs.[36] Scientific protocols are categorized based on the severity of pain inflicted on the animal by the said scientific procedure (Table 11.2).

The project leader or principal investigator should adhere to the approved protocol. For protocol amendments, they must submit the revised parts of the

TABLE 11.2 Categorization of Scientific Procedure Based on Pain and Distress as Stated in AO40[18]

Category of Scientific Procedure	Level of Pain and Distress	Reviewers	Examples of Scientific Procedures
Category 1	Mild	IACUC Chair	Venipuncture Skin irritation test with mild irritating property Superficial tissue biopsy Peripheral blood vessel cannulation
Category 2	Moderate	IACUC Chair + 1–2 members	Efficacy and pharmacologic tests Routine toxicity test avoiding lethal endpoints Animal surgeries
Category 3	Severe	Quorum of IACUC	Antimicrobial and vaccine tests Animal disease and surgical modeling Toxicity test with significant morbidity and mortality endpoint

protocol to the IACUC chairperson for evaluation and approval. The IACUC members are informed through an official memorandum by the IACUC chairperson of the action taken on the protocol regardless whether it underwent an expedited or full review.

Noncompliance with the approved ACUP and/or protocol may lead to revocation of the approval to conduct the animal scientific procedure at the institutional level. Consequently, the IO in consultation with the IACUC shall review the reasons for revocation and take appropriate corrective action.

Another vital role of the IACUC is to evaluate, approve, and monitor the ACUP through regular facility inspection conducted twice a year. This ensures that facility manager and personnel are adhering to the set forth guidelines on proper care, use of animals, and proper maintenance of the facility. The COP and the National Research Council Guide for the Care and Use of Laboratory Animals[57] are used as references for ACUP review and facility inspections.

The AO40[18] provides the ACUP inspection checklist, which includes daily, holiday, and emergency veterinary care; facility personnel qualifications and training; animal environment, housing, and management such as cage type and density, temperature, humidity, ventilation, illumination, noise control, environmental enrichment, food, water, and sanitation; as well as veterinary medical care to address pain and distress, disease prevention, analgesia, anesthesia, and euthanasia.

After the thorough evaluation, deficiencies are classified as either minor or significant, which can pose minor or major threat to animal health or safety, respectively. The facility manager and/or the IO are informed of these deficiencies to make necessary corrective measures in an agreed appropriate time frame. If significant deficiencies are not resolved by the set correction dates, the IACUC must notify the AHWD-BAI within 30 working days. Actions taken for the identified deficiencies are included in the required annual report submitted to BAI, which then issues a Certificate of Approval.

Animal Care and Use Areas Addressed by the Regulations

Veterinary Care

The AO40[18] briefly outlines the requirements for veterinary care in an ACUP. The veterinarian may be employed part-time or as a consultant, but the frequency and duration of visits must be described in the ACUP. Weekend, holiday, and emergency veterinary care should be available. The ACUP should describe quarantine and acclimation, disease surveillance, preventive medicine, analgesia, anesthesia, and euthanasia. The IACUC has a responsibility to provide necessary technical recommendations and guidelines/references (e.g., recommended doses for anesthetics and analgesics) to animal users. Finally, as discussed previously, the COP[19] provides a more comprehensive reference for various aspects of veterinary care. For example, it lists various pathogens found on laboratory animals based on animal species, zoonotic and/or opportunistic potential, and the potential to confound research.

Euthanasia

A significant portion of the AO40[18] lists euthanasia methods approved for various laboratory animal species. These include the use of carbon dioxide (for rodents placed in an uncrowded chamber), inhalant anesthetics, barbiturates (most recommended for dogs and cats), stunning, cervical dislocation (for mice, rats weighing less than 250 g, and rabbits weighing less than 1 kg), decapitation, and exsanguination (performed under anesthesia). Agents such as ether and chloroform are also listed as unsatisfactory and are discouraged. Whatever method is employed, euthanasia should always be done away from public view.

Transport of Animals

Under the Animal Welfare Act,[17] three separate DA Administrative Orders, for the transportation of animals were issued, AO43 Series of 1999, transport of animals by Air,[58] AO19 Series of 2006 for transport of animals by land,[59] and AO02 Series of 2002 transport of animals by Sea[60] were promulgated to promote, ensure, and protect the welfare of all animals during transportation. These Orders take into considerations the 5Fs[35] in the transport of live animals. In realizing this objective, it is recognized that there are certain limitations during transport that may restrict the enjoyment of the 5Fs, provided that such procedures adopted for transportation are not contrary to law.

These Administrative Orders state that the management of any air, land, and sea transport utility has the responsibility to provide animals with appropriate, clean, and sanitary facilities for the safe conveyance and delivery to their final destination. Brokers and handlers should ensure that sufficient food and water for such animals are made available whenever necessary while in carriage for more than 12 hours. Animals transported by air should conform to the rules and regulations of the IATA[45] in terms of caging or crates with appropriate ventilation. Airport authorities should provide necessary assistance and appropriate holding area for the safety and comfort of the animals. All animals being transported by land, sea, or air should have appropriate government documents/certificates issued by the Department of Environment and Natural Resources (DENR) or BAI-DA such as Livestock Handlers License, Veterinary Certificate, rabies vaccination for dogs and cats, shipping permits, and in case of wildlife, Wildlife Transport Permit issued by DENR. Others items include identification of animals, ownership of animals pursuant to PD533 otherwise known as "Anti Cattle Rustling Law," and negative Coggins test for horses. They should be also accompanied by trained animal caretakers/handlers during transport. Complete guidelines and considerations during travel/transport of animals are shown in AO19 for the Transport of Animals by Land.[59]

Highlights of RA.10631, an Act Amending Certain Sections of RA 8485, Otherwise Known as the Animal Welfare Act of 1998[17]

Section 1 of the Amended Act clearly states that the purpose of the law is to protect and promote the welfare of all terrestrial, aquatic, and marine animals in the Philippines by supervising and regulating the establishment and operations of all facilities utilized for breeding, maintaining, keeping, treating, or training of all animals either as objects of trade or as household pets. For purposes of this Act, pet animal shall include birds.

The same Section of the Act considers animal welfare to pertain to the physical and psychological well-being of animals. It includes, but is not limited to, the avoidance of abuse, maltreatment, cruelty, and exploitation of animals by humans by maintaining appropriate standards of accommodation, feeding and general care, the prevention and treatment of disease, and the assurance of freedom from fear, distress, harassment, and unnecessary discomfort and pain and allowing animals to express normal behavior.

In Section 3 of the amended Act, a new section was inserted as Section 7, which states that it shall be unlawful for any person who has custody of an animal to abandon the animal.

Section 7 states that if any person being the owner or having charge or control of any animal shall without reasonable cause or excuse abandon it, whether permanently or not, without providing for the care of that animal, such act shall constitute maltreatment under section 9. If the animal is left in circumstances likely to cause the animal any unnecessary suffering, or if its abandonment results in the death of the animal, the person liable shall suffer the maximum penalty of fine and imprisonment on conviction by final judgment.

Among the highlights of the amended Animal Welfare Act is the imposition of stiffer penalty by setting the maximum imprisonment of 2 years and 1 day to 3 years and/or maximum fine not exceeding 25,000 pesos.

In Section 5 of the amended Act, a new section was inserted as Section 10, which states that the Secretary of the DA shall deputize Animal Welfare Enforcement Officers (AWEOs) from recognized nongovernment organization, citizens group, community organization, and other volunteers who have completed and passed the training requirements for the AWEO. The Secretary shall also create a mechanism for the monitoring, supervision, and reporting of these enforcement officers. The AWEO shall have the authority to seize and rescue illegally traded and maltreated animals and arrest violators of this Act subject to the existing laws, guidelines, rules, and regulation, on arrest and detention.

SINGAPORE

Background and Current Situation

Research involving the use of animals is regulated to safeguard laboratory animal welfare. Under the Animals and Birds (Care and Use of Animals for Scientific Purposes) Rules 2004,[22] any research facility that intends to keep or use animals for scientific purposes must obtain a license from the Agri-Food & Veterinary Authority of Singapore (AVA).

The National Advisory Committee for Laboratory Animal Research (NACLAR) was established in 2003 with the aim of promoting responsible and humane care and use of animals for research. Members included representatives from academia, research organizations, the AVA, veterinarians, as well as legal and ethical specialists. In 2004, NACLAR developed national guidelines known as the NACLAR Guidelines.[23] Research facilities are required to comply with the guidelines as part of licensing requirements.

The NACLAR Guidelines are modeled closely after best practices in countries such as Australia, Canada, New Zealand, the United States, and documents such as the old CIOMS International Guiding Principles[61] and the European

Convention for the Protection of Vertebrate Animals Used for Experimental and Other Scientific Purposes.[62] Consultation with local research communities, associations, government agencies, religious groups, animal welfare groups, and the general public was conducted in its formulation. Today, the NACLAR Guidelines set out responsibilities for all parties involved in the care and use of animals for scientific purposes, in accordance with widely accepted scientific, ethical, and legal principles.

In Singapore, self-regulation is enforced. Research facilities are licensed, but not the individual researchers or the projects. However, requirements imposed through licensing make each research facility accountable for researchers and research projects under it. Each research facility has to put in place a program and a system of checks to ensure that animals are used responsibly. In this regard, the appointed IACUC of each facility acts as an internal approval and auditing body to evaluate all proposed use of animals for scientific purposes in compliance with the NACLAR Guidelines. The AVA maintains regulatory oversight and audits IACUCs to check for effective self-regulation.

Summary of Regulations or Legal Requirements

Under the Animals and Birds Act, Chapter 7,[21] the Animals and Birds (Care and Use of Animals for Scientific Purposes) Rules 2004,[22] governs the care and use of animals for scientific purposes. The Rules commenced in effect from November 15, 2004. The NACLAR Guidelines[23] establish best practices in the care and use of animals for research.

The NACLAR Guidelines are presented in three parts:

a. The Guiding Principles for the Care and Use of Animals for Scientific Purposes (Guiding Principles).
b. The IACUC Guidelines detail how the IACUC should operate in carrying out its responsibilities.
c. The Training Guidelines provide the training scope and requirements for all personnel involved in the care and use of animals for scientific purposes.

Scope

The Guiding Principles of the NACLAR Guidelines[23] cover all aspects of the care and use of animals for scientific purposes performed to acquire, develop, or demonstrate knowledge or techniques in any scientific discipline. This includes the purpose of teaching, field trials, environmental studies, research, diagnosis, product testing, and the production of biological products.

In relation, animals are defined under the Animals and Birds (Care and Use of Animals for Scientific Purposes) Rules 2004[22] and the NACLAR Guidelines[23] as any live vertebrate, including any fish, amphibian, reptile, bird, and nonhuman mammal.

Applicability

Research facility licensing requirements and the NACLAR Guidelines[23] apply to all research, which is conducted in the facility. The IACUC of the facility maintains overall responsibilities in ensuring that all animal research complies with relevant legislation as well as the NACLAR Guidelines.

The Principles

The 3Rs principles of replacement, reduction, and refinement[36] are integral to the NACLAR Guidelines[23] and form the basis of the Guiding Principles. Chapter 2: General Principles for the Care and Use of Animals for Scientific Purposes recommends an approach to animal use based on the 3Rs.

Proposals submitted to the IACUC must contain sufficient information to justify the proposed use of animals. Researchers must demonstrate in their proposals that procedures will avoid or minimize discomfort, distress, and pain to the animals and that alternatives have been considered. They must also ensure that experiments are not unnecessarily duplicated. The rationale for involving animals and the appropriateness of the species and approximate numbers of animals used must also be explained. In all experiments, animal welfare must be observed and maintained at all times.

Animal research facilities are further required to submit an annual report by the Chief Executive Officer (CEO) per calendar year. Assurance must be provided that all researchers and personnel comply with the Guiding Principles and that any exceptions to this are explained, approved by IACUC and recorded officially. Assurance must also be given that professionally acceptable standards governing the care and use of animals were maintained, with appropriate use of anesthetics, analgesics, and tranquilizers where necessary, and that each researcher has considered alternatives to any procedure, which can cause pain or distress to animals.

Ethical Review

The IACUC is the institutional oversight body appointed by the CEO of each animal research facility. The IACUC reports directly to the CEO, who is then responsible for acting on the IACUC's recommendations. The IACUC advises the CEO of the institution's compliance, establishes plans and schedules for correction of deficiencies, and makes recommendations regarding any aspects of the animal program, facilities, or personnel training. This system ensures that management is fully apprised of the facility's care and use of animals for scientific purposes.

As such, the IACUC must be provided with adequate and appropriate facilities, powers, and resources to carry out its duty. It reviews and approves or rejects proposals involving animal use, taking into consideration ethical and welfare aspects as well as scientific or educational value. It is mandatory to review the animal and care use program semiannually and inspect the housing and procedural rooms of the research facility annually.

The IACUC considers these areas when reviewing proposals involving animal use:

- Species and numbers of animals to be used
- Rationale for using animals and justification of the appropriateness of the species and numbers to be used
- Complete description of the propose use of the animals
- Whether proposals avoid or minimize stress, distress, or pain
- Whether the 3Rs have been complied with, i.e., replacement using nonliving systems, reduction of numbers, or refinement of techniques
- That there is no unnecessary duplication of previous experiments
- That animals are not used repeatedly unless justified
- Adequate description and justification of humane endpoints
- Appropriate use of drugs (sedatives, analgesics or anesthetics) and techniques in consultation with the AV
- That paralytics will not be used unless when combined with anesthesia
- Qualifications and training of personnel conducting procedures on animals
- Adequate and appropriate care and husbandry of animals
- Adequate and appropriate presurgical, surgical, and postsurgical procedures, techniques, and care
- Euthanasia methods

The IACUC also reviews and investigates complaints about animal care and use, monitors compliance of approved projects involving animals, ensures animal users and caretakers are adequately trained, and ensures occupational health and safety standards are complied with. Additionally, the IACUC serves a key role in advising of potential security risks and vulnerabilities and in conducting appropriate risk assessments to develop an emergency preparedness and disaster plan to deal with any potential or real threats.

The IACUC comprises at least five members, including one separate member from each of the following four categories:

a. A veterinarian with training or experience in laboratory animal science and medicine, and who has experience in the routine care of the species of animals used in the research facility
b. A person with appropriate experience in the use of animals for scientific purposes
c. A person who is not affiliated in any way with the research facility and is not a user of animals for scientific purposes
d. A person whose primary concerns or interests is in nonscientific areas.

The AV must be appointed to the IACUC, but the CEO or equivalent cannot be appointed to the IACUC. Additionally, a maximum of three members are allowed from the same unit or department within the institution to avoid influencing IACUC decisions.

Certain IACUC decisions, such as approvals or suspensions/withdrawals of projects, require a quorum. In Singapore, a quorum is defined as more than 50%

of the IACUC members with at least one representative from the "nonscientific" or "nonaffiliate" category.

Institutional and Designated Personnel Responsibilities and Qualifications

CEO: Defined in the NACLAR Guidelines[23] as the Chief Executive Officer (or person of similar standing) of an Institution, which is in the position to grant resources to the IACUC and to enforce the IACUC's recommendations. The CEO cannot be appointed to the IACUC.

AV: Defined in the NACLAR Guidelines as a veterinarian engaged formally on a full-time or part-time basis, to maintain oversight of the ACUP and to provide adequate veterinary care. All veterinarians including the AV must have qualifications in veterinary science and are licensed by the AVA.

IACUC: Defined in the NACLAR Guidelines as the Institutional Animal Care and Use Committee constituted by institutions.

Investigator: Defined in the NACLAR Guidelines as a person who proposes or has approval to conduct a project involving the use of animals.

Staff: Defined in the NACLAR Guidelines as all persons involved in the housing, feeding, and general care of the animals or who otherwise assist investigators. Staff should have appropriate veterinary or animal care qualifications or experience, as recommended in the Training Guidelines.

Personnel Training, Education, and Competency Requirement

Personnel training and education are described in the IACUC Guidelines, Chapter 2, NACLAR Guidelines[23]: Oversight of the Animal Care and Use Program. The research facility has a responsibility to ensure that all groups of personnel involved in the care and use of animals for scientific purposes (researchers, research technicians, animal technicians, and other personnel) receive appropriate training and are adequately trained to recognize and alleviate signs of pain and distress, as well as conduct animal procedures without causing unnecessary harm. Training should also be extended to temporary staff such as students or visiting researchers, where necessary.

The NACLAR Guidelines recommend that training programs should include the knowledge of humane methods of animal handling and experimentation; basic needs of different animal species; proper handling and care; proper preprocedural, procedural, and postprocedural techniques and care; aseptic surgical methods and procedures; and proper use of anesthetics, analgesics, and tranquilizers. Occupational health and safety is also mentioned as a component. The NACLAR Guidelines encourage institutions to support training programs and to encourage active participation by personnel, especially where training is not mandatory in a facility. Training records are advised to be maintained, either by the individuals, IACUC, departments, or training coordinators.

Specific training requirements and recommendations are highlighted in the Training Guidelines for animal facility staff (caretakers, technicians, managers, and veterinarians), researchers, IACUC members, service personnel, and teachers at Tertiary Institutions (training for teachers at non-tertiary level is to be determined by the Ministry of Education). There is also a recommendation that training be provided for any biohazard-related work involving genetically modified organisms, radiations, chemicals, and general hazards.

IACUC members are required to undergo the Responsible Care and Use of Laboratory Animals Course, which provides an understanding of basic animal experimentation requirements and teaches basic handling skills and manipulation techniques. This course is compulsory for all IACUC members within the first 12 months of their appointment to the IACUC.

At least 50% of IACUC members are required to receive formal IACUC training; nevertheless, all IACUC members are encouraged to undertake this training to better understand their individual roles and responsibilities within the various categories of IACUC membership. IACUC training introduces members specifically to the mandates, compositions, and functions of the IACUC as well as relevant regulations, policies, and operations. IACUC training must be undertaken within the first 12 months of appointment to the IACUC.

Animal Care and Use Areas Addressed by the Regulations

Animal Transportation

Animal transportation is addressed in the Guiding Principles of the NACLAR Guidelines,[23] Chapter 4: Procurement and Transport of Animals. The chapter lists factors that can cause distress during transportation, such as confinement, movement, noise, environmental conditions, personnel, duration and mode of transportation, food and water restrictions, and stocking density.

Animals must be transported under species-appropriate conditions, which meet welfare standards or international transport standards, with potential sources of distress identified and minimized. Sufficient shelter, food, water, nesting or bedding material, and space must be provided to the animal to express normal behaviors. At the same time, animals should be protected from sudden movements, extremes of temperature and humidity, and be prevented from escaping. Delivery and receipt of animals must be carried out appropriately and responsibly, with all animals accounted for.

Physical Plant and Environmental Conditions

Physical plant and environmental conditions are detailed in the Guiding Principles of the NACLAR Guidelines,[23] Chapter 3: Animal Housing and Management. Information is also provided in the IACUC Guidelines, Chapter 2: Oversight of the Animal Care and Use Program. Animals should be provided with environmental conditions that suit their behavioral and biological needs where possible, with appropriate air exchange, temperature, humidity, noise, light intensity, light cycles, and ventilation. Recommended temperature and humidity ranges, ventilation, and lighting for common laboratory animal species are further listed in the NACLAR Guidelines, Appendix IID.[23]

Environmental conditions should be monitored daily and recorded, with prompt investigation of any deviations or abnormalities. It is recognized that outdoors-housed animals face a different set of environmental conditions, especially in a tropical climate like Singapore's. Nevertheless, the conditions should be suitable for the species of animals.

Housing or Enclosures

Chapter 3: Animal Housing and Management of the Guiding Principles of the NACLAR Guidelines[23] and Chapter 2: Oversight of the Animal Care and Use Program of the IACUC Guidelines (of the NACLAR Guidelines) recommend that animal housing facilities should be appropriately staffed, designed, constructed, equipped, and maintained to provide a high standard of animal well-being. Both outdoor and indoor housing facilities should suit the needs of the particular species, with pens, cages, and containers designed, constructed, and maintained to meet species-specific behavioral and environmental requirements.

Generally, pens, cages, and containers should be of durable materials, be maintained in a clean and good condition, provide safety and shelter to animals, and allow for expression of species-specific behaviors. At the same time, they should not allow or cause injury to the animals and also prevent animals from escaping. Bedding or nesting materials should be provided where appropriate. Housing or caging systems should allow for easy observation of animals with minimal disturbance.

The Guiding Principles, Appendix II, Standards for Housing and Environmental Conditions, provides cage size recommendations for the most common laboratory animal species. Recommendations for floor area and height of cages are based on individual animal body weights and are similar to the *Guide for the Care and Use of Laboratory Animals*, NRC 1996.[8]

The Guiding Principles specifically advise that rodents are not to be housed on wire-floor cages, unless absolutely necessary, and if then, only for short periods. A solid resting area should be provided for wire-floor cages. Social housing is encouraged as far as possible. Animals that must be housed individually should receive environmental enrichment or have the impact of social isolation minimized.

Housing facilities should be disinfected and cleaned appropriately, and there should be a pest control program in place. Adequate contingency plans must be in place to cover emergencies such as flooding or fire, or the breakdown of lighting, heating, cooling, or ventilation.

Daily Care Activities

Chapter 3: Animal Housing and Management of the Guiding Principles details guidelines for daily care activities, as does Chapter 2: Oversight of the Animal Care and Use Program of the IACUC Guidelines (NACLAR Guidelines).[23] Animals should receive sufficient species-appropriate, uncontaminated, palatable, and nutritionally adequate food to meet their needs for growth or maintenance. Food should be stored in appropriate storage areas off the floor to minimize deterioration and contamination with regular cleaning of feeding equipment performed.

Animals should have access to a constant and reliable source of drinking water, which should be clean and uncontaminated. Any variation to food or drinking requirements in nutrition-based trials must be evaluated and approved by the IACUC and should be gradual.

Adequate and effective hygiene and sanitation practices should be maintained and monitored on a regular basis. This should include cleaning and disinfection of all individual pens, cages or enclosures, animal housing areas, procedural areas, feed and equipment storage areas, etc. A pest control program should be in place.

Behavioral and Environmental Management Enrichment

Enrichment and environmental complexities are also detailed in the Guiding Principles, Chapter 3: Animal Housing and Management and the IACUC Guidelines and Chapter 2: Oversight of the Animal Care and Use Program (NACLAR Guidelines).[23] Wherever possible, animals should be provided with environmental enrichment or complexities that promote expression of normal behaviors.

Group housing is recommended for social animals as far as possible, in socially compatible settings. Where animals must be individually housed, the effects of physical isolation should be minimized by the use of noncontact communication and reflection or by increasing environmental complexity as far as possible. A variety of enrichment types can be provided apart from social enrichment, such as occupational enrichment, physical enrichment, sensory enrichment, or nutritional enrichment.

Cultural Aspects in the Use of Animals

Research in Singapore is generally carried out by researchers of diverse nationalities, and cultural aspects have not tended to influence choices of animal species. Animals are generally chosen according to their suitability for the study. Commonly used animal species include rodents (rats, mice), guinea pigs, rabbits, pigs, goats, NHPs, poultry, and fish.

Animal use should be undertaken only after due consideration of the project's value to human or animal health, or the advancement of knowledge, weighed against the potential effects on animal welfare. Animals used must be of an appropriate species and quality for the scientific purpose concerned. While a minimum number may be required to obtain scientifically valid results, this should not result in greater suffering of individual animals in a project or study. Animal use projects must be as brief as possible, with death as an endpoint avoided if at all possible.

Occupational Health and Safety

In general, the facility's occupational health and safety program must conform to relevant biosafety standards set out by the Ministry of Health, Singapore, and the occupational health and safety legislation set out by the Ministry of Manpower, Singapore. Occupational health and safety is an important component of the facility's ACUP (Chapter 2: Oversight of the Animal Care and Use Program of the IACUC Guidelines, NACLAR Guidelines).[23] The IACUC participates in the facility's occupational health and safety program by incorporating reviews of research activities that may be hazardous to personnel into the IACUC protocol review process. The IACUC additionally should ensure that personnel have received appropriate training.

Veterinary Care

A dedicated chapter of the Guiding Principles, Chapter 6: Veterinary Care describes standards of veterinary care required. The specific role of the veterinarian is detailed in Chapter 2: Oversight of the Animal Care and Use Program of the IACUC Guidelines (NACLAR Guidelines).[23] Each facility must formally engage an AV who is licensed by the AVA, with the appropriate authority to oversee the ACUP and to provide veterinary care to the animals. The AV can be employed on a full-time or part-time basis with a formal schedule of regular visits to the facility. Interim arrangements must be made to ensure constant and ready access to veterinary care when the AV cannot be present.

Components of veterinary care include the following:

- Availability of appropriate facilities, personnel, equipment, and services, which comply with the Guiding Principles
- Use of appropriate methods to prevent and control diseases and diagnose and treat diseases and injuries
- 24-h emergency, weekend, and holiday care
- Monitoring of animals' health and well-being
- Provision of training and guidance to researchers or other personnel involved in the care and use of animals in handling, immobilization, anesthesia, analgesia, tranquilization, surgical manipulations, and euthanasia
- Adequate preprocedural, surgical, and postprocedural care
- Preventative medicine programs
- Attention to both physical health and psychological well-being
- Provision of advice to researchers on the design and implementation of study proposals.

There must be effective communication channels between the staff and the veterinarian, such that the veterinarian is notified of any case requiring his/her attention. The veterinarian must be able to provide prompt care and treatment and to keep the staff informed of any further monitoring to be made, or daily treatments to be carried out.

Euthanasia

Euthanasia is described under Chapter 8, Responsibilities of Investigators in the Guiding Principles, and under Chapter 3, Review of Proposals of the IACUC Guidelines (NACLAR Guidelines).[23] Examples of acceptable, conditionally acceptable (requires IACUC approval of scientific justification), and unacceptable methods are listed in Chapter 3: Review of Proposals of the IACUC Guidelines. There are three references on euthanasia (Appendix I): *Euthanasia of Animals Used for Scientific Purposes* by the Australian and New Zealand Council for the Care of Animals in Research and Teaching,[24] the Canadian Council *Guide to the Care and Use of Experimental Animals*, Chapter XII,[25] and The AVMA *Report of the AVMA Panel on Euthanasia.*[26]

Euthanasia of animals is recognized as necessary to relieve pain, distress, or suffering of animals when humane endpoints have been reached. The choice of euthanasia method is dependent on factors such as species, age, availability of restraint, personnel skill, and others. Any method used must avoid distress, be reliable, and produce rapid loss of consciousness without pain till death occurs.

Euthanasia should be performed in a quiet, clean environment, preferably away from other animals, with death determined prior to carcass disposal. Any dependent neonates must have alternative care provided, or also be euthanized, where applicable. Only personnel who have been designated by a veterinarian, or demonstrated to a veterinarian their competency, should perform euthanasia on animals.

Generally, IACUC should review and approve the choices of euthanasia methods based on minimum pain, distress, anxiety, or apprehension caused; minimum delay until unconsciousness; reliability and irreversibility; personnel safety; emotional effect on personnel; compatibility with research goals, requirements and purposes; compatibility with species, age, and health status; and drug availability and human abuse potential.

Special Considerations

Nonhuman Primates

Special considerations for the care and use of NHPs for scientific purposes are indicated in Appendix III, Additional Information on the Care and Use of Nonhuman Primates for Scientific Purposes of the Guiding Principles of NACLAR Guidelines.[23] NHPs are regarded as being similar to humans, with highly developed mental and emotional capacities. As such, environmental enrichment is especially important. Group housing is preferred due to their complex social hierarchy and interactions, but the potential for problems such as dominance hierarchies, wounding, and disease transmission should be considered. Housing enclosures should be designed in ways such that NHPs can express specific behaviors such as gathering food, perching, swinging, and vertical flight. Appendix III also advises that while interaction is encouraged, it should not be forced; direct physical contact should be minimized to prevent bilateral disease transmission, as well as to avoid emotional attachment.

Genetically Altered Animals

The Genetic Modification Advisory Committee (GMAC) oversees and advises on issues relating to genetic modification and genetically modified organisms. The GMAC Guidelines[63] detail general considerations in transport of transgenic animals, with two critical principles:

- The animals are prevented from escaping.
- The need for arrival to the intended delivery, with proper identification processes and accounting for all animals by a competent biologist.

The aim of both principles is to ensure that transgenic animals do not escape into the environment and potentially interbreed with feral populations. It may also be necessary for the Institutional Biosafety Committee of the facility to make additional rules or conditions and inspect transport arrangements to ensure compliance with the above two principles. In addition to transportation of animals adhering to transportation guidelines detailed in the Guiding Principles of the NACLAR Guidelines,[23] the GMAC Guidelines also stipulate that the transport carriers should comply with IATA regulations.[45] They must be escape-proof and allow for easy inspection without the carriers being opened.

Special considerations for the use of transgenic animals are described in Chapter 3: Review of Proposals of the IACUC Guidelines. The IACUC has a responsibility in determining if any mutant gene can result in a severely debilitating phenotype and mitigating measures, which can be put into place to address this. For example, modified husbandry measures or housing conditions can be of use.

General criteria for humane endpoints should be included in the project proposal. Specific clinical abnormalities, which are known or suspected to occur in the development of a new mutant model, should either be included at the start or made known to the IACUC when the information becomes available. It is recognized that experimental endpoints may differ from mutant animals as compared to "normal" animals where the phenotype involves clinical abnormalities.

Personnel performing genetic manipulations should be qualified and trained in these procedures as well as related procedures such as aseptic surgery.

Reuse of Animals

Reuse of animals is addressed in the Guiding Principles of the NACLAR Guidelines[23] under Chapter 8: Responsibilities of Investigators. Generally, animals should not be used in more than one experiment (whether in the same project or a different one) without evaluation and approval by the IACUC. While reuse of animals may reduce the total number of animals used in a project, the general principles are that animals should not be used in unnecessarily repeated or duplicated scientific activities and that a reduction of animals used should not result in greater suffering of individual animals thereafter.

Animal reuse should not be advocated as a reduction strategy, and reduction should not be a rationale for reusing animals. Thus, refinement and reduction goals should be balanced on a case-by-case basis. In all circumstances, humane endpoints should be observed to determine when animals can be removed from the study, treated or euthanized.

The IACUC is advised to consider if experiments involving reuse of animals contain procedures, which cause animals' pain or distress; whether there are potential long-term or cumulative effects of pain or distress; the total time an animal will be used; the amount of pain, distress, or biological stress likely to be caused by next and subsequent procedures; and whether animals are allowed to recover fully from the first study prior to being used in subsequent studies. If the IACUC approves animal reuse, it may then decide to require that the previously approved protocol be amended to include the approved changes, or a separate protocol be developed to cover the reuse of animals in another study. Animals are specifically not permitted for use in multiple major survival surgeries (MMSSs) unless scientifically justified by the researcher in writing, such as if the MMSS is a related component of the same study; or if it is required by the AV as a procedure for the animal's health and well-being; or unless there are any other special circumstances, which have been duly evaluated and approved by the IACUC.

Annual Reports

Each research facility is required to prepare and submit an annual report per calendar year, i.e., January 1 to December 31. The report must be signed off and certified by the CEO of the facility. The categories of information required in the report fall under the following sections.

Assurance

Assurance should be provided that the facility has adhered to the NACLAR Guidelines.[23] Any exceptions must be explained by the researcher and approved by the IACUC. Each researcher must have considered alternatives to painful procedures, and professionally acceptable standards must have been applied to the care and use of animals.

Background Information and Statistics

The composition of the IACUC and the name of the AV have to be stated. Locations of all facilities where animals were housed or used should be indicated. Species types and numbers of animals used in different pain categories should also be listed. *The different pain categories are as follows*:

a. Scientific purposes involving no pain or distress, or with the use of pain-relieving drugs
b. Scientific purposes involving accompanying pain or distress for which appropriate anesthetics, analgesics, or tranquilizers were used
c. Scientific purposes involving accompanying pain or distress for which appropriate anesthetics, analgesics, or tranquilizers were not used due to potential adverse effects on the procedures or results and which an explanation is provided by the researcher and accepted by IACUC
d. Animals being bred, conditioned, or held for scientific purposes but not yet used for such purposes

Self-regulation

Dates of the semiannual ACUP review, and the annual facility inspection review are to be indicated. Deficiencies should be identified as significant or minor with relation to compromise of animal welfare, with corrective actions planned, and a time frame for rectification scheduled.

Care and Use of Fish

Appendix IV, Additional Information on the Care and Use of Fish for Scientific Purposes of the Guiding Principles (NACLAR Guidelines),[23] describes special considerations for fish in terms of choice of species used; procurement; transportation and handling; acclimatization; quarantine; design and construction of facilities; water quality and temperature; illumination; stocking density and water flow; diet and feeding; health program and disease control; analgesia, anesthesia, and invasive procedures; euthanasia; and dangerous species and zoonoses.

THAILAND

Background Information and Current Situation

Animals have been used in research for almost 100 years. Laboratory animals such as mice and rats were introduced for biomedical research around 1960 at the Department of Medical Sciences and medical schools in Thailand, but the beginning of laboratory animal science was virtually unknown.

In 1961, the Southeast Asia Treaty Organization (SEATO) Cholera Research Project set up a laboratory animal unit for conducting biomedical research on tropical diseases. In 1960, the laboratory joined the Armed Forces Research

Institute of Medical Sciences (AFRIMS). AFRIMS operates as a joint Royal Thai and United States Army medical research venture composed of the Royal Thai Army Component and the US Army Medical Component (USAMC, the current name is US Army Medical Directorate, USAMD). The laboratory ACUP at the USAMC-AFRIMS has followed the US Animal Welfare Law and all other relevant regulations. It has earned and maintained AAALAC International accreditation since 1999.

The needs for laboratory animals for biomedical research in Thailand did not come to light until 1971 after series of ad hoc committee meetings, which led to a memorandum of understanding to set up the National Laboratory Animal Center (NLAC) under Mahidol University's responsibility for breeding and supplying small laboratory animals to users in the country. At that time, Chulalongkorn University and Mahidol University responded by including laboratory animal science subject in their curriculum at the Faculty of Veterinary Science and the Faculty of Science, respectively. The development of laboratory animal care and use in Thailand has since slowly progressed and has the following landmarks:

- In 1973, the World Health Organization (WHO)/United Nations development program agreed to provide technical support for the Thai laboratory animal project and assigned Dr Stian Erichsen (the International Council for Laboratory Animal Science's (ICLAS's) Secretary-General), to Mahidol University as a consultant to establish the NLAC.
- In 1974, the National Economic and Social Development Board approved the project proposed by Mahidol University.
- In 1975, the NLAC was established at Salaya Campus, Mahidol University.
- In 1978, the NLAC commenced its supply of rats and mice nationwide.
- In 1988, Thailand hosted the 9th ICLAS International Symposium in Bangkok.
- In 1999, the National Research Council of Thailand (NRCT) issued the *Ethical Principles and Guidelines for the Use of Animals for Scientific Purposes* (the *Ethical Guidelines*)[28] based on the old CIOMS International Guiding Principles for Biomedical Research involving animals[61] and appointed a national committee to promote and regulate the issues relevant to animal users through the IACUC.
- In 2001, the Thai Government Cabinet approved the NRCT's proposal for the National Committee for Research Animal Development (NCRAD). The NCRAD has organized several training and seminars as well as the annual symposia on laboratory animal science for members of IACUC, researchers, veterinarians, technicians, and animal caretakers.
- In 2002, the Thai Association of Laboratory Animal Science (TALAS) was founded. Since 2007, the TALAS has organized an annual laboratory animal science conference focusing on IACUC training, continuing education for personnel involving in animal care and use, and current issues in laboratory animal science.

- In 2006, the *Ethical Guidelines* were revised and the English version was established.
- In 2007, the Cabinet approved the first National Strategic Plans for the Care and Use of Animals for Scientific Purposes and appointed the new NCRAD under the Secretariat Office of the NRCT (SONCRAD) to oversee the strategic plans. The NRCT proposed Animal for Scientific Purposes Bill to the Government Cabinet. The Standard and Guideline for Care and Use of Laboratory Animals was drafted and the final document is pending.
- In 2009, the NRCT organized the International Symposium on Laboratory Animal Sciences 2009 (ILAS2009) in Bangkok.
- In 2012, TALAS organized the 5th Asian Federation of Laboratory Animal Science Associations (AFLAS) Congress in Bangkok.

The Standard for Institutional Animal Care and Use Committee[29] was issued by the NRCT.

- In 2015, the *Animals for Scientific Purposes Act, B.E. 2558* (A.D. 2014),[27] was approved and published in the Royal Thai Government Gazette on March 13, 2015, specifically controlling the care and use of animals for scientific purposes. The protected animals include all nonhuman vertebrates and some invertebrates such as cephalopods, decapods, and arthropods. All procedures on animals must be performed by licensed holders and be complied with the *Ethical Guidelines.*[28] The Act is enforced by both IACUC and SONCRAD.
- In 2015, the Institute of Animal for Scientific Purposes Development (IAD) has replaced the SONCRAD and the IAD's missions are stated in the *Act*. The *Ethical Guidelines* was changed to the "*Ethics in Procedures on Animals for Scientific Purposes (Ethics)*"[30] and published in the Royal Thai Government Gazette on March 14, 2016.

There are at least 30 Thai governmental institutions with over 70 laboratory animal units throughout the country using laboratory animals for research, teaching, production of biological materials, and testing. AAALAC International has accredited five additional Thai institutions since 2010. A few more laboratory animal units may seek accreditation within the next 2 years. The NLAC supplies approximately 300,000 rodents, guinea pigs, and rabbits each year. About 90% of rodents are mice and the others are rats. Outbred rodents are kept under strict hygienic conventional system, and inbred animals are under specified pathogen-free condition. Most laboratory animal facilities in Thailand are generally set up for biomedical research, and the animals on demand are mice, rats, guinea pigs, rabbits, and hamsters. Rats, mice, and rabbits have been occasionally used for drug testing and biological products development. Dogs, cats, pigs, and monkeys are seldom used due to lack of sources and facilities. Recently, a private company was founded as a joint venture with Mahidol University to provide an alternative source of specific pathogen-free (SPF) laboratory rodents. All animal users are required by the

Ethical Guidelines to prepare protocols for using animals, and the protocols must be approved by the IACUC before using the animals.

Summary of Regulations, Guidelines, or Legal Requirements

Before 2015, Thailand had no laws concerning animal welfare and the care and use of laboratory animals for scientific purposes, but had issued the NRCT's *Ethical Guidelines* in 1999.[28] Animal users in Thailand are expected to use animals only for acquiring new knowledge in biomedical sciences for improving the well-being of both humans and animals without offending the public by carefully selecting only quality animals, providing high standards of animal care and management, and using refined techniques to reduce animals' pain and distress.

In 2009, the NRCT issued the *Fundamental Principles on Designing Animal Care and Use Facilities for Laboratory Animals*.[31] The principles provide recommendations on the following areas including location, accessibility, building characteristics, functional and support areas, facility design criteria, air ventilation and management, air temperature, relative humidity and air pressure control and management, and safety standard.

Scope

The *Ethical Guidelines*[28] define that "an animal" means any vertebrate and some kind of invertebrate animals and cover educational institutions, state enterprises, and governmental and private sectors.

The Principles

The 3Rs principles[36] are integral to the *Ethical Guidelines*.[28] The *Ethical Guidelines* emphasize the following recommendations:

1. Animal users should be aware of the value of life of animals. Animals must be used only for specific purposes, which have been carefully considered as beneficial and most necessary for the development of the quality of both human and animal lives and/or the progress of science, and when it has been established that there is no other available equivalent option.
2. Animal users must be aware of the accuracy of research outcome using the minimal number of animals. Animal users must be aware that the objective and goal of the project can be achieved accurately with the least number of animals depending on the genetics and health quality of the animals and the appropriate planning and techniques used in the experimental design.
3. The use of wild animals must not violate laws or policies for wildlife conservation. The use of wildlife is restricted to scientific research that cannot be replaced by any other kind of animals, and it is to abide by the laws and policies for wildlife conservation.

4. Animal users need to be aware that animals are just as humans are living beings. Animal users have to be aware that animals experience a sense of pain and respond to their surroundings in the same way that humans do. Animals are treated with caution to avoid stress, pain, and suffering by providing optimal conditions for transportation, animal husbandry, environmental enrichment, prevention of diseases, and appropriate experimental techniques.
5. Animal users must keep detailed data and records of animal experiments. Animal users must strictly follow the protocol described in their proposals, and all the details of the experiments are recorded in full and made available for public release or investigation at all times.

Ethical Review

From 2007 to 2015 the NCRAD, which was appointed by the Government Cabinet, was responsible for monitoring and promoting the ethical use of animals in research, testing, production of biological materials, and teaching, through the *Ethical Guidelines*,[28] and the amended *Ethical Guidelines* as needed. The NCRAD promoted and reinforced that the use of animals in Thailand. Both governmental and private institutions follow the *Ethical Guidelines*. The care and use of laboratory animals in all institutions are overseen by the SONCRAD. The Committee has the authority to investigate the internal affairs of an institute, which is subject to public complaints or accusation by the people, public media, published documents, academic publications, and research funding organizations for its violation of the *Ethical Guidelines*. The Committee cultivates awareness in the Budget Bureau and institutional budgeting units to support the necessity of the *Ethical Guidelines* and to endorse requests from institutions for sufficient funding and coordinates with research funding organizations to render their supports to the projects endorsed by the Institutional Committee.

Every institution that uses animals in research, testing, production of biological materials, and teaching is obliged to establish at least one committee to manage and be accountable for the use of animals, so that the *Ethical Guidelines* are followed. The committee should be pluralistic; and its membership should be diverse including members of the management of the institution, researchers, and lay people. The responsibilities of the Institutional Committee are as follows:

- To set up standard operational procedures according to the *Ethical Guidelines*.
- To review every project that involves the use of animals in research, testing, production of biological materials, and teaching at or outside the institution, and to forward it to the executive board of the institute for approval. Only projects that follow the *Ethical Guidelines* should be authorized to proceed or continue.
- To monitor the use of animals so that experimentation follows the *Ethical Guidelines* and that described in the approved protocol.
- To manage the animal unit of the institute so that it complies with the standards as set forth in the *Ethical Guidelines*.

- To support and ensure that the animal unit of the institute is sufficiently funded for facilitating the standard set up in the *Ethical Guidelines.*
- To provide personnel with continuing education and training to increase the knowledge in laboratory animal science.

In addition, the *Ethical Guidelines* recommend that editorial boards of academic journals should request the author(s) of the submitted research paper to provide detailed information concerning the genetic background and the number of animals used, animal care provided, and experimental protocols including the certificate of approval for the research project issued by the authorized Institutional Committee. The manuscript should be rejected unless all the above requirements are fulfilled.

The *Ethical Guidelines* recommend that before using animals, the users should submit their protocols in detail, including the following information and considerations:

1. Research steps and plans, including the objectives and expected benefit to uplift the quality of life of humans or animals and/or the progress of science, and the accumulation of academic information.
2. The user should support the need to use animals with evidence and reasons that there is no other alternative. Animal users should reserve the use of animals for situations when there is unavoidable necessity or when there is no other available option.
3. The users should carefully review all information and reported documents related to their research, and they should utilize all the information to make the most out of the use of the animals in their research.
4. Animal users should select the species and the breed of animal that fits in with the objectives and goals of the research.
5. A statistical method should be employed for estimating the least number of animals required while ensuring the most accurate and acceptable results.
6. Animal users should include in the protocol the setup of the experiment; the techniques and the materials to be used; and the care and management of the animals before, during, and after experimentation. Animal users should employ only the most appropriate techniques, technology, and statistical methods in planning and evaluating the project.
7. Animal users and animal caretakers should treat animals with kindness and should avoid any procedure or process that causes pain and stress to the animals. In case the situation is unavoidable, they should explicitly state the reasons that should be entirely based on academic grounds.
8. The experiment should be concluded before the animal will die of extreme pain. At the end of each experiment, the users are responsible to euthanize all animals. In case the animals are to survive, the users must provide the reasons for such necessity in their proposals and must be responsible for taking care of the animals under conditions appropriate for the species. The animals should neither be released to nature, nor should they be abandoned at the animal unit without appropriate care.

Animal use protocols are to be reviewed by the IACUC in accordance with criteria determined in the "*Guidelines on Animal Use Protocol Review*"[64] that was published by the NRCT in 2014.

The NRCT issued the *Standard for* IACUC in 2012.[29] Any institution maintaining and using animals for scientific purposes must appoint one committee (IACUC) and clearly show the IACUC's reporting line of command and its relationship to other sections in its organizational chart. The standard determines responsibilities of the IACUC to follow the *Ethical Guidelines* and describes IACUC composition and functions.

The IACUC should include at least the following seven members, and one member may have more than one role:

- An Institutional Administrator or his/her representative as the Chair of the Committee
- The Chief of the laboratory animal unit as a member and secretary
- The AV
- At least two scientists with experiences in research using laboratory animals
- One member who is nonscientific discipline and not affiliated with the institution
- An expert or professional in others areas, i.e., a statistician, a lawyer, or a librarian
- Must have at least one nonaffiliated member

The IACUC has the following functions:

- Establish an annual budget plan to improve the ACUP
- Submit the annual budget plan to the IO
- Review animal use protocols
- Perform an oversight of the animal use protocol procedures
- Personnel training and education
- Communicate continuously to the public and distribute knowledge
- Establish standard operating procedures (SOPs) and practices in the care and use of animals at the institution
- Make a routine annual institutional report and submit it to the IO

Accreditation of the Standard for IACUC based on policies and implementation of the institution and functions of IACUC has been given to the three institutions since 2015. The NRCT believes that competency of IACUC is the most crucial for monitoring and improving ethics in laboratory animals.

Animals for Scientific Purposes Act, B.E. 2558 (A.D. 2015)

Although the *Ethical Guidelines* has been issued and recommended to all institutions, its implementation is still loose. The NRCT, therefore, has made an enduring effort to propose Animals for Scientific Purposes Bill to the Cabinet since 2007. After several attempts, the act known as "Animals for Scientific Purposes Act, B.E. 2558 (A.D. 2015)"[27] was approved by the National Legislative Assembly in December 2014. But not until March 2015, the Act was published in the Royal Thai Government Gazette as an official announcement. However, it was enforced as of September 14, 2015, after a 180-day grace period for compliance, which is usual for many Thai laws.

The Act has 56 sections, which are described in 7 chapters including the Supervision and Promotion Committee, Ethics, Regulations on the Care and Use of Animals, Competent Officials, Non-compliances, Penalty, and Transitory Provision. This Act is a complement of the Animal Welfare and Prevention of Torture of Animal Act, B.E. 2557 (A.D. 2014),[65] which protects primarily pets and other live animals.

Protected Animals and Animal Procedures

The animals protected by the Animals for Scientific Purposes Act are living nonhuman vertebrates and their fetuses or embryonated eggs from halfway through development and other organisms recognized to have pain sensation (to be prescribed in the Ministerial regulation) that are handled/treated for scientific purposes. The procedures involving the protected animals include research, testing, production of biomaterials, and teaching.

Regulatory Bodies

Two national committees are described in the Act, i.e., the Supervision and Promotion for Procedures on Animals for Scientific Purposes Committee (Supervision and Promotion Committee) and the Ethics Committee. The former has 10 ex-officio members and 12–14 qualified members that are stakeholder representatives. The qualified members have a term of 3 years and cannot serve for more than two consecutive terms. This committee plays a major role in making policies, ethics, standards, and procedures for animal care and uses, disseminating and promoting information to all interested parties including the public, considerations of issues arising from the operation of the Act including appeals of the decision made by the licensor and the Ethics Committee, and reporting the operation of the Act to the Cabinet.

On the other hand, the Ethics Committee is consisted of at least seven members appointed by the Supervision and Promotion Committee from qualified experts in animal sciences. The committee has a term of 3 years for not more than two consecutive terms. The primary role of this committee is to determine and make appropriate decisions on noncompliances to the Ethics. The penalty could be an admonition, probation, suspension, or revocation of the license depending on the severity of the breach of Ethics.

As prescribed by the Act, an IAD is established as a division of NRCT serving as a secretariat office for enforcing the Act. This office has the following important functions:

- Make a survey, study, and analysis of academic information necessary for formulating goals, policies, work plans, programs, and measures in promoting the procedures on animals for scientific purposes
- Develop and issue the standards on the procedures on animals that are compatible with international standards
- Promote, develop, and monitor animal facilities and the procedures on animals to be complied with the national standard
- Set up data base on animals for scientific purposes and disseminate the information on policies, work plans, programs, and ethics to the interest groups and public
- Coordinate and cooperate with, and give consultation to, local and international agencies concerning the development of the national standards
- Promote and support studies on research and technology development and offer training courses in the procedures on animals
- Prepare an annual report on the activities of all institutions to the Cabinet

Ethics and Regulations on the Care and Use of Animals

The Ethics in Procedures on Animals for Scientific Purposes (*Ethics for Animal Procedures*)[30] approved by the Supervision and Promotion Committee is a slight modification of the NRCT's *Ethical Guidelines* described above.[28] The *Ethics for Animal Procedures* are, however, more extensive to include all persons involved in the care and use of animals such as the CEO or the owner of the facilities, the head of the animal unit, the AV, the animal caretakers, and the students.

Any institution that raises or maintains, and/or uses the protected animals, must register its facilities to the NRCT. The registered animal facilities must have a location, surrounding, physical structure, and essential equipment as prescribed in the Ministerial regulation. The facilities must be staffed with an AV and the head of the unit. In addition, an IACUC must be appointed to review and monitor animal use protocols so as to be in line with the *Ethics for Animal Procedures*. The number and qualification of members of IACUC is specified in the Ministerial regulation. The IACUC has to make a semiannual report on the institution's activity to NRCT. The head of the animal unit is responsible for monitoring the use of animals in terms of number, species, and procedures approved in the protocol, keeping the facilities clean, safe, and fit in with the national standard, and preparing statistics of the animals used.

Any change of the objectives of animal uses, renovation of the physical structure or construction of new building, species of animals, AV or head of the animal unit, name or location of the facilities, or closedown of the facilities must be reported to NRCT. Before closing down, the CEO or owner must follow the guidelines specified in the Ministerial regulation.

A license is required for anyone who uses or produces the protected animals. Exception, however, is applied to students, who enrolled in the course work or participants of the workshop that the procedure on animals is an integral part of the program, and animal caretakers. Although the animal caretakers do not need to have licenses, they have to work under a license holder. The license is valid for 4 years and renewable. A license holder whose work involves with improvement of breeds, breeding, animal production, stem cells, genetic modification, or cloning must prepare and submit a semiannual report to the NRCT. The report is classified as strictly confidential. In addition, any person who trades, import, export, or transit the animals protected by this Act must inform and follow the guidelines described in the Ministerial regulation. The Supervision and Promotion Committee is also responsible to issue guidelines for animal transfer and transportation, euthanasia, and waste management.

Inspections and Competent Officials

Enforcement of the Act is authorized by the Prime Minister (PM), who appoints Competent Officials (COs), mostly personnel of the NRCT and qualified experts, to implement it. The PM also approves all Ministerial regulations.

The CO may enter any establishment during daytime to examine the facilities, performances, documents, and data, which may indicate noncompliance. If there is reason to believe that an offense under the Act has been committed, places or vehicles may be searched by CO without a search warrant but with a prior approval of the Secretary-General of NRCT. Similarly, individuals may be summoned to give statements. The CO must report to the Secretary-General within a week. In case there is a seizure of animals as evidence, the CO may allow the owner or others to take care of the animals until the end of litigation.

Noncompliances and Penalties

Noncompliances to Ministerial regulations regarding the standard of animal facilities, IACUC, and specific procedures such as improvement of breeds, breeding, animal production, stem cells, genetic modification, or cloning are subjected to rectification, a halt, or termination of the activity as instructed by the Secretary-General. Noncompliances to the *Ethics for Animal Procedures*,[30] on the other hand, are primarily monitored by the IACUC, who would approve or not approve the protocols or stop the project. If the case is reported to the Ethics Committee, the license holder may lose his/her license as mentioned above.

Offenses to the Secretary-General's orders under this Act are liable to a fine ranging from Baht 10,000–300,000. The most serious penalty is an imprisonment for up to 1 year for those who do not have license, yet perform procedures on the protected animals.

Conclusion

Although it takes so long for Thailand to establish and implement the guidelines and law on ethics in animals for scientific purposes, prior to this, most Thai scientists are aware of it and have adopted both the national and international ethical guidelines and guidelines in their practices. The essence of the Act is the control on the welfare of animals in scientific work and the ethics for animal users and breeders that incorporates the globally accepted 3Rs principles. Thus, implementation of the Act further ensures that animal users and breeders would follow the consensus agreed on by the international agencies. It is hoped that the Act would promote high quality of work and products from animals in scientific discipline.

VIETNAM

Background and Current Situation

Modern medical research in Vietnam was marked with the foundation of Pasteur Institute in Ho Chi Minh City in 1891, followed by the establishment of Pasteur Institute in Nha Trang in 1895 and the National Institute of Hygiene and Epidemiology in 1926 in Hanoi. Each of these institutions established an experimental animal unit for tropical diseases and vaccine research, control, and testing. It can be said that experimental animals have been playing important roles in protection of public health in Vietnam.

Modern medical education started in Vietnam in 1902 with the Foundation of Indochina Medical College (l'Ecole de Médecine de l'Indochine) that is also the first modern university in Vietnam (the current name is Hanoi Medical University). The Faculty of Veterinary Medicine established 2 years later (in 1904) was affiliated to the this college and then became a unit of the Agricultural University in Hanoi in 1956.

The country is considered as a development success story, thanks to the political and economic reforms (Đổi Mới) launched in 1986. The main indicators of population health and education have been improved, and higher biomedical education has made progresses for decades. Dozens of medical colleges and veterinary faculties in agricultural universities have been founded since 1956 to meet the requirement of manpower for human and animal health protection in the era of international integration. These biomedical education and research bodies are planning to gain international recognition and reputation through international communication including scientific publications. Therefore, good laboratory animal facilities and practices are required. Despite the long history of medical education and research, the implementation of good laboratory animal management and usage has not made adequate progress yet.

The first symposium on "Laboratory Animals in Research and Application" organized by Vietnam National University of Agriculture in October 2012 attracted participants from a number of medical and agricultural institutions.

After the meeting, the Vietnam Laboratory Animal Network was established and provided an exchange platform for laboratory animal researchers and users in the country.

Nowadays, animal welfare issues are gaining more attention from educational and research institutions as well as more public concern. Good practical protocols and SOPs have been applied in laboratory animal management and use by many institutions and medical companies. Laboratory animal and animal welfare subjects have been included in veterinary medicine curricula. In August 2016, a meeting of scientists from medical and veterinary institutions was organized to discuss about the need of IACUC for research using animals.

Nafovanny Vietnam (Vietnam Primate Breeding and Development Corporation) is the only center accredited by AAALAC International in Vietnam since 2009. Other laboratory animal units should be on their ways to this target.

Summary of Regulations or Legal Requirements

Laboratory Animals

Major institutions and companies have their own protocols for laboratory animal research and usage following the guidelines and criteria of Ministry of Health[33] and WHO Laboratory Biosafety Manual.[34] Academic institutions follow the OIE *Terrestrial Animal Health Code* (Section 7 Animal Welfare)[2] in teaching, vocational training, and practices.

Wild Animals

Two panel codes, two laws, six decrees, three circulars and several decisions demonstrate regulations on wild animal protection and biodiversity conservation. Different ministries and governmental organizations are involved animal protection tasks. Some regulations (among others) are listed below.

Decree 32/2006/ND-CP[66] indicates two groups of endangered, precious, and rare animals. The exploitation and use for commercial purposes is strictly prohibited, with a permit required for scientific research and conservation purposes to group IB (endangered and critically endangered species). A permit is required for all purposes including scientific research, conservation, and commercial exploitation for group IIB (threatened and rare species).

Decree 157/2013/NĐ-CP[67] assigns levels of punishments toward violations related to wildlife. The handling is also based on the status of the animals: alive, dead, in parts, domestic, or introduced species.

If the animal belonging to group IB is dead or in parts, it is (1) transferred to a scientific institution, a training institution, an environmental education institution, a specialized museum, a specialized management agency, or a medical facility for research or for medicine and (2) destructed in cases of infection and/or cannot be handled by the above measures.

Decree 160/2013/ND-CP[68] addresses the management of endangered, precious, and rare species of animals.

The New Veterinary Law

The new Veterinary Law[32] has come into effect in July 2016 marking a progress in animal welfare in the country. The article 21 of the law states the following:

- Organizations and individuals who keep and use animals take the following responsibilities: (1) to provide species-appropriate care, feed and transport and (2) to minimize pain and fear of animals during transport, slaughter, euthanasia, medical treatment, and scientific research. This covers all animals, both terrestrial and aquatic animals.
- Organizations and individuals who keep animals as pets and animals for conservation purposes take responsibilities to provide appropriate care, disease prevention, and veterinary treatment.

Compared to Veterinary Ordinance 2004[69] and 2007[70] and Decree 32 (2006)[66] where the use of animals for experimentation is allowed and no regulations on how to protect animals from suffering can be found, the new Veterinary Law is making a significant step toward.

Near Future Steps

The first symposium on laboratory animals in Hanoi in 2012 and the first animal welfare conference in Ho Chi Minh City in 2014 as well as the increasing number of seminars throughout the country on the two subjects have made positive effects among scientific community. The "5Fs"[35] and the "3Rs"[36] have become familiar to biomedical scientists and students. Encouragingly the government supports the moves at Association of Southeast Asian Nations (ASEAN) to bring its legislation into line with European Union legislation regarding the use of animals in experiments for cosmetic purposes. It is believed that the change may happen in the near future.

The Vietnamese Association for Laboratory Animal Science will be soon founded for fostering the development and the role of laboratory animal science in the country.

REFERENCES

1. The World Organisation for Animal Health (OIE). In: *Terrestrial animal health code.* 21st ed. vols. 1–2. Available at: http://www.oie.int/en/international-standard-setting/terrestrial-code/access-online/. Accessed November 13, 2016.
2. The World Organisation for Animal Health (OIE). Section 7. Animal Welfare. Available at: http://www.oie.int/index.php?id=169&L=0&htmfile=titre_1.7.htm. Accessed December 1, 2016.
3. *CIOMS-ICLAS international guiding principles for biomedical research involving animals.* Council for International Organizations of Medical Sciences and the International Council for laboratory Animal Science; 2012. Available at: http://aaalac.org/accreditation/RefResources/IGP2012.pdf.
4. Law of the Republic of Indonesia No. 18. 2009. [Chapter VI], Veterinary Public Health and Animal Welfare; Part 2, Animal Welfare; Articles 66–67 (Undang-Undang Republik Indonesia Nomor 18 Tahun 2009. Bab VI. Kesehatanmasyarakat Veteriner. Bagiankedua: Kesejahteraan Hewan, Pasal 66–67).

5. Law of the Republic of Indonesia No.41. 2014. [Chapter VI], Veterinary Public Health and Animal Welfare; Part 2, Animal Welfare; Articles 66A (Undang-Undang Republik Indonesia Nomor 41 Tahun 2014. Bab VI. Kesehatanmasyarakat Veteriner. Bagiankedua: Kesejahteraan Hewan, Pasal 66A).
6. *The national guidelines on health research ethics*. Republic of Indonesia: Health Research Ethics Committee; 2011. The Ministry of Health.
7. *Teaching guide book for ethics on health research*. Republic of Indonesia: Health Research Ethics Committee; 2011. The Ministry of Health.
8. National Research Council (NRC). *Guide for the care and use of laboratory animals*. Washington: National Academies Press; 1996.
9. American Veterinary Medical Association (AVMA). *AVMA guidelines on euthanasia*. Schaumburg, IL: AVMA; 2007.
10. Conservation Law No. 5. 1990. The Ministry of Forestry, Republic of Indonesia. (Undang-Undang Konservasi Hayati No. 5 Tahun 1990. Kementrian Kehutanan Republik Indonesia).
11. AWA2015 [Animal Welfare Act 2015]. *Laws of Malaysia. Putrajaya*. Malaysia: The Attorney General's Chambers; 2015. Available at: http://www.mvc.gov.my/doc_downloads/AnimalWelfareAct_AKTA%20772_BI_akta%20772-BI.pdf.
12. AA2013 [Animals (Ammendment) Act 2013]. *Laws of Malaysia. Putrajaya*. Malaysia: The Attorney General's Chambers; 2013. Available at: http://www.mvc.gov.my/doc_downloads/Animal_Amendment_Act_2013.pdf.
13. *The Malaysian Code of Practice for the Care and Use of Animals for Scientific Purposes (Draft)*. The Laboratory Animal Science Association of Malaysia (LASAM). Available at: http://www.rmc.upm.edu.my/dokumen/PTPPY1_92272_upm_code_of_practice.pdf. Accessed November 13, 2016.
14. American Veterinary Medical Association (AVMA). *AVMA guidelines for the euthanasia of animals*. 2013 ed. Schaumburg, IL: AVMA; 2013. Available at: http://avma.org/KB/Policies/Documents/euthanasia.pdf.
15. WCA2010 [The Wildlife Conservation Act 2010], Laws of Malaysia Act 716. Department of Wildlife and National Parks. Available at: https://www.unodc.org/res/cld/document/wildlife-conservation-act-2010_html/Wildlife_Conservation_Act_2010.pdf. Accessed November 13, 2016.
16. The Republic Act 8485. *An act to promote animal welfare in the philippines, otherwise known as the animal welfare act of 1998*. Available at: www.angelfire.com/ok2/animalwelfare/welfa-react.html. Accessed November 13, 2016.
17. Republic Act No 10631. *An act amending certain section of republic act 8485, otherwise known as the animal welfare act of 1998"*. 2013. Available at: http://www.bai.da.gov.ph/index.php/laws-issuances/republic-act.
18. Administrative Order No. 40. Series of 1999. *Rules and regulations on the conduct of scientific procedures using animals*. Philippines Department of Agriculture. Available at: http://www.bai.da.gov.ph/index.php/laws-issuances/administrative-order/category/79-ao-1999. Accessed November 13, 2016.
19. Philippine Association for Laboratory Animal Science (PALAS). *Code of practice for the care and use of laboratory animals in the Philippines*. 2nd ed. 2002.
20. Revised Implementing Rules and Regulations of Republic Act 8485 otherwise known as "The Animals Welfare Act of 1998 as amended by Republic Act 10631". Available at: http://www.bai.da.gov.ph/index.php/laws-issuances/republic-act. Accessed November 13, 2016.
21. *Animals and birds act*. Agri-Food and Veterinary Authority of Singapore; 2002. [Chapter 7]. Available at: http://www.ava.gov.sg/docs/default-source/legislation/animals-and-birds-act/7web_abact.

22. *Animals and birds (care and use of animals for scientific purposes) rules*. Agri-Food and Veterinary Authority of Singapore; 2004. Available at: http://www.ava.gov.sg/docs/default-source/legislation/animals-and-birds-act/19web_ab_careanduseofanimalsforscientificpurposesr.pdf.
23. *NACLAR guidelines (national advisory committee for laboratory animal research guidelines on the care and use of animals for scientific purposes)*. Agri-Food and Veterinary Authority of Singapore; 2004. Available at: http://www.ava.gov.sg/docs/default-source/tools-and-resources/resources-for-businesses/Attach3_AnimalsforScientificPurposes.PDF.
24. ANZCCART (the Australia, New Zealand Council for the Care of Animals in Research and Teaching). In: Reilly JS, editor. *Euthanasia of animals used for scientific purposes: A monograph*. 1993. SA: Glen Osmond.
25. CCAC (Canadian Council on Animal Care). 2nd ed. *Guide to the care and use of experimental animals*, vol. 1. Albert St., Ottawa, Ontario, Canada, K1P5G4: CCAC; 1993. . [Chapter XII]. Available at: http://www.ccac.ca/Documents/Standards/Guidelines/Experimental_ Animals_Vol1.pdf.
26. AVMA (American Veterinary Medical Association). Report of the AVMA panel on euthanasia. *J Am Vet Med Assoc* 2001;**218**(5):669–96.
27. *Animals for scientific purposes act, B.E. 2558 (A.D. 2015)*. 2015. Available at: http://labanimals.net/images/Download/Laws/ASP_-_English.pdf.
28. *Ethical principles and guidelines for the use of animals for scientific purposes*. National Research Council of Thailand (NRCT); 1999. Available at: http://labanimals.net/images/Download/Ethics/Ethics%20Animal%20Eng.pdf.
29. *Standard for institutional animal care and use committee*. National Research Council of Thailand (NRCT); 2012. Available in Thai at: http://labanimals.net/images/Download/IACUC/iacuc%20standard.pdf.
30. *The ethics in procedures on animals for scientific purposes*. Committee for Supervision and Promotion of Procedures on Animals for Scientific Purposes. National Research Council of Thailand; 2015. Available in Thai at: http://labanimals.net/images/Law2015/Committee/02.PDF.
31. *Fundamental principles on designing animal care and use facilities for scientific work*. National Research Council of Thailand (NRCT); 2009. Available in Thai at: http://labanimals.net/images/Download/Ethics/Ethics%20Animal%20Eng.pdf.
32. Veterinary law, No. 79/2015/QH13 signed by the Chairman of National Assembly of Vietnam on June 19, 2015. Available in Vietnamese at: http://vanban.chinhphu.vn/portal/page/portal/chinhphu/hethongvanban?class_id=1&mode=detail&document_id=180584. Accessed October 21. 2016
33. Decision, No. 371/BYT-QĐ signed by the Minister of the Ministry of Health, the Socialist Republic of Vietnam on March 12, 1996. Available in Vietnamese at: http://moj.gov.vn/vbpq/lists/vn%20bn%20php%20lut/view_detail.aspx?itemid=9243. Accessed November 13.2016.
34. World Health Organization. *Laboratory biosafety manual*. 3rd ed. 2004. Available at: http://www.who.int/csr/resources/publications/biosafety/en/Biosafety7.pdf.
35. *Five freedoms*. The UK Farm Animal Welfare Council; 1965. Available at: http://webarchive.nationalarchives.gov.uk/20121007104210/http:/www.fawc.org.uk/freedoms.htm.
36. Russel WMS, Burch RL. *The principles of humane experimental technique*. London: Methuen and Co; 1959. Reissues: 1992, Universities Federation of Animal Welfare, Herts, UK.
37. Statsblad Law No. 432. 1912. Government Act (Ordonantie) Regarding Legal Review on Regulations Related to Government's Supervision on Animal and Veterinary Inspector (Ordonansi tentang Peninjauan Kembali Ketentuan-Ketentuantentang Pengawasan Pemerintahdalam Bidang Kehewanandan Polisi Kehewanan 1912 No. 432) (Herziening van de Bepalingen Omtrent het Veeartsenijkundige Staatstoezicht en de Veeartsenijkundige Politie, Staatsblad 1912 No. 432).

38. Law of the Republic of Indonesia No. 6. 1967. [Chapter 22]; Article 22 (Undang-Undang Republik Indonesia No. 6 Tahun 1967 Pasal 22).
39. Law of the Republic of Indonesia No. 18. 2009. [Chapter VII], Veterinary Authority; Article 74 (Undang-Undang Republik Indonesia Nomor 18 Tahun 2009 Bab VII—Otoritas Veteriner, Pasal 74).
40. Decree M. *Ministry of health of Republic of Indonesia*. 1030/MENKES/SK/VII/2005. 2005.
41. Decree M. *Ministry of health of Republic of Indonesia*. 562/MENKES/SK/V/2007. 2007.
42. *National bioethics committee report*. Republic of Indonesia: The Ministry of Health; 2008.
43. Government of Republic of Indonesia's Regulation No. 95. 2012. Veterinary Public Health and Animal Welfare (Peraturan Pemerintah Republik Indonesia. Kesehatan Masyarakat Veterinerdan Kesejahteraan Hewan Nomer 95 Tahun 2012).
44. Law of the Republic of Indonesia No.41. 2014. [Chapter XIII]. Criminal Provisions (Undang-Undang Republik Indonesia Nomor 41 Tahun 2014. Bab XIII. Ketentuan Pidana pasal 91B).
45. IATA (International Air transport Association). *Live Animals Regulations*. 40th ed., 199–206. Available at: http://www.iata.org/publications/store/Pages/live-animals-regulation.aspx. Accessed November 13, 2016.
46. Law of the Republic of Indonesia No.21. 2004. Protocol on Biosafety to the Convention on Biological Diversity (Undang-Undang Republik Indonesia Nomor 21 Tahun 2004. Protocol Cartagena tentang Keamanan Hayati atas Konvensi tentang Keanekaragaman Hayati).
47. *National Biosafety Framework of the Republic of Indonesia Ministry of Environment of the Republic of Indonesia Cooperating with UNEP-GEF Project for the Development of National Biosafety Framework in Indonesia*. 2004.
48. FA 1985(Fisheries Act 1985). *Laws of Malaysia. Putrajaya*. Malaysia: The Attorney General's Chambers; 1985. Available at: http://www.agc.gov.my/agcportal/uploads/files/Publications/LOM/EN/Act%20317%20-%20Fisheries%20Act%201985.pdf.
49. WCE 1997(Wildlife Conservation Enactment 1997). *Sabah Lawnet*. Sabah, Malaysia: Kota Kinabalu: State Attorney General's Chambers; 1997. Available at: http://www.lawnet.sabah.gov.my/Lawnet/SabahLaws/StateLaws/viewdoc.aspx?document=WildlifeConservationEnactment1997.pdf.
50. WPO 1998(Wildlife Protection Ordinance 1998). *Sarawak Lawnet*. Sarawak, Malaysia: Kuching: State Attorney General's Chambers; 1998. Available at: http://lawnet.sarawak.gov.my/lawnet_file/Subsidiary/SUB_Swk.%20%20L.N.%20%2094_98%20watermark.pdf.
51. IFAE 2003(Inland Fisheries, Aquaculture Enactment 2003). *Sabah Lawnet*. Sabah, Malaysia: Kota Kinabalu : State Attorney General's Chambers; 2003. Available at: http://www.lawnet.sabah.gov.my/Lawnet/SabahLaws/StateLaws/viewdoc.aspx?document=SabahInlandFisheriesAquacultureEnactment2003.pdf.
52. MEA 1996(Malaysian Education Act 1996). *Laws of Malaysia*. Malaysia: Putrajaya: The Attorney General's Chambers; 1996. Available at: http://www.agc.gov.my/agcportal/uploads/files/Publications/LOM/EN/Act%20550.pdf.
53. *Principles and guidelines for the use of animals in Precollege education*. USA: Institute of Laboratory Animal Resources Commission on Life Sciences, National Research Council; 2004. Available at: https://www.nabt.org/websites/institution/File/Principles%20and%20Guidelines%20for%20the%20Use%20of%20Animals%20in%20Precollege%20Education.pdf.
54. NMHRC [National Health, Medical Research Council]. *Australian code for the care and use of animals for scientific purposes*. 7th ed. Australia: Canberra: National Health and Medical Research Council; 2004. Available at: https://www.nhmrc.gov.au/guidelines-publications/ea16.

55. NMHRC [National Health, Medical Research Council]. *Australian code for the care and use of animals for scientific purposes.* 8th ed. Australia: Canberra: National Health and Medical Research Council; 2013. Available at: https://www.nhmrc.gov.au/guidelines-publications/ea28.
56. Masangkay JS. The status of laboratory animal research in the Philippines. In: *Proceedings of the 3rd AFLAS Congress and the 8th CALAS annual meeting in Beijing.* September 27–29, 2008. p. 13–6. China.
57. National Research Council (NRC). *Guide for the care and use of laboratory animals.* Washington: National Academies Press; 2011.
58. Administrative Order No.43 Series of 1999, *Rules and Regulations on Animal Air Transport.* Philippine Department of Agriculture. Available at: http://www.bai.da.gov.ph/index.php/laws-issuances/administrative-order/category/79-ao-1999. Accessed November 13, 2016.
59. Administrative Order No.19 Series of 2006, *Rules and Regulations on the Transport of Live Animals by Land.* Philippine Department of Agriculture. Available at: http://www.bai.da.gov.ph/index.php/laws-issuances/administrative-order/category/85-ao-2006. Accessed November 13, 2016.
60. Administrative Order No.02 Series of 2002, *Rules and Regulations on the Transport of Animal by Sea.* Philippine Department of Agriculture. Available at: http://www.bai.da.gov.ph/index.php/laws-issuances/administrative-order/category/82-ao-2002. Accessed November 13, 2016.
61. CIOMS (Council for International Organizations of Medical Sciences). *International guiding principles for biomedical research involving animals.* 1985. Available at: http://www.cioms.ch/images/stories/CIOMS/guidelines/1985_texts_of_guidelines.htm.
62. *European convention for the protection of vertebrate animals used for experimental and other scientific purposes.* 1986. Available at: http://conventions.coe.int/treaty/en/treaties/html/123.htm.
63. GMAC Guidelines (The Genetic Modification Advisory Committee Guidelines). *The Singapore biosafety guidelines for research on GMOs.* Singapore: GMAC - Genetic Modification Advisory Committee; 2006. Available at:.
64. *Guidelines on animal use protocol review.* National Research Council of Thailand; 2014. Available in Thai at. http://www.labanimals.net/images/Protocol/Criterion.pdf.
65. The Animal Welfare and Prevention of Torture of Animal Act, B.E. 2557 (A.D. 2014). Available in Thai at: http://library2.parliament.go.th/giventake/content_nla2557/law87-261257-4.pdf. Assessed November 13, 2016.
66. The government decree 32/2006/ND-CP on the management of endangered, precious and rare species of wild plants and animals signed by the Prime Minister of the Socialist Republic of Vietnam on March 30, 2006. Available in Vietnamese at: http://www.moj.gov.vn/vbpq/lists/vn%20bn%20php%20lut/view_detail.aspx?itemid=16378. Accessed November 13, 2016.
67. The government decree 157/2013/ND-CP on the management and protection of forest and its products signed by Prime Minister of the Socialist Republic of Vietnam on November 11, 2013. Available in Vietnamese at: http://vanban.chinhphu.vn/portal/page/portal/chinhphu/hethongvanban?class_id=1&mode=detail&document_id=170847.
68. The government decree 160/2013/ND-CP on the management of endangered, precious and rare species of animals signed by the Prime Minister of the Socialist Republic of Vietnam on November 12, 2013. Available in Vietnamese at: http://www.chinhphu.vn/portal/page/portal/chinhphu/hethongvanban?class_id=1&mode=detail&document_id=170893.
69. Veterinary Ordinance, No. 18/2004/PL-UBTVQH11 signed by Chairman of National Assembly of Vietnam on April 29, 2004. Available in Vietnamese at: http://vanban.chinhphu.vn/portal/page/portal/chinhphu/hethongvanban?class_id=1&mode=detail&document_id=84971.
70. Veterinary Ordinance, No. 05/2007/NĐ-CP signed by the Prime Minister of the Socialist Republic of Vietnam on January 09, 2007. Available in Vietnamese at: http://vanban.chinhphu.vn/portal/page/portal/chinhphu/hethongvanban?class_id=1&mode=detail&document_id=19635.

Chapter 12

Laboratory Animals Regulations and Recommendations: Australia and New Zealand

Denise Noonan[1], Virginia Williams[2]
[1]*The University of Adelaide, Adelaide, Australia;* [2]*Virginia Williams Consulting, North Canterbury, New Zealand*

GENERAL FRAMEWORK—AUSTRALIA

Regulatory Framework

The Australian framework for the humane use of animals for research is underpinned by the principles of prevention of cruelty, duty of care, and Russell and Burch's "Three Rs" principles of humane animal use (Replacement, Reduction, and Refinement[1]). The Commonwealth of Australia is a federation of six states and two territories, with three tiers of government. Under the authority of the Australian constitution, the Australian Government controls national and international matters including those relating to gene technology, Australian biosecurity (quarantine), wild animal management on Commonwealth lands, live animal exports and imports, export-registered slaughter establishments, and trade in animal products. Although the Australian Government is responsible for representing the nation's position on animal welfare with regard to international treaties and trade, it is the Australian states and territories that have the primary authority and responsibility for animal welfare within Australia. Matters relating to scientific use of animals and animal welfare are therefore subject to the animal welfare legislation and regulatory authority of the particular Australian state or territory. Local governments have responsibility for management of companion animals, and some aspects of public health.[2]

The responsible government minister in each state or territory appoints an Animal Welfare Advisory Committee, an expert panel drawn from members of the community, that provides the minister with advice on animal welfare issues and on associated legislation, guidelines, and Codes of Practice. The government department responsible for regulation of animal welfare in each jurisdiction varies, but is principally aligned with either Primary Industries

Laboratory Animals. http://dx.doi.org/10.1016/B978-0-12-849880-4.00012-X
Copyright © 2018 Elsevier Inc. All rights reserved.

and Agriculture, or Environment, or Local Government, as summarized in Table 12.1. Historically, both Australian and New Zealand legislation have been strongly influenced by the UK animal welfare legislation. More recently each Australian jurisdiction has developed its legislation in slightly different ways. The state of New South Wales (NSW) has a general *Prevention of*

TABLE 12.1 Summary of Principal Australian Legislation for Scientific Use of Animals[3]

State	Animal Welfare Legislation	Regulator
Australian Capital Territory (ACT)	Animal Welfare Act 1992 Animal Welfare Regulation 2001	ACT Government—Transport Canberra and City Services
New South Wales (NSW)	Animal Research Act 1985 Animal Research Regulation 2010	NSW Government—Department of Primary Industries—Animal Welfare Branch
Northern Territory (NT)	Animal Welfare Act Animal Welfare Regulations	Northern Territory Government of Australia—Department of Primary Industry and Fisheries - Animal Welfare Authority
Queensland (Qld)	Animal Care and Protection Act 2001 Animal Care and Protection Regulations 2012	Queensland Government—Department of Agriculture and Fisheries (DAF)
South Australia (SA)	Animal Welfare Act 1985 Animal Welfare Regulations 2012	Department of Environment, Water and Natural Resources-Conservation and Land Management
Tasmania (Tas)	Animal Welfare Act 1993 Animal Welfare (General) Regulations 2015	Department of Primary Industries, Parks, Water and Environment- Animal Biosecurity and Welfare Branch
Victoria (Vic)	Prevention of Cruelty to Animals Act 1986 Prevention of Cruelty to Animals Regulations 2008	Victorian Government—Department of Economic Development, Jobs, Transport and Resources (Agriculture Victoria)
Western Australia (WA)	Animal Welfare Act 2002 Animal Welfare (Scientific Purposes) Regulations 2003	Western Australian Government—Department of Agriculture and Food

Cruelty to Animals Act 1979 and also a separate *Animal Research Act 1985* that specifically regulates aspects of the scientific use of animals.[4] Other Australian states instead have specific sections referring to scientific use of animals within their general animal welfare legislation, summarized in Table 12.1: *Prevention of Cruelty to Animals Act*,[5] *Animal Welfare Act*,[6–10] or *Animal Care and Protection Act*[11].

Although the specific legislation differs between the states and territories,[3] all require the following: (1) prior licensing or registration of the scientific establishment that uses animals, and/or premises where animals are bred, held, or used for scientific purposes; (2) prior ethical review and approval, and ongoing oversight, of the animal care and use by an appropriately constituted Animal Ethics Committee (AEC); and (3) adherence to the principles and guidelines set out in the Guideline document entitled the *Australian Code for the care and use of animals for scientific purposes 2013* (eighth edition; "Australian Code").[12] Details of authorized animal welfare inspectors and compulsory reporting requirements are included in the legislation for each jurisdiction.

Scope in Brief

Scientific uses or "Scientific Purposes" are defined in state and territory legislation, and also in a national animal ethics and welfare Code of Practice. The *Australian Code for the care and use of animals for scientific purposes (2013)* defines scientific purposes as: "*all activities conducted with the aim of acquiring, developing or demonstrating knowledge or techniques in all areas of science, including teaching, field trials, environmental studies, research (including the creation and breeding of a new animal line where the impact on animal wellbeing is unknown or uncertain), diagnosis, product testing, and the production of biological products*."[12]

The species of animals protected by animal welfare legislation differ slightly between the states and territories; however, the Australian Code defines an animal as: "*any live non-human vertebrate, that is, fish, amphibians, reptiles, birds and mammals, encompassing domestic animals, purpose-bred animals, livestock, wildlife, and cephalopods*."[12] In the states and territories of Victoria, Queensland, The Australian Capital Territory, and The Northern Territory some decapod species (such as crabs and lobsters) are also included as protected animals under legislation, following on from public concerns and recent literature suggesting sentience in these species.[5]

State and territory animal welfare legislation makes provision for prohibition and restriction of particular animal procedures and uses. In accordance with the former Australian National Consultative Committee on Animal Welfare (NCCAW) 1993 guidelines, scientific use of the Draize test and the Lethal Dose 50 test was restricted, and in general requires approval from both the institutional AEC and the government minister with responsibility for the animal welfare legislation.[13]

National Harmonization

Despite differences between the animal welfare legislation in each jurisdiction, there are also mechanisms to harmonize and to seek consensus with regard to animal welfare. The most important of these with regard to laboratory animals is the *Australian Code for the care and use of animals for scientific purposes,* which is recognized in each jurisdiction and included within the regulatory framework that applies to scientific use of animals. The Australian Code is endorsed by the National Health and Medical Research Council (NHMRC), Australian Research Council, Commonwealth Scientific and Industrial Research Organisation, and Universities Australia.[12]

Harmonization of animal welfare legislation: as discussed earlier, regulatory responsibility for animal welfare in Australia is principally a state and territory matter, with the Australian Government playing a facilitating role. In the past, the Council of Australian Governments Standing and Select Councils framework was used for coordination and harmonization of state and territory matters relating to primary industries and animal welfare. The Australian government minister with responsibility for Primary Industries, and the Primary Industries ministers for each jurisdiction, would meet as a Council at regular intervals and as required. Model Codes of Practice for the Welfare of Animals were developed for livestock industries under the former Primary Industries Ministerial Council and Standing Council on Primary Industries. These voluntary Model Codes for animal welfare are being progressively replaced by regulated standards and guidelines under the authority of legislation in each jurisdiction.[14] The process of developing Australian Animal Welfare Standards and Guidelines commenced under the impetus of the Australian Animal Welfare Strategy (2005–13).[15,16] The current framework for harmonization consists of the Agriculture Ministers Forum, Agriculture Senior Officials Committee, and the Animal Welfare Task Group.[17] Apart from livestock industries, development of Animal Welfare Standards and Guidelines have been undertaken by working groups from a range of animal sectors and industries.

The Australian Code and Other Guidance Documents

The Australian Code has been prepared by a joint working party of the NHMRC, Commonwealth Scientific and Industrial Research Organisation (CSIRO), Australian Research Council, Universities Australia, representatives of Australian government departments, state and territory governments of Australia, and Animal welfare organisations (Royal Society for the Prevention of Cruelty to Animals and Animals Australia).[12] The Australian Code is regularly revised and the eighth edition was endorsed and released in 2013. Authorship and review of the Australian Code is coordinated and overseen by the main funding body, the NHMRC. Representatives of the regulatory agencies in each state or territory participate in the revision process via the Jurisdictional Reference Group (previously known as the Code Reference Group).

The NHMRC has produced supplementary policies and guidelines intended to be read in conjunction with the Australian Code. Compliance with these additional documents is particularly relevant to institutions receiving NHMRC grant funding.[18]

- Principles and guidelines for the care and use of non-human primates for scientific purposes (2016).[30]
- A guide to the care and use of Australian native mammals in research and teaching (2014).[18]
- NHMRC Guidelines on the Care of Cats Used for Scientific Purposes (2009).[18]
- NHMRC Guidelines on the Care of Dogs Used for Scientific Purposes (2009).[18]
- Guidelines on the use of animals for training interventional medical practitioners and demonstrating new medical equipment and techniques (2009).[18]
- Guidelines to promote the wellbeing of animals used for scientific purposes: The assessment and alleviation of pain and distress in research animals (2008).[18]
- NHMRC Guidelines for monoclonal antibody production (2008).[18]
- Guidelines for the generation, breeding, care and use of genetically modified and cloned animals for scientific purposes (2007).[19]
- Australian code for the responsible conduct of research (2007).[18]

In addition to the Australian Code, a large number of national Model Codes for animal welfare have been developed for livestock species.[20,21] State and territory animal welfare codes of practice have been developed from animal welfare legislation and Australian Model Codes. These are currently under review as part of the *Australian Animal Welfare Strategy*, and will be incorporated as standards and guidelines in Regulations under the animal welfare legislation in each state and territory. The number of these is extensive; however, a select list of Codes of Practice and guidance documents relating specifically to animals used for scientific purposes is as follows:

1. *National Consultative Committee of Animal Welfare Guidelines*[13] (N.B. NCCAW has been disbanded):
   a. Genetic manipulation of animals;
   b. LD50 and Draize tests;
   c. Pound sourced animals.
2. *New South Wales*:
   a. Animal Research Review Panel Guidelines for the Care and Housing of Dogs in Scientific Institutions;
   b. Animal Research Review Panel Guidelines for radio tracking in wildlife research;
   c. Animal Research Review Panel Animal care guidelines for wildlife surveys;

   d. Animal Research Review Panel Guidelines on opportunistic research on free living wildlife;
   e. Animal Research Review Panel Guidelines on the use of feral animals in research;
   f. Animal Research Review Panel Guidelines on collection of voucher specimens;
   g. Animal Research Review Panel Guidelines on the use of pitfall traps;
   h. Animal Research Review Panel Guidelines for the production of monoclonal antibodies;
   i. Animal Research Review Panel guidelines on supply of dogs and cats for use in research;
   j. Several other New South Wales Animal Research Review Panel Guidelines for care and housing of specific laboratory animals;[22]
   k. Animal Welfare Branch Fact Sheet 21: Supply of dogs and cats for use in research.[23]
3. *Victoria*:
   a. Code of Practice for the Housing and Care of Laboratory Mice, Rats, Guinea Pigs and Rabbits;
   b. Code of practice for the use of animals from municipal pounds in scientific procedures.[24]
4. *South Australia—Wildlife Ethics Committee policies and guidelines*:
   a. Policy on the euthanasia of research animals in the field;
   b. Guidelines for the collection of hair and feather samples;
   c. Guidelines for the use of live traps to capture terrestrial vertebrates;
   d. Policy on the use of microchips for marking wildlife;
   e. Guidelines for the use of tracking tunnels;
   f. Guidelines for the transportation of live animals;
   g. Policy on the collection of voucher specimens;
   h. Policy on the collection of blood from wildlife.[25]

Where there is use of a species not specifically covered by one of the above research context-specific Guidelines, then Codes or Standards pertinent to the care and use of that species in another context are used to provide general guidance on the minimum standards expected for the care of those species being held and used in a research environment. The Australian and New Zealand Council for the Care of Animals in Research and Teaching (ANZCCART) provides useful resource and reference material, which is available at the ANZCCART website.[26]

GENERAL FRAMEWORK—NEW ZEALAND

In New Zealand (NZ), the Animal Welfare Act 1999 (the Act)[27] and the Animal Welfare Amendment Act (No 2) 2015 (the Amendment Act)[28] provide a legislative basis under which the welfare of animals, including those used in research, testing, and teaching (RTT), is regulated. The Act[27] requires owners of animals, and persons in charge of animals, to attend properly to the welfare of those animals and specifies conduct that is or is not permissible in relation to any animal or class of animals. It also provides a process for approving the use of animals in RTT. In recognition that animals that are used in this way may be subject to impacts of varying degrees on their welfare, a special section—Part 6—of the Act[27] applies specifically to this use of animals.

In addition, the Act[27] establishes the National Animal Ethics Advisory Committee (NAEAC) whose primary functions include advising the Minister for Primary Industries on ethical issues and animal welfare issues arising from RTT; providing advice on the development and review of codes of ethical conduct (CECs); making recommendations on the approval, amendment, suspension, or revocation of a CEC; and providing information to AECs. The National Animal Welfare Advisory Committee (NAWAC) provides advice on any matters relating to the welfare of animals apart from those involved with their use in RTT. NAWAC also develops codes of welfare for animal species and uses.

Under the Amendment Act,[28] a new Section (84A) prohibits the use of animals in any RTT for the purpose of developing, making, or testing a cosmetic or an ingredient that is intended exclusively for use in a cosmetic.

THE PRINCIPLES—AUSTRALIA

Justification, Responsibilities, and the "Three Rs"

Although the specific pieces of legislation relevant to animal use for scientific purposes differ between Australian states and territories, the key principles and framework are similar because they all adopt the *Australian Code for the care and use of animals for scientific purposes 2013 (eighth edition;* "*Australian Code*") into their respective legislation.[12] The way in which the guidance in this document is incorporated differs in subtle ways between states and territories; in some, mandatory compliance with the entire document is specified in the legislation; while in others, the key principles and terminology are embedded in the legislation itself. A detailed description and comparison of these differences are beyond the scope of this chapter, therefore, for simplicity the focus here is mainly on the requirements detailed in the Australian Code. In its introduction, the Australian Code states that its purpose is "*to promote the ethical, humane and responsible care and use animals for scientific purposes.*" The Australian Code details the responsibilities of all those involved in the care and use of animals. This embraces a duty of care, an obligation to respect animals, and support for the well-being of the animals.[12]

The Australian Code was the first, and remains the principal Australian guideline to mandate consideration of the principles of Russell and Burch's "Three Rs"[1] in the context of animal use for scientific purposes. The "Three Rs" must be applied at every stage of animal use, starting from the time of breeding or acquisition of the animals, for the duration of animal holding and use, and also including the fate of the animals at the end of scientific use.[12]

According to the Australian Code, a scientific or teaching activity can only be approved by an AEC if the proposed activity has sufficient justification and that the work is essential for progress toward one of the following purposes: (1) to obtain and establish significant information relevant to the understanding of humans and/or animals; (2) for the maintenance and improvement of human and/or animal health and welfare; (3) for the improvement of animal management or production; or (4) to obtain and establish significant information relevant to the understanding, maintenance, or improvement of the natural environment; or (5) for the achievement of educational outcomes in science, as specified in the relevant curriculum or competency requirements. This prior approval pertains to all aspects of the proposed animal use, including the animals being used (includes the specification and justification of the requested species, strain, gender, age, number, etc.), the procedures to be performed (including maximum permissible severity and intervention/endpoints), the investigators/personnel involved (including review of past experience and competence), and the institution and premises where the work will occur.[12]

Ultimately, before approving the use of animals for scientific purposes the AEC must be convinced that the use is (1) essential; and (2) has been appropriately justified by weighing the predicted scientific or educational value of the project against the potential effects on the welfare of the animals; and (3) that all ethical and animal welfare issues have been considered and the "Three Rs" have been adequately applied.

THE PRINCIPLES—NEW ZEALAND

The Amendment Act[28] recognizes in (a) (i) of the long title that animals are sentient and that their welfare must be properly attended to. Section 80 of the Act[27] lays down the principles under which the use of animals in RTT is allowed, confining it to cases in which there is a good reason to believe:

- **i.** "That the findings of RTT will enhance:
  - **a.** the understanding of human beings, animals, or ecosystems; or
  - **b.** the maintenance or protection of human or animal health or welfare; or
  - **c.** the management, protection, or control of ecosystems, plants, animals, or native fauna; or
  - **d.** the production and productivity of animals; or
  - **e.** the achievement of educational objectives; and
- **ii.** that the benefits derived from the use of animals in RTT are not outweighed by the likely harm to the animals."

The other principles require animal users:

"i. to ensure that

a. all reasonable steps are taken to ensure the physical, health and behavioural needs of those animals are met in accordance with both good practice and scientific knowledge; and

b. where animals used in RTT are ill or injured, they receive, where practicable, treatment that alleviates any unreasonable or unnecessary pain or distress; and

c. where, because of the nature of the work, the needs cannot be provided, any degree of pain or distress is reduced to the minimum possible in the circumstances.

ii. to promote efforts

a. to reduce the number of animals used in research, testing, and teaching to the minimum necessary:

b. to refine techniques used in any research, testing, and teaching so that the harm caused to the animals is minimised and the benefits are maximised:

c. to replace the use of animals in teaching by substituting for animals, where appropriate, non-sentient or non-living alternatives or by imparting the information in another way."[27]

SCOPE/APPLICABILITY—AUSTRALIA

State or territory animal welfare legislation applies to all animals in that jurisdiction. The general definition of an animal is a living nonhuman vertebrate; however, there are inclusions and exceptions of animal species that vary between the states and territories. Additional requirements and protection, as detailed in the legislation, apply to all animals used for RTT purposes; however, common laboratory animal species such as rats, mice, guinea pigs, rabbits, and nonhuman primates (NHPs) may be specifically addressed. As the *Australian Code for the care and use of animals for scientific purposes eighth edition* 2013 ("Australian Code") is adopted under state and territory legislation, the definition of an animal used in the Australian Code applies to all scientific use of animals nationally.[21] Terminology used in state legislation also varies; however, the Australian Code defines the scope and terminology, thus:

Animal: "*any live non-human vertebrate (that is, fish, amphibians, reptiles, birds and mammals, encompassing domestic animals, purpose-bred animals, livestock, wildlife) and cephalopods.*" And "*embryos, fetuses and larval forms* [that] *have progressed beyond half the gestation or incubation period of the relevant species, or they* [have] *become capable of independent feeding.*"

Scientific purposes: "*all activities conducted with the aim of acquiring, developing or demonstrating knowledge or techniques in all areas of science, including teaching, field trials, environmental studies, research (including the creation and breeding of a new animal line where the impact on animal wellbeing is unknown or uncertain), diagnosis, product testing, and the production of biological products.*"

Teaching: "*developing, imparting or demonstrating knowledge or techniques to achieve an educational outcome in any area of science as specified in the relevant curriculum or competency requirements.*"

Scope as stated in the Australian Code:

> *"The Code encompasses all aspects of the care and use of animals for scientific purposes in medicine, biology, agriculture, veterinary and other animal sciences, industry and teaching with an educational outcome in science. It includes the use of animals in research, teaching, field trials, product testing, diagnosis, the production of biological products and environmental studies.*
>
> *The Code provides general principles for the care and use of animals, specifies the responsibilities of investigators, teachers and institutions, and details the terms of reference, membership and operation of AECs [Animal Ethics Committees]. It also provides guidelines for the humane conduct of scientific activities, and for the acquisition of animals and their care, including their environmental needs.*
>
> *The Code applies throughout the animal's involvement in activities and projects, including acquisition, transport, breeding, housing, husbandry, the use of the animal in a project, and the provisions for the animal at the completion of their use.*
>
> *The Code covers all live non-human vertebrates and cephalopods. Institutions are responsible for determining when the use of an animal species not covered by the Code requires approval from an AEC, taking into account emerging evidence of sentience and the ability to experience pain and distress. Animals at early stages in their development-that is, in their embryonic, fetal and larval forms-can experience pain and distress, but this occurs at different stages of development in different species. Thus decisions as to their welfare should, where possible, be based on evidence of their neurobiological development. As a guide, when embryos, fetuses and larval forms have progressed beyond half the gestation or incubation period of the relevant species, or they become capable of independent feeding, the potential for the experience of pain or distress should be taken into account."*[12]

All scientific RTT uses of animals must have obtained prior ethical approval from an AEC before animal use can commence. There is no expedited review process for certain categories of animal use. All new animal use proposals or project applications must be considered at a quorate face-to-face committee meeting; however, minor modifications to approved projects may be considered by an executive subcommittee of the AEC. Decisions made by the executive subcommittee must be presented for ratification by the AEC at the next meeting.

In the ethical review process, a cost-benefit analysis approach is used as the tool for an AEC to consider and determine, by consensus, whether the likely or anticipated scientific or educational value outweighs the potential impact on the animals; whether the principles of the "Three Rs" have been applied in all respects of the care and use of the animals.

SCOPE/APPLICABILITY—NEW ZEALAND

There are a number of definitions that underpin the applicability of New Zealand's legislation pertaining to the use of animals in RTT. These include the definition of "animal," "research, testing and teaching," "manipulation," and "physical, health and behavioural needs."

According to the Animal Welfare Act[27] and Animal Welfare Amendment Act[28]:

Animal (the Act[27] s2)

"**a.** means any live member of the animal kingdom that is—

i. a mammal; or

ii. a bird; or

iii. a reptile; or

iv. an amphibian; or

v. a fish (bony or cartilaginous); or

vi. any octopus, squid, crab, lobster, or crayfish (including freshwater crayfish); or

vii. any other member of the animal kingdom which is declared from time to time by the Governor-General…to be an animal for the purposes of this Act; and

b. includes any mammalian foetus, or any avian or reptilian pre-hatched young, that is in the last half of its period of gestation or development; and

c. includes any marsupial pouch young."[27]

Research, Testing, and Teaching (the Act[27] s5)

"**1.**

a. any work (being investigative work or experimental work or diagnostic work or toxicity testing work or potency testing work) that involves the manipulation of any animal; or

b. any work that—

i. is carried out for the purpose of producing antisera or other biological products; and

ii. involves the manipulation of any animal; or

c. any teaching that involves the manipulation of any animal.

i. Addition included in the Amendment Act[28] (2[3])

d. Any routine breeding of animals that may result in the birth or production of an animal that is more susceptible to, or at greater risk of, pain and distress during its life, being breeding for the purpose of carrying out any work or teaching of a type specified in paragraphs (a) to (c) on any offspring.

i. In addition, the Amendment Act[28] includes in Section 5(1A): The term defined by subsection (1) (a) or (b) carried out on the body or tissues of an animal after the animal was killed for the purpose, if the killing of the animal was a manipulation under Section 3(1A).

2. The term defined by subsection (1) does not include any manipulation that is carried out on any animal that is in the immediate care of a veterinarian, if—
   a. the veterinarian believes on reasonable grounds that the manipulation will not cause the animal unreasonable or unnecessary pain or distress, or lasting harm; and
   b. the manipulation is—
      i. for clinical purposes in order to diagnose any disease in the animal or any associated animal; or
      ii. for clinical purposes in order to assess the effectiveness of a proposed treatment regime for the animal or any associated animal; or
      iii. for the purposes of assessing the characteristics of the animal with a view to maximising the productivity of the animal or any associated animal."[27,28]

Manipulation (the Act[27] s3)

"1. In this Act, unless the context otherwise requires, the term manipulation, in relation to an animal, means interfering with the normal physiological, behavioural, or anatomical integrity of the animal by deliberately—
   a. subjecting it to a procedure which is unusual or abnormal when compared with that to which animals of that type would be subjected under normal management or practice and which involves—
      i. exposing the animal to any parasite, micro-organism, drug, chemical, biological product, radiation, electrical stimulation, or environmental condition; or
      ii. enforced activity, restraint, nutrition, or surgical intervention; or
   b. depriving the animal of usual care."[27]

Additions included in the Amendment Act[28] (s3)

"1A. The term defined by subsection (1) includes the killing of an animal (other than an animal in a wild state) for the purpose of interfering with the animal's body or its tissues in a manner specified in that subsection.
1B. The term defined by subsection (1) includes the breeding or production of an animal using any breeding technique (including genetic modification) that may result in the birth or production of an animal that is more susceptible to, or at greater risk of pain or distress during its life as a result of the breeding or production."

Physical Health and Behavioral Needs (s4)

"In this Act, unless the context otherwise requires, the term physical, health, and behavioural needs, in relation to an animal, includes

a. proper and sufficient food;
b. Proper and sufficient water;
c. adequate shelter;
d. opportunity to display normal patterns of behaviour;
e. physical handling in a manner which minimises the likelihood of unreasonable or unnecessary pain or distress;
f. protection from, and rapid diagnosis of, any significant injury or disease being a need which, in each case, is appropriate to the species, environment, and circumstances of the animal."[27,28]

AUTHORIZATION OF USER-BREEDING INSTITUTIONS/ INSPECTION/PENALTIES BY COMPETENT AUTHORITIES—AUSTRALIA

Scientific institutions that use animals for scientific purposes (RTT) are authorized to do so by licenses issued under the state or territory animal welfare or animal research legislation.[4–11] The breeding and supply of particular animal species for scientific use may also require authorization or licensing. The minister or government department with responsibilities for the animal welfare legislation issues these licenses and administers any conditions imposed. Additional requirements and conditions usually apply to dogs and cats sourced from private owners and municipal pounds for scientific purposes.[12,18,23] Examples of additional state Codes and Guideline documents have been listed earlier in this chapter, in the section General Framework.

Inspections of animal facilities and the breeding and use of the animals are responsibilities of both the AEC and the relevant government department in most jurisdictions. Each licensed institution or breeding establishment is required to comply with the license conditions, as well as the responsibilities, procedures, and practices stated in the *Australian Code for the care and use of animals for scientific purposes, eighth edition* 2013 ("Australian Code").[12] The AEC listed on the license must oversee the sourcing, supply, breeding, transport, and husbandry of animals intended for scientific use. The external review of institutions, a requirement under the Australian Code, usually incorporates inspection of animal teaching and research areas, and animal holding and breeding facilities.[12]

The government regulator receives annual statistics reports on animal use for scientific purposes from each licensed breeding establishment and research and teaching institution. The details of the statistical reporting are included in the relevant legislation, or in guidelines promulgated by the regulator.[4–11]

AUTHORIZATION OF USER-BREEDING INSTITUTIONS/ INSPECTION/PENALTIES BY COMPETENT AUTHORITIES—NEW ZEALAND

Under Part 6 of the Animal Welfare Act (the Act),[27] any person or institution that wishes to use animals in RTT must do so under a CEC approved by the

Director-General for Primary Industries. CECs set out policies and procedures[29] to be followed by institutions and by the AECs established under those CECs. Approval is personal to the institution and is not transferable without prior consent of the Director-General. The contents of a code are specified in Section 88 of the Act[27]:

1. "Each code of ethical conduct must contain provisions that set out, in relation to the carrying out of the research, testing, or teaching to which the code relates, the policies to be adopted and the procedures to be followed:
   a. by the code holder; and
   b. by an animal ethics committee appointed by the code holder.
2. The policies and procedures must:
   a. enable the animal ethics committee to carry out its functions effectively; and
   b. enable persons who are members of the animal ethics committee but who are not employed by the code holder to have an effective input into the working of the committee; and
   c. make provision for adequate monitoring of compliance with the conditions of project approvals to be carried out; and
   d. make provision for the code holder to collect the information and to maintain the records required by regulations made under this Act; and
   e. specify animal management practices and facilities that are such as to enable the purposes of this Part to be met adequately; and
   f. be such as to ensure that where any member of the animal ethics committee makes a complaint, that complaint may be dealt with fairly and promptly by the animal ethics committee or the code holder."

Institutions must be reviewed by an accredited reviewer, within 5 years of their previous approval. In addition, an institution that has been approved for the first time must undergo an independent review within 2 years of being granted that approval. The purpose of the periodic reviews is to establish compliance by the code holder and by the institution's AEC with the institution's CEC and with the requirements of the legislation. The independent review report is evaluated by the NAEAC and the Ministry of Primary Industries.[27]

An institution (or person) that carries out RTT using animals without the requisite approved CEC is subject, in the case of an individual to imprisonment for a term not exceeding 6 months or to a fine not exceeding $25,000 or to both, while, in the case of a body corporate, to a fine not exceeding $125,000.

NONHUMAN PRIMATES: SPECIAL CONSIDERATIONS AND RESTRICTIONS—AUSTRALIA

NHPs are not indigenous to Australia. There are international treaties and Commonwealth, state and territory legislative restrictions upon their importation. Biocontainment holding and exotic animal licensing and permit conditions are enforced by government regulatory authorities.[30] Additional ethical and animal welfare concerns apply to use of NHP, in general, and great apes in particular. These are addressed in animal welfare legislation, the *Australian Code for the care and use of animals for scientific purposes, eighth edition* 2013 ("Australian Code"), and additional guidelines such as the NHMRC *Principles and guidelines for the care and use of non-human primates for scientific purposes* (2016; recently revised).[12,30] Currently great apes are not used for biomedical research or testing in Australia.

NHMRC Statement on the Use of Nonhuman Primates Including Great Apes

"**1.** Non-human primates must not be used for scientific purposes except when:
   i. no alternative to the use of non-human primates is suitable to achieve the stated aims of the project, and
   ii. the potential effects on the non-human primates are justified by the potential benefits.

2. As the closest species to human beings with the most advanced social and behavioural skills, the use of great apes (gorilla, orangutan, chimpanzee and bonobo) raises even greater ethical concerns than that of other non-human primates.

3. No great apes are held in Australia specifically for scientific purposes. The only great apes held in Australia are in zoological collections for conservation breeding purposes.

4. Great apes must not be imported from overseas for use for scientific purposes.

5. The use of great apes for scientific purposes in Australia is permitted only when their use:
   i. will not have any appreciable negative impact on the animals involved, e.g., observational studies, activities already being undertaken for management or veterinary purposes
   ii. will potentially benefit the individual animal and/or their species.

6. No other use of great apes in Australia is permitted under this guideline.

7. The institution should ensure that the Animal Welfare Committee (AWC) of NHMRC is notified of projects involving the use of great apes for scientific purposes approved by the institutional animal ethics committee (AEC), before the project commences. For NHMRC-funded research, this requirement is mandatory."[30]

Breeding, Sourcing, and Supply of Nonhuman Primates

The NHMRC supports Australian Breeding Colonies for macaques (*Macaca nemestrina* and *Macaca fascicularis*), marmosets (*Callithrix jacchus*), and baboons (*Papio hamadryas*). These colonies have been established to centralize breeding, provide a consistently high standard of animal care and management, and to allow access to NHPs for research.[30]

Breeding of Nonhuman Primates

The NHMRC states that:

"**12.** New breeding colonies of non-human primates must not be created outside of the established Australian breeding colonies unless as an integral component of the research project, and must be approved by an institutional AEC and relevant regulatory authorities. The institution should ensure that the AWC of NHMRC is notified before a new breeding colony is established. For breeding colonies funded by NHMRC, this requirement is mandatory.

13. Imported species of non-human primates must not be used to establish a breeding colony outside the established Australian breeding colonies.

14. Non-human primates must always be accompanied by a documented complete and current life history that includes source; housing, care and management at both source facility and subsequent facility (facilities); clinical history; health status; behaviour; and transport.

16. When non-human primates are supplied to a project approved by an institutional AEC, the animals must be obtained from an established Australian breeding colony unless another source is approved by the AEC.

17. When importation of non-human primates is proposed, the investigator must provide the institutional AEC with all available information regarding animal welfare at the source facility. This information should include housing conditions, social grouping, environmental enrichment, diet, health status and breeding performance of animals. Information provided should also include arrangements and conditions for transport between the source colony and the institution.

18. Non-human primates that are imported from overseas must be captive bred and must be accompanied by documentation to certify their captive-bred status.

21. The institution should ensure that the AWC of NHMRC is notified of the importation of non-human primates after approval from the institutional AEC has been obtained. For NHMRC-funded activities, this requirement is mandatory."[30]

Provisions at the Conclusion of Their Use

The NHMRC states that:

"**23.** Provisions for non-human primates at the conclusion of their use must take into account their long-term welfare. Retirement must be considered as an option if suitable in terms of the health and temperament of the animal, and space and resources are available at a facility that can meet their species-specific physical, social and behavioural needs.

24. When non-human primates are humanely killed for any reason, all possible attempts should be made to inform other interested investigators of the availability of tissues and organs and the disease status of the animals. This may be achieved effectively through the managers of the established Australian breeding colonies, who can disseminate the information to interested investigators.

25. Non-human primates originally sourced from another country must not be exported from Australia.

26. Australian-bred non-human primates must not be exported unless:

- **i.** for a specific purpose. Examples of specific purposes would be maintenance of genetic diversity or provision of overseas researchers with a model of a primate disease
- **ii.** prior approval has been obtained from the institutional AEC by those responsible for the export
- **iii.** export is to an institution with animal welfare standards that comply with the Code at a minimum
- **iv.** there are adequate animal protection laws in the destination country, and an active organisation with the legal authority to enforce animal protection laws
- **v.** appropriate Commonwealth and state or territory government permits, licences and approvals are in place.

27. The institution should ensure that the AWC of NHMRC is notified of the export of non-human primates after approval from the institutional AEC has been obtained. For NHMRC-funded institutions, this requirement is mandatory."[30]

Projects Conducted in Other Countries

The NHMRC states that:

"**35.** If a project involving the use of non-human primates is to be conducted in another country, Clauses 2.6.9–2.6.14 of the [Australian] Code must be upheld. The conduct of a project in another country should not be used as a mechanism for avoiding compliance with Code."[30]

Further details regarding the special needs of NHP are also addressed in the NHMRC Guidelines.

NONHUMAN PRIMATES: SPECIAL CONSIDERATIONS AND RESTRICTIONS—NEW ZEALAND

The Animal Welfare Act[27] prohibits the use of nonhuman hominids unless approved by the Director-General, and only if the research is in the best interest of the animal, or in the interests of the species to which it belongs. These are defined to include the gorilla, chimpanzee, bonobo, or orangutan. There are no nonhuman hominids in research institutions in New Zealand.

GENETICALLY ALTERED ANIMALS—AUSTRALIA

The Commonwealth of Australia *Gene Technology Act* 2000[31] and *Gene Technology Regulations* 2001[32] regulate genetically modified (altered) animals and other organisms (genetically modified organisms, GMOs) in Australia. The aim of the legislation is to protect the health and safety of humans and the environment by identifying and managing risks posed by, or resulting from, gene technology. Biocontainment of GMOs is required, with the details of the physical containment (PC) determined by legislation and guidelines. This involves government accreditation of organizations generating or holding GMOs, certification of facilities, licensing of use of GMOs ("dealings"), monitoring and compliance activities, and maintenance of a "Record of GMO and GM Product Dealings" as required by the *Gene Technology Act 2000*.[31] The *Guidelines for the transport, storage and disposal of GMOs* (2011) applies to all genetically modified animals and other organisms regardless of their commercial, domestic, or scientific use.[33]

The *Australian Code for the care and use of animals for scientific purposes, eighth edition* 2013 ("Australian Code")[12] and the NHMRC Animal Welfare Committee's *Guidelines for the generation, breeding, care and use of genetically modified and cloned animals for scientific purposes* (2007)[19] address the animal welfare and ethical issues and requirements relating to genetically modified and cloned animals used for scientific purposes. These guidelines provide a brief overview of issues such as the potential for adverse animal welfare outcomes, the uncertainties associated with the genetic alteration on the resultant genetic constitution (genotype) and biological characteristics or phenotype, and the large number of animals generated to obtain a small number with the desired genotype. The roles and responsibilities of the AEC, the research scientist, and the animal carer are detailed and cross-referenced with the requirements of the Australian Code.

The Australian NCCAW *Position statement on genetic manipulation of animals* (1991) states that "All stages of producing genetically modified animals, including developmental stages such as breeding-up prior to release, should be: Regarded as experimental under the *Australian Code of Practice for the Care and Use of Animals for Scientific Purposes*, and Covered by Animal Experimentation Ethics Committees (AEECs)."[13]

GENETICALLY ALTERED ANIMALS—NEW ZEALAND

The Hazardous Substances and New Organisms Act 1996[34] and the Biosecurity Act 1993[35] provide for controls on the use of genetically altered animals. The controls on the housing and use of genetically altered animals are detailed in the *Biosecurity Authority Standard 154.03.03 Containment Facilities for Vertebrate Laboratory Animals*.[36] These controls require the maintenance of registers, which record the numbers of genetically modified animals held in containment. The registers must be updated each month. Movement from secure containment in the animal facility to a research laboratory must also be documented. The transfer of genetically modified animals requires authorization by the Ministry for Primary Industries. Government officials may audit RTT institutions that hold genetically altered animals at any time to ensure compliance with the Standard.

INSTITUTIONAL AND DESIGNATED PERSONNEL RESPONSIBILITIES—AUSTRALIA

The state and territory animal welfare legislation determines the licensing requirements and conditions that apply to scientific establishments that use animals for scientific purposes.[4–11] These conditions include, generally, the requirement to appoint one or more AEC, to report annually to the government regulator, and to fulfill the responsibilities detailed in the *Australian Code for the care and use of animals for scientific purposes, eighth edition* 2013 ("Australian Code").[12] The state or territory legislation also usually stipulates that a particular individual, nominated by the institution, is named on the license for the scientific institution or establishment. This person is usually a senior manager with responsibilities that include oversight of the RTT conducted by the institution, and with the authority to represent the institution on these matters to the government regulator.[4–11]

The Australian Code (Section 2.1) details institutional responsibilities and states that matters of compliance with relevant legislation, and the Australian Code, lie with "*the governing body of the institution.*" The governing body of the institution ("the institution"), or its delegate, is held responsible and accountable for the activities of the personnel under their authority who use animals for scientific purposes, as well as other personnel and resources controlled by them on behalf of the institution. The institution is responsible for the effective function of the AECs listed on the license, is required to seek comments and reports from the AEC, and to resource it adequately so that it can fulfill its responsibilities and thereby assist the institution. In effect, the AEC performs a significant compliance management role on behalf of the institution.[12]

The Chairperson of the AEC plays a pivotal role in the success of this arrangement, as communication between the AEC and the institution must be both frequent and effective. The Australian Code (Section 2.2.2) recommends thus:

"Institutions should consider appointing a chairperson who holds a senior position in the institution. If the chairperson is an external appointee, institutions must provide the chairperson with the necessary support and authority to carry out the role."[12]

The responsibilities incumbent upon scientific researchers and teachers are similar to those detailed in the following New Zealand section. The responsibilities of institutions are detailed in Australian Code Section 2.1, and include:

- Establishing or accessing one or more AEC, to ensure that all scientific use of animals complies with the relevant legislation, Codes, and guidelines;
- Provision of education, training, workshops, and other information so that all animal users, animal care staff, and AEC members are aware of their responsibilities under the Code and legislation;
- Seeking comment from the AEC on all matters that may affect animal welfare, including building or modification of animal facilities;
- Seeking AEC approval for all institutional animal care and use guidelines and emergency planning;
- Responding promptly to recommendations of the AEC;
- Establishing mechanisms to respond to enquiries or complaints concerning the use of animals within the institution;
- Ensuring adequate numbers of trained animal care staff and veterinary services are available;
- Establishing procedures for resolution of disputes between members of the AEC, animal users, and the institution;
- Ensuring that the AEC operation and report are reviewed annually by the institution, and at least every 4 years by an external review committee.[12]

INSTITUTIONAL AND DESIGNATED PERSONNEL RESPONSIBILITIES—NEW ZEALAND

The responsibilities of the facility manager are described in the *Good Practice Guide for the Use of Animals in Research, Testing and Teaching*[37] as follows:

"The process of animal acquisition and the management of breeding and holding facilities should be supervised by persons with appropriate veterinary or animal care qualifications or experience with the species involved. The facility manager should ensure that adequate veterinary care is provided for all animals held for breeding and for experimental manipulations with ready access to institutional or consultant veterinary services seven days a week.

The facility manager should:

- be responsible for the management of the day-to-day care of the animals in holding and breeding facilities;
- supervise the work of other personnel in the facility;
- act as liaison between investigators and teachers and facility personnel;
- communicate with the AEC [Animal Ethics Committee] on management of the facility and any adverse incidents;
- contribute to the development and maintenance of the institution's animal care policies and procedures;

- be knowledgeable regarding signs of pain, distress and illness specific to each species housed, and should ensure that the well-being of all animals is regularly assessed. (After animals are allocated to an approved project, the investigator has primary responsibility for ensuring adequate monitoring of the animals' wellbeing);
- ensure that ill or injured animals which are not assigned to approved projects are treated promptly and that the cause of death is investigated for animals which die unexpectedly;
- ensure that personnel are provided with appropriate protective clothing, maintain high standards of personal hygiene and do not eat, drink or smoke in animal areas;
- document procedures used in the management of small animal holding and breeding facilities. These procedures should be approved as Standard Operating Procedures (SOPs) by the AEC and should take into account the requirements of the species held, the experiments being conducted, and the health and safety of the personnel, and include transport, quarantine and disposal of animals, routine husbandry, prevention, diagnosis and treatment of disease, monitoring of health status and genetic constitution, and physical environmental factors. These procedures should be made known to all personnel involved in the care and use of the animals and should be reviewed regularly. For large animals, transport and routine husbandry procedures as recommended in the relevant codes of welfare should be followed unless alternative procedures have been approved by the animal ethics committee;
- for small animal colonies, maintain a regular schedule of cage, equipment and facility sanitisation to ensure that potential pathogens are kept at minimum levels in the environment;
- ensure that adequate records are maintained of:
  - the source, care, allocation, movement between locations, use and disposal of all animals, and of any diseases developed;
  - the fertility, fecundity, morbidity and mortality in animal breeding groups, in order to monitor the management of the groups, and to assist detection of the origin and spread of disease; and
  - the health status, genetic constitution and the physical environment of the animals, when definition of these is required.

Records maintained by the facility manager must be made available to investigators. The facility manager should ensure that investigators and teachers are informed of any changes to the conditions under which animals are held and which may affect their work."[37]

OVERSIGHT AND ETHICAL REVIEW PROCESS—AUSTRALIA

The mainstay of the Australian system is the AEC, whose terms of reference, role, membership, and mode of operation are comprehensively outlined in

Section 2.2 of the *Australian Code for the care and use of animals for scientific purposes, eighth edition* 2013 ("Australian Code").[12]

The AEC comprises a diverse membership consisting of at least one of each of the following categories of persons: a qualified veterinarian with experience relevant to the activities of the institution (Category "A"); a suitably qualified person with substantial recent experience in the use of animals in scientific or teaching activities (Category "B"); a person with demonstrable commitment to, and established experience in, furthering the welfare of animals, who is not employed by, or otherwise associated with the institution, and who is not involved in the care and use of animals for scientific purposes (Category "C"); and a person who is both independent of the institution and has never been involved in the use of animals in scientific or teaching activities, either in employment or beyond their undergraduate education (Category "D"). The Australian Code recommends, but does not currently mandate the appointment of persons responsible for the routine care of animals within the institution (e.g., Animal Facility Managers) as AEC members. Invitation of persons with specific expertise to provide advice as required is also advised. Finally, to ensure a fair balance between the numbers of persons who are independent of the institution (Category C and D) and those who may be considered to not be, it is stated that where an AEC has more than four members, the Categories C and D should represent no less than a third of the members.

The primary role of the AEC is to ensure that the use of animals is justified, provides for the welfare of those animals, and incorporates the principles of "Three Rs."[1] However, in reality, the Australian Code gives the AEC a relatively large role in that it is responsible, on behalf of the institution, not only for the ethical review and approval of proposed projects, but also for overseeing and ensuring that all aspects of animal work (including acquisition, use, husbandry, housing, care, welfare, use, and fate) are acceptable, and for ongoing monitoring of these parameters and the implementation of the "Three Rs." On the surface, this may appear to be an impossibly large task for the AEC, as a small group of persons, at least a third of whom are generally serving on a voluntary and casual basis; however, in reality, by law, the onus of compliance and implementation of the "Three Rs" actually rest directly on the institution and on the persons involved directly in animal use and animal care. Nevertheless by entrusting AECs with the oversight and monitoring of animal use and care to the AEC, the Australian Code better provides them with the necessary authority to assist the institutions in humane use by intervening, suspending, and making recommendations for change as necessary.

Postapproval monitoring of animal use is an AEC responsibility that is detailed in Section 2.3 of the Australian Code. The AEC members monitor animals by:

(1) performing inspections of animal facilities where animals are housed and laboratories where animal use occurs; (2) inspecting animal monitoring records maintained by researchers and teachers; (3) reviewing written annual, progress, and completion reports on projects; (4) considering reports from investigators on unexpected adverse events.[12]

OVERSIGHT AND ETHICAL REVIEW PROCESS—NEW ZEALAND

In general, no RTT activities involving animals can proceed without prior approval by an AEC, properly constituted under an approved CEC.[29] Section 118 of the Animal Welfare Act[27] does allow for the Minister to authorize research or testing that is in the national interest without the approval of an AEC.

The functions and powers of the AEC are laid out in Section 99 of the Act[27] as:

1. "to consider and determine on behalf of the code holder applications for the approval of projects;
2. to set, vary, and revoke conditions of project approvals;
3. to monitor compliance with conditions of project approvals;
4. to monitor animal management practices and facilities to ensure compliance with the terms of the code of ethical conduct;
5. to consider and determine applications for the renewal of project approvals;
6. to suspend or revoke, where necessary, project approvals;
7. to recommend to the code holder amendments to the code of ethical conduct."[27]

In considering any application for the approval of a project and in setting, varying, or revoking conditions of the approval of a project, s100 of the Act,[27] states that every AEC must have regard to such of the following matters as are relevant:

1. "the purposes of this part of the Act and any matters that the committee is required to consider by regulations made under this Act; and
2. the scientific or educational objectives of the project; and
3. the harm to, or the distress felt by, the animals as a result of the manipulation, and the extent to which that harm or distress can be alleviated by any means (including, where the pain or distress cannot be held within reasonable levels, the abandonment of the manipulation or the humane destruction of animals), but this paragraph does not apply to the killing of animals for the purpose of any project where RTT are to be performed on their bodies or tissues; and
4. whether the design of the experiment or demonstration is such that it is reasonable to expect that the objectives of the experiment or demonstration will be met; and
5. the factors that have been taken into account in the choice of animal species; and
6. the extent to which there has been
   i. assessment of the suitability of using non-sentient or non-living alternatives in the project; and
   ii. replacement of animals as subjects with suitable non-sentient or non-living alternatives

7. whether the number of animals to be used is the minimum necessary to ensure a meaningful interpretation of the findings and the statistical validity of the findings; and
8. whether adequate measures will be taken to ensure the general health and welfare of animals before, during, and after manipulation; and
9. whether suitably qualified persons will be engaged in supervising and undertaking the research, testing, or teaching; and
10. whether any duplication of an experiment is proposed and, if so, whether any such duplication will be undertaken only if the original experiment—
    - is flawed in a way that was not able to be predicted; or
    - needs to be duplicated for the purpose of confirming a result that was unexpected or has far-reaching implications; and
11. whether the same animals are to be used repeatedly in successive projects, and, if so, the cumulative effect of the successive projects on the welfare of the animals; and
12. whether there is a commitment to ensuring that findings of any experiment will be adequately used, promoted, or published; and
13. any other matters that the committee considers relevant."[27]

Membership of the AEC is defined in s101 of the Act[27] as follows:

1. "Each animal ethics committee is to consist of at least 4 members.
2. If the code holder is an organisation, the members of the animal ethics committee must be appointed by the chief executive of the organisation or his or her nominee.
3. One member must be—
   a. the code holder; or
   b. if the code holder is an organisation, a senior member of the organisation appointed by the chief executive to be a member of the committee.
4. Any senior member of an organisation who is appointed under subsection (3) (b) must be a person who is capable of evaluating
   a. each proposal for a project; and
   b. the qualifications and skills of the proposer of a project; and
   c. the scientific value or the teaching value, as the case may require, of a project.
5. One member must be a veterinarian (not being a veterinarian who is an employee of, or is otherwise associated with, the code holder) appointed by the code holder on the nomination of the New Zealand Veterinary Association or a similar national body of veterinarians.
6. One member must be a person appointed by the code holder on the nomination of an approved organisation (defined as having the promotion of animal welfare as its principal purpose).
7. The person appointed under subsection (6) must not be:
   a. a person who is in the employ of, or is otherwise associated with, the code holder; or
   b. a person who is involved in the use of animals for research, testing, or teaching.

8. One member must be a person appointed by the code holder on the nomination of a territorial authority or regional council.
9. The person appointed under subsection (8) must not be:
   a. a person who is in the employ of, or is otherwise associated with, the code holder; or
   b. a person who is associated with the scientific community or an animal welfare agency."[27]

REUSE: REQUIREMENTS—AUSTRALIA

Repeated use of individual animals is permitted, subject to requirements detailed in the *Australian Code for the care and use of animals for scientific purposes, eighth edition* 2013 ("Australian Code").[12] Section 2.3.15 states that:

"Individual animals must not be used in more than one scientific activity, either in the same or different projects, without Animal Ethics Committee (AEC) approval. However, appropriate reuse of animals may reduce the total number of animals used in a project, result in better experimental design, reduce distress or avoid pain to other animals. When considering approval for the reuse of animals, the AEC must take into account:

i. the pain or distress, and any potential long-term or cumulative effects, caused by any previous activities and conditions;
ii. the time allowed for recovery of the animals between activities;
iii. whether an animal has recovered fully from the previous activities;
iv. the pain and distress likely to be caused by the next and subsequent activities; and
v. the total time over which an animal will be used."

And Section 5.1.6: "Reuse of individual animals requires AEC approval (see 3.3.11). However, the nature of wildlife field studies may require, or result in, recapture of individual animals. Measures should be put in place to minimize the effects of this."[12]

REUSE: REQUIREMENTS—NEW ZEALAND

Section 100 (k) of the Animal Welfare Act[27] requires the AEC to determine whether the same animals are to be used repeatedly in successive projects, and, if so, the cumulative effect on the welfare of the animals.

SETTING FREE/REHOMING REQUIREMENTS—AUSTRALIA

Setting free or release of genetically altered, feral, domestic, exotic, or pest animal species into the natural environment is often prohibited by state and territory legislation.[12] There are several concerns, including: the welfare of the particular animal, the potential effects upon the natural environment or ecosystem, and biosecurity concerns relating to the risk of spread of disease from released animals to wild animal populations. Rehoming of suitable domestic animal species is a matter subject to AEC policy decision, or on a case-by-case basis. Native wildlife release requirements are detailed in Section 3.3.39 of the *Australian Code for the care and use of animals for scientific purposes, eighth edition* 2013 ("Australian Code"), but may also be subject to other legislation (e.g., Gene Technology Act 2000; Agricultural and Veterinary Chemicals Act 1994; Biosecurity Act 2015; environmental pest animals). It is outside the scope of this chapter to discuss these; however, references to relevant regulatory restrictions are detailed and listed in the Australian Code and related NHMRC guidelines, and Commonwealth, state, and territory government websites.[3,12]

The Australian Code requires the capture, holding, transportation, handling; and release of wildlife must be in accordance with the principles stated in Sections 3.3.33–3.3.46. Requirements for release of wildlife:

"**i.** the duration of captivity must be minimized and consistent with the purpose and aims of the project. If animals are to be released they must be handled in a manner where the risk of disease transmission and environment disturbance is minimized. All possible steps must be taken to avoid animals becoming habituated to human activity;
ii. release occurs at the site of capture, unless otherwise approved by the AEC;
iii. the timing of the release coincides with the period of usual activity for the species. Animals are protected from injury and predation at the time of their release; and
iv. animals that have been sedated or anaesthetized have recovered to full consciousness before their release."[12]

SETTING FREE/REHOMING: REQUIREMENTS—NEW ZEALAND

Most species of indigenous fauna are protected by law. The Department of Conservation must be consulted when these species are required. Permits are usually necessary to collect, keep, release, or kill protected fauna, and further permits are usually required to import or export such species. Any conditions imposed on permits must be observed. Native fauna apart, institutions are free to develop their own policies on rehoming of animals used in research.

Under the Hazardous Substances and New Organisms Act 1996,[34] any approval to develop or field test new organisms, including genetically modified animals, must include controls that address particular matters including facility and access requirements aimed at limiting the likelihood of any accidental release, monitoring and phytosanitary requirements, eradication plans for escaped organisms, and inspection and monitoring of facilities.

OCCUPATIONAL HEALTH AND SAFETY—AUSTRALIA

Health and safety in the workplace is the shared responsibility of employers and employees and is subject to state and territory occupational health and safety

(OHS) legislation.[3] A detailed description and comparison of the legislation in each jurisdiction is beyond the scope of this chapter, therefore the focus here is mainly on the requirements detailed in the *Australian Code for the care and use of animals for scientific purposes, eighth edition* 2013 ("Australian Code").[12] The Australian Code requires the AEC to seek information from animal users concerning the potential OHS impact of their animal work upon other personnel and other animals in the animal facility, and also information concerning related discussions with the Institutional Biosafety and OHS Committees.[12] The purpose is not to duplicate the functions of other Committees or authorities, but rather for a better understanding of the impact of the experiment upon the animals and personnel, and to ensure that all other compliance matters have been addressed.

OCCUPATIONAL HEALTH AND SAFETY—NEW ZEALAND

Management of health and safety in the workplace is governed by the Health and Safety at Work Act 2015.[38] The main purpose of this legislation is to provide for a balanced framework to secure the health and safety of workers and workplaces.

EDUCATION, TRAINING, AND COMPETENCE OF PERSONNEL—AUSTRALIA

The *Australian Code for the care and use of animals for scientific purposes, eighth edition* 2013 ("Australian Code"), requires scientific institutions, AECs, teachers, and the animal users themselves to accept responsibility for ensuring education, professional training, and competence of all persons involved with animal use.[12] Section 1.29 requires that all people caring for and using animals must be competent for the procedure they perform, or under the direct supervision of a person who is competent to perform the procedure. The Institution and the AEC are charged with the responsibility of determining whether investigators have the required education, training, and competence to perform the techniques required for the particular activity. Teachers and supervisors must ensure that prior to using animals, the students receive appropriate instruction in their ethical and legal responsibilities as well as the appropriate methods of animal care and use [Sections 4.12 and 4.13].

Several states also require that researchers and members of AECs are provided with an understanding of the ethical and humane care and use of animals. Recognizing that the skills and knowledge of those involved are essential to achieve high standards of animal welfare and scientific outcomes, courses for people using animals in research and teaching have been developed.[39–41] In Victoria a program to develop Standard Operating Procedures for registered training authorities for the use of animals in training is underway. Specific training for researchers and teachers is currently undertaken at the institutional level.

Some training and education for AEC members has been identified as being provided by the ANZCCART, and the Animal Welfare Units in Queensland, NSW, and Victoria.[26,39–44] An on-line training program in NSW is currently being revised and updated (see Animal Ethics Infolink website below for links to international courses and Caring for Animals in Research and Education course). Many universities and other research institutions require that new research, teaching, and postgraduate students undertake an induction training program, which includes ethics and regulation of animals used for scientific purposes.

Publically available on-line courses and recorded public seminars, webinars, and lectures hosted by Australian institutions:

- NSW Animal Research Review Panel & NSW Dept. Primary Industries Animal Welfare Branch: Animal Ethics Infolink[39]
- University of Melbourne[42]
- Animal Welfare Science Centre[43]
- Australian Animal Welfare Strategy[44]

Animal Care Personnel

The Australian Code requires that institutions should ensure that animal care staff are appropriately trained and instructed (Sections 2.1.2 and 2.1.5).[12] In many cases, these staff will have attended or be attending appropriate tertiary education institutions providing training in animal technology, veterinary nursing, and other animal care courses. It would be expected that these staff have had training so that they:

- understand the ethical principles, which govern the care and uses of animals as detailed in the Code;
- are familiar with the laws, guidelines, and codes of practice, which govern the acquisition, care, breeding, handling, and use of animals in their workplace;
- and are able to provide and maintain animals in accordance with the relevant codes to meet species-specific needs and to ensure animals are maintained in an optimum state of well-being.

Education, Training, and Competency With Regard to Nonhuman Primates

The NHMRC provides the principal funding, and the key policy and guidelines, for scientific use of NHPs in Australia.[12,30] The *Principles and guidelines for the care and use of non-human primates for scientific purposes* (2016) states:

"**33.** All persons involved with the care and use of non-human primates must have a sound knowledge of the species in their charge. Education and training for new and inexperienced investigators and animal carers must be provided.

34. Investigators and animal carers involved with the care and use of non-human primates must ensure that procedures are performed competently, and
    i. be competent for the procedure they perform and the species with which they are working, or
    ii. be under the direct supervision of a person who is competent to perform the procedure and with the particular non-human primate species involved."[30]

And

"8.2 Investigators and animal carers must be familiar with all aspects of the care and use of nonhuman primates, before obtaining animals or commencing projects. When preparing an application to the institutional AEC [Animal Ethics Committee] for a research project involving the use of non-human primates, investigators should consult with the managers of the Australian breeding colonies as a source of helpful information. The application to the institutional AEC must document evidence of the competence of all investigators with the species to be used.

8.3 When the research project involves extended interaction with non-human primates, training for new or inexperienced animal carers and investigators must be arranged in consultation with the managers of the Australian breeding colonies. Training through the Australian breeding colonies will facilitate familiarity between the animal and the investigator and the transition of the nonhuman primates to a new environment."[30]

EDUCATION, TRAINING, AND COMPETENCE OF PERSONNEL—NEW ZEALAND

The Animal Welfare Act[27] requires that the AEC determines the qualifications of personnel using animals in RTT, as detailed in Section 100: "(i) whether suitably qualified persons will be engaged in supervising and undertaking the research, testing, or teaching."

The *Good Practice Guide for the Use of Animals in Research, Testing and Teaching*[37] covers training in Section 5.1.2 *Personnel:*

- "*The most important factor ensuring high standards of animal care is a sufficient number of well-trained, committed staff. Personnel working with animals in a holding facility should be appropriately instructed in the care and maintenance of those animals, and in how their actions may affect the animals' well-being and the outcome of experiments.*
- *Institutions should encourage and promote formal training in animal science or technology (a list of courses is given as an appendix).*
- *Personnel employed in the care of animals should be trained to recognise at an early stage changes in animal behaviour, performance and appearance.*

- *New personnel who will care for animals should be appropriately instructed in their duties and in institutional policy and procedures.*
- *Personnel should be informed of allergy hazards and the important zoonotic diseases of animals under their care and of precautions that should be taken. Regular health checks and appropriate immunisation (e.g., against tetanus and other zoonoses) of all personnel who handle animals are recommended in the interests of both personnel and animals."*

TRANSPORT—AUSTRALIA

Importation of animals into Australia poses a biosecurity risk, which is managed and administered by the Australian Government Department for Agriculture and Water Resources and their Biosecurity Import Conditions System (formerly Australian Quarantine Inspection Service). Importation of exotic and wild animal species is subject to Commonwealth of Australia quarantine legislation and international treaty such as the Convention on International Trade in Endangered Species of Wild Fauna and Flora (CITES).[45,46]

International and domestic air transport of exotic and genetically modified animals is subject to Commonwealth, state, and territory government biosecurity regulations,[45] Office of the Gene Technology Regulator transportation guidelines for GMOs,[33] and International Air Transport Association (IATA) Live Animal Regulations 2012 guidelines.[47] Domestic transport of domestic animals, native animals, and animal tissues within Australia are subject to state and territory regulatory requirements and restrictions.

The *Australian Code for the care and use of animals for scientific purposes, eighth edition* 2013 ("Australian Code"), Section 3.2[12] provides guidance on transport of animals used for scientific purposes, and stipulates that domestic animal and livestock transport must comply with the relevant Commonwealth, state, and territory animal transportation legislation and standards. These include the 2012 Australian Animal Welfare Standards and Guidelines for the Land Transport of Livestock.[21] IATA guidelines apply to air transport of animals within Australia.

Specific National Health and Medical Research Council Importation Requirements for Nonhuman Primates

The NHMRC *Principles and guidelines for the care and use of non-human primates for scientific purposes* (2016) states:

"**19.** Importation of non-human primates for scientific purposes must comply with relevant legislation and must not occur unless appropriate Commonwealth and state or territory government permits, licences and approvals are in place.

20. Ordering and dispatch of non-human primates must be accompanied by evidence of an institutional AEC [Animal Ethics Committee] approval for the project.
21. The institution should ensure that the AWC [Animal Welfare Committee] of NHMRC is notified of the importation of non-human primates after approval from the institutional AEC has been obtained. For NHMRC funded activities, this requirement is mandatory."[30]

TRANSPORT—NEW ZEALAND

The *Good Practice Guide for the Use of Animals in Research, Testing and Teaching*[37] (the Guide) covers transport in Section 3.3 and refers to requirements under the Animal Welfare Act[27]:

- Section 22 "requires that the person in charge of any vehicle or aircraft, and the master, or where there is no master the person-in-charge, of every ship, shall ensure that the welfare of any animal conveyed therein is attended to, and that the animal is provided with reasonably comfortable and secure accommodation. The consignor is responsible for arranging for the supply of food and water *en route*."[37]
- Under Section 23, "it is an offence to confine or transport an animal in a manner or position that causes the animal unreasonable or unnecessary pain or distress, or to transport it if its condition or health is such as to render it unfit to be transported."[37]

The Guide[37] also references the requirements in the Animal Welfare (Transport within New Zealand) Code of Welfare 2011.[48]

HOUSING AND ENRICHMENT—AUSTRALIA

Care of all animals is addressed in state and territory animal welfare legislation,[4–11] and other state and territory animal welfare codes of practice. In general, the owner or person in charge of any animal must provide the animal with adequate and appropriate food, water, shelter, exercise and veterinary attention, provide a suitable living environment and take all reasonable steps to prevent animal harm. These obligations apply equally to the owners of companion animals, livestock, or animals used for scientific purposes. In the case of domestic livestock, the states and territories have based their animal welfare codes of practice upon the Primary Industries Standing Committee Model Animal Welfare Codes of Practice, which provide guidance on animal housing, husbandry, transport, euthanasia and other relevant issues.[20] The Model Codes are currently under review as part of the *Australian Animal Welfare Strategy* and will be replaced by regulated standards under the authority of state and territory legislation.[21]

Section 3 of the *Australian Code for the care and use of animals for scientific purposes eighth edition* 2013 ("Australian Code"), discusses care of animals in production and holding facilities in scientific establishments.[12] This section provides an outline of the issues that need to be satisfactorily addressed, including: species-suitable accommodation (buildings and pens or cages); building maintenance and repair; cleaning and hygiene of buildings and animal pens or cages; vermin exclusion and control; water supply and drainage; environmental controls (temperature, humidity, ventilation, light, noise, ammonia, dust, odors); contingency planning for emergencies. Routine husbandry procedures on livestock that are not part of a project must comply with relevant legislation and animal welfare codes of practice.

The Australian Code does not prescribe cage, pen or engineering standards. Guidance on environmental enrichment strategies is provided in Part III Fact Sheet E from the NHMRC's *Guidelines to promote the wellbeing of animals used for scientific purposes: The assessment and alleviation of pain and distress in research animals* (2008).[49]

In the state of Victoria, *The Prevention of Cruelty to Animals Act, 1986* mandates compliance with the *Code of Practice for the Housing and Care of Laboratory Mice, Rats, Guinea Pigs and Rabbits*.[24] This code applies to all institutions breeding or using laboratory rats, mice, guinea pigs and rabbits for scientific purposes. The primary objective of this code is to establish minimum standards and promote more detailed recommendations for the housing and care of these species than is currently provided in the Australian Code. The particular aspects this Housing and Care Code include are: nutrition, animal enclosures, climate and environment, behavior and environmental enrichment, hygiene, handling and basic procedures, health monitoring, transportation, euthanasia, and monitoring and record keeping. It specifies minimum standards for the housing, and recommends monitoring sheets, welfare assessment sheets, other record keeping sheets and a phenotype report for genetically modified animals.

In the state of NSW, similar standards apply. These are detailed in six species-specific Animal Research Review Panel *Guidelines* for the housing and care of dogs, rats, rabbits, guinea pigs, sheep and mice in scientific establishments.[22]

HOUSING AND ENRICHMENT—NEW ZEALAND

The *Good Practice Guide for the Use of Animals in Research, Testing and Teaching*[37] provides general recommendations for housing conditions and enrichment. New Zealand has elected to avoid the use of engineering standards for cage sizes. Instead, performance standards are in place.

"Animal accommodation should be designed and managed to meet species-specific needs. Pens, cages, and containers should be designed, constructed, and maintained to ensure the comfort and wellbeing of the animals. Any variations to these requirements as part of a project must receive prior AEC [Animal Ethics Committee] approval. The following factors should be taken into account:

- species-specific behavioural requirements, including the availability and design of space to allow free movement and activity, sleeping, enclosed spaces, contact with others of the same species, and the opportunity to perform a species-specific behavioural repertoire;
- provision of single housing for animals when appropriate for the species and, if necessary, for the purpose of the experiment, e.g., during recovery from surgery or collection of samples;
- the need to provide ready access to food and water;
- the need to clean the pen, cage or container;
- protection from spread of pests and disease;
- requirements of the project; and
- the need to observe the animals readily.

Pens, cages and containers should:

- be constructed of durable, impervious materials;
- be kept clean;
- be maintained in good repair;
- be secure and escape-proof;
- protect the animals from climatic extremes;
- not cause injury to the animals;
- be large enough to ensure the wellbeing of the animal or animals, with adequate space to allow them to stretch out when recumbent and to stand upright.

Wire floor cages for rodents should not be used unless essential to the research protocol and then only for brief periods. Animals should have a solid resting area when housed on wire floors.

The population density of animals within cages, pens or containers and the placement of these in rooms should be such that acceptable social and environmental conditions for the species can be maintained. Where it is necessary to individually house animals of a species, which are normally kept in a social group, the conditions should be managed to minimize the impact of social isolation. Animals should be housed in these circumstances for the minimum time necessary.

Bedding, litter or other environmental provisions should be provided, if appropriate, to the species, and should be comfortable, absorbent, dust-free, non-palatable, nontoxic, able to be sterilized if needed, and suitable for the particular research purpose. Pregnant animals must be provided with nesting materials where appropriate."[37]

Section 4.7 of The Guide[37] states:

"Most animals used in research, testing and teaching are housed in unnatural environments. Wherever possible such animals should be provided with an environment that can accommodate the behavioural and physiological needs of the species.

Almost all the species of animals used in research, testing and teaching have well defined social structures and prefer to live in groups, although care must be taken to ensure that animals are socially compatible. Individual housing is stressful for social animals, and social isolation should be avoided whenever possible and limited to meet the specific research objectives as approved by an AEC. The effects of physical isolation should be minimized where possible by the use of noncontact communication, whether visual, auditory or olfactory. Judicious use of mirrors can also be helpful, as can an environment of increased complexity.

The living areas of the animals should be set up and provisioned with the means that will enable them to perform a behavioural repertoire appropriate to the species."[37]

HUSBANDRY AND ENVIRONMENT—AUSTRALIA

Section 3 of the *Australian Code for the care and use of animals for scientific purposes, eighth edition* 2013 ("Australian Code"), discusses care of animals in production and holding facilities in scientific establishments.[12] The Victorian and NSW government regulators also provide minimum standards and recommendations, which have been discussed above and are similar to those that apply in New Zealand. Additional detail on suggested enrichment is provided for each particular species in the NHMRC *Guidelines to promote the wellbeing of animals used for scientific purposes: The assessment and alleviation of pain and distress in research animals (2008)* and the NHMRC, Victorian, and NSW state animal housing and care guidelines listed in the section General Framework.[22,49]

Codes and guidelines for meeting the needs of domestic livestock species, captive native species, and exotic zoo animal species may apply in some situations. National and international guidelines for husbandry and environment of captive wildlife, zoo, and aquatic species exist; however, a detailed listing of these is outside the scope of this chapter. References to Commonwealth of Australia Model Codes for animal welfare and Australian animal welfare standards and guidelines are provided earlier in the section General Framework, and other references are provided in the Australian Code and at the Australian Animal Welfare Strategy website.[12,14,15,20,21]

HUSBANDRY AND ENVIRONMENT—NEW ZEALAND

The *Good Practice Guide for the Use of Animals in Research, Testing and Teaching*[37] requires that, in terms of environment, species-specific requirements such as lighting, temperature, air quality, appropriate day/night cycles, and protection from excessive noise and vibrations should be taken into account.

It also recommends the following in Section 5.2:

- "Animals should receive appropriate, clean and nutritionally adequate food according to accepted requirements for the species. The food should be in sufficient quantity and of appropriate composition to maintain normal growth of immature animals or normal weight of adult animals and to provide for the requirements of pregnancy or lactation.
- Uneaten perishable food should be removed promptly unless contrary to the needs of the species. Where possible, alteration to dietary regimes should be gradual.
- When animals are fed in groups, there should be sufficient trough space or feeding points to avoid undue competition for food, especially if feed is restricted. Feeding space is determined by the size and number of animals that must eat at one time.
- Drinking water should be constantly and reliably available, and be clean, fresh and uncontaminated.
- Routine husbandry procedures should comply with any code of welfare for the species involved and must be performed by competent personnel. Variations to normal procedure as part of an experimental project must receive prior AEC [Animal Ethics Committee] approval. Procedures applied to the maintenance of breeding stock and supply of animals are viewed as routine husbandry and fall outside the definition of manipulation. When special breeding requirements are integral to a research or teaching project such as in the creation of a genetically modified animal, then procedures applicable to breeding must be regarded as a manipulation and should be included in the proposal to the AEC."[37]

VETERINARY CARE—AUSTRALIA

Rather uniquely, compared to the rest of the world, there is currently no legislative requirement for a dedicated veterinary surgeon who is laboratory animal science trained or laboratory animal medicine trained to be appointed by institutions to assist with proactive implementation of initiatives consistent with the "Three Rs"[1] principles in the use, care, and welfare of the animals. Instead there is a more basic requirement in the *Australian Code for the care and use of animals for scientific purposes, eighth edition* 2013 ("Australian Code"), for institutions to ensure "availability and access to veterinary advice for the management and oversight of a program of veterinary care, quality management and project design to safeguard animal wellbeing."[12] However, the current Australian Code does advise institutions to consider appointing an officer with veterinary, or other appropriate, qualifications who is authorized by the AEC to ensure that projects are proceeding in compliance with the Code and the decisions of the AEC. Furthermore, many institutions do appoint a veterinary officer or animal welfare officer to fulfill the duties and responsibilities often undertaken by veterinarians appointed as, for example, "designated veterinarians" in the European Union, "clinical or attending veterinarians" in the United States, "named veterinary surgeons" in the United Kingdom, and "animal welfare officers" in Canada, the Netherlands.

VETERINARY CARE—NEW ZEALAND

The provision of veterinary care is addressed indirectly in Section 9 of the Animal Welfare Act,[27] which:

- "requires owners of animals, and persons in charge of animals, to take all reasonable steps to ensure that the physical, health, and behavioural needs of the animals are met in accordance with both good practice and scientific knowledge"; and
- "requires owners of ill or injured animals, and persons in charge of such animals, to ensure that the animals receive, where practicable, treatment that alleviates any unreasonable or unnecessary pain or distress from which the animals are suffering."[27]

The *Good Practice Guide for the Use of Animals in Research, Testing and Teaching*[37] requires, in Section 5.1, that facility managers "should ensure that adequate veterinary care is provided for all animals held for breeding and for experimental manipulations with ready access to institutional or consultant veterinary services seven days a week." It also requires (Section 6.1) that investigators "should consult with veterinarians whenever unexpected adverse effects occur in order that standard veterinary care treatment regimes are immediately implemented. This responsibility parallels the public's duty of care to seek veterinary management of any sick animals in their charge."[37]

CONDUCT OF EXPERIMENTAL PROCEDURES—AUSTRALIA

Planning to Maximize Well-Being and Minimize Pain and Distress in Animals

The *Australian Code for the care and use of animals for scientific purposes, eighth edition* 2013 ("Australian Code"), is based upon the general principles of prevention of cruelty, duty of care, and Russell and Burch's "Three Rs"[1] principles of humane animal use.[12] Investigators are required to treat animals with respect and to consider their welfare when planning and conducting scientific RTT. Implementation of the "Three Rs" at the planning stage is a process that involves both the investigators and the AEC, assisted by the structured animal use Proposal or Application documentation, and the AEC assessment process [Australian Code Sections 2 and 3]. The Australian Code (Sections 2.4 and 3) places emphasis in the need for the animal user or investigator to anticipate the animal welfare impacts of all the scientific procedures that comprise a particular experiment, study or use, and to prepare a strategy that will promote animal well-being and limit animal pain and distress (Section 3).[12] Ensuring technical skills and competence of all personnel involved in animal care and use is a responsibility of the scientific investigator or teacher. Both anticipated and unanticipated (or accidental) adverse impacts need to be incorporated in the strategy. Often the AEC requires an animal use flowchart or chart that outlines step-by-step what will happen (and when) to each experimental group of animals. A period of conditioning or acclimatization of the animal to the project environment, personnel, and procedures prior to commencing the animal use is recommended to be included in the plan. Familiarity with the species-specific signs of pain and distress, and incorporation of these in a record of animal welfare monitoring observations is usually required, and is a most useful tool if tailored to the particular animal use proposed. The monitoring plan usually incorporates details of the frequency of animal welfare observations as determined by the animal use protocol and the animal's condition, and provides specific criteria for interventions that maximize well-being and minimize pain and distress. Interventions include withdrawal of the animal from the experiment, use of analgesia, provision of veterinary treatments, nursing and palliative care, and humane killing to limit pain and distress (euthanasia).[12] Further details and suggestions concerning the importance of the planning process and ways in which well-being can be promoted are detailed in the NHMRC *Guidelines to promote the wellbeing of animals used for scientific purposes: The assessment and alleviation of pain and distress in research animals, Part II (2008).*[49]

Detecting and Limiting Pain and Distress

The Australian Code Section 3 (ii) states:

"Pain and distress may be difficult to evaluate in animals. Unless there is evidence to the contrary, it must assume that procedures and conditions that would cause pain and distress in humans cause pain and distress in animals. Decisions regarding the possible impact of procedures or conditions on an animal's wellbeing must be made in consideration of an animal's capacity to experience pain and distress." and:

Section 2.4.18 (x): "Use of pharmacological agents such as anaesthetics, analgesics and sedatives must be appropriate to the species, the individual animal (e.g., age, physiological status) and the scientific aims, and must be consistent with current veterinary or medical practice. Anaesthesia must be used for procedures that are likely to cause pain of a kind and degree for which anaesthesia would normally be used in veterinary or medical practice."[12]

Further details and suggestions concerning the importance of the detecting and limiting animal pain and distress are detailed in the NHMRC *Guidelines to promote the wellbeing of animals used for scientific purposes: The assessment and alleviation of pain and distress in research animals, Parts II and III Fact sheets- Pain management: anaesthesia, analgesia and anxiolytics, and Humane killing and euthanasia (2008).*[49]

Monitoring, Reporting, and Reviewing Research and Teaching Use

Reporting annually to the AEC on progress and outcomes of animal use projects is a requirement of the Australian Code and the responsibility of the scientific investigator or teacher. The investigator is also responsible for reporting promptly to the AEC any unexpected adverse events. The AEC is required to perform inspections of animal facilities and animal use projects and records to ensure that welfare of animals is monitored as agreed. The AEC reports at least annually to the governing body (or delegate) of the scientific institution. The AEC report includes a summary of its activities, and the animal use conducted by the institution.[12] The scientific institution reports on scientific use of animals to the government regulator, in accordance with state and territory legislative requirements.[4–11]

Currently neither the Australian Code nor the state/territory legislation provides any guidelines on severity classifications of projects or procedures. However, in the state of Victoria, the "impact" (severity) of procedures are required to be specified as part of the annual use statistics reporting requirement (required by the Regulations associated with the Victorian *Prevention of Cruelty to Animals Act*).[5] The categories used are as follows: observational, animal unconscious without recovery, minor conscious intervention no anesthesia, minor surgery with recovery, minor physiological challenge, surgery with recovery, physiological challenge, and "death as an end point." Note that the latter lethality study is a special category, which requires special approval from the Minister of Agriculture.[5]

CONDUCT OF EXPERIMENTAL PROCEDURES—NEW ZEALAND

Section 6.4 of the *Good Practice Guide for the Use of Animals in Research, Testing and Teaching*[37] (the Guide) outlines responsibilities of investigators in the conduct of experiments, including the following:

- Limiting pain and distress
- Animal welfare monitoring of pain or distress
- Study endpoints
- Repeated use of animals in experiments
- Duration of experiments
- Handling and restraining animals
- Completion of projects
- Euthanasia
- Autopsy
- Preoperative planning
- Surgery
- Postoperative care
- Implanted devices
- Organ and tissue transplantation

- Neuromuscular paralysis
- Electroimmobilization
- Animal models of disease
- Modifying animal behavior
- Toxicological experiments
- Experiments involving hazards to humans or other animals
- Experimental manipulation of animals' genetic material
- Cloning of animals
- Experimental induction of neoplasia
- Production of monoclonal antibodies
- Lesions of the central nervous system
- Withholding food or water
- Fetal experimentation
- Research on pain mechanisms and the relief of pain
- Animal welfare and animal health research

Investigators are ultimately responsible for all matters relating to the welfare of animals under their control. However, the Guide[37] recognizes that in many institutions the duty of managing routine animal husbandry is delegated to professional animal care personnel on a daily basis. Therefore, strategies must be in place for effective communication between facility managers and investigators regarding animal welfare and research concerns.

The Guide[37] addresses the management of painful procedures in Appendix 1. The basis of the Guide's recommendations on pain management is the Swiss Academy of Medical Sciences *Ethical Principles and Guidelines for Experiments on Animals*,[50] which states in Section 4:

"Experiments on animals shall be carried out in accordance with the latest developments. Known prophylactic, diagnostic and therapeutic processes shall be taken into account and the scientific guidelines provided by international expert bodies shall be observed."

"If pain, suffering or stress are inevitable concomitants of an experiment, their duration and intensity must be limited to the minimum. To this end, the animals shall be monitored by specially trained personnel in accordance with predefined criteria and at predefined times and measures necessary to alleviate suffering shall be taken insofar as this is compatible with the objective of the experiment. The animal must be able to express its sensations and where possible avoid painful stimuli. Hence, the use of substances that induce paralysis without loss of consciousness and analgesic effects is unauthorized."[50]

As part of the annual collection of statistics on the use of animals in RTT, investigators must grade the impact on animal welfare of the required procedures. The five grades used are as follows:

Grade A "No impact or virtually no impact"
Grade B "Little impact"
Grade C "Moderate impact"

Grade D "High impact"
Grade E "Very high impact"

A detailed description of the different grades can be found in *Animal Use Statistics 2013*.[51]

EUTHANASIA—AUSTRALIA

The *Australian Code for the care and use of animals for scientific purposes, eighth edition* 2013 ("Australian Code"), requires animals to be killed humanely in circumstances where animal welfare is compromised and animal pain and distress is unable to be alleviated promptly (Section 3.1).[12] The Australian Code Section 3.3.45 states:

"The method and procedures used for killing an animal must be humane and:

- **i.** avoid pain or distress and produce rapid loss of consciousness until death occurs;
- **ii.** be compatible with the scientific and educational aims of the project or activity;
- **iii.** be appropriate to the species, age, developmental stage and health of the animal;
- **iv.** require minimum restraint of the animal;
- **v.** be reliable, reproducible and irreversible;
- **vi.** ensure that animals are killed in a quiet, clean environment away from other animals;
- **vii.** ensure that death is established before disposal of the carcass, fetuses, embryos and fertilized eggs."[12]

The most commonly relied upon local documents to provide guidance on methods of euthanasia are the ANZCCART Guidelines on Euthanasia (2001, currently under revision)[52] and the NHMRC *Guidelines to promote the well-being of animals used for scientific purposes: The assessment and alleviation of pain and distress in research animals, Parts II and III Fact sheets- Pain management: anaesthesia, analgesia and anxiolytics, and Humane killing and euthanasia (2008)*.[49] In the state of Victoria the mandatory *Code of Practice for the Housing and Care of Laboratory Mice, Rats, Guinea Pigs and Rabbits* contains guidelines on euthanasia methods in those species.[24] Likewise, the NSW Animal Research Review Panel and other states' animal care guidelines also include suitable humane species-specific methodologies.[22] All of these guidelines and methods are similar to the American and Veterinary Medical Association's *AVMA Guidelines for the euthanasia of animals: 2013 edition.*[53]

As mentioned earlier in this chapter, the *Australian Animal Welfare Strategy* promotes humane animal protocols for a wide range of wild and domestic species and circumstances,[15] and Regulated and Model Animal Welfare Codes include details of acceptable euthanasia techniques for domestic livestock species.[14,21,49]

EUTHANASIA—NEW ZEALAND

The *Good Practice Guide for the Use of Animals in Research, Testing and Teaching*[37] provides the following general guidance on the principles of euthanasia:

> *"When it is necessary to kill an animal, humane procedures must be used. These procedures should fulfil the following requirements:*

- *They should be minimally stressful.*
- *They should be reliable.*
- *They should produce rapid loss of consciousness without pain until death occurs.*
- *The appropriate means must be readily at hand.*
- *The procedures should be compatible with the aims of the experiments.*
- *The procedures should be performed only by persons who have demonstrated to a veterinarian or other qualified person that they are competent in the methods to be used.*
- *Animals should be killed in a quiet, clean environment, and normally away from other animals. There should be no disposal of the carcass until death is established.*
- *Dependent neonates of animals being killed must also be killed or provision made for their care.*
- *Methods of killing must be appropriate to the developmental stage of the animal. Disposal of fertilised eggs, fetuses and embryos must not occur until death is confirmed.*[37]

> *Investigators are referred to the monograph* Euthanasia of Animals Used for Scientific Purposes[52] *published by the Australian and New Zealand Council for the Care of Animals in Research and Teaching (ANZCCART) or the* AVMA Guidelines for Euthanasia[53] *for specific recommendations for euthanasia of the various species of laboratory animals."*

EQUIPMENT AND FACILITIES—AUSTRALIA

As discussed earlier in this chapter, the *Australian Code for the care and use of animals for scientific purposes, eighth edition* 2013 ("Australian Code"), does not prescribe cage, pen, or engineering standards; however, in some cases Commonwealth, state, and territory legislation and animal welfare Codes of Practice stipulate general standards or provide specific guidance.[4–11,14,20–22,24]

Regulatory requirements apply in circumstances where facilities are used for holding or transporting animals or biological materials subject to Australian Department of Agriculture and Water Resources biosecurity, and/or Office of the Gene Technology Regulator Physical Containment conditions for genetically modified animals and organisms.[33,45,47]

As imported and genetically modified laboratory rodents are commonly used in biomedical research, many animal facilities are constructed or modified to accommodate animals requiring PC2 and Department of Agriculture and Water Resources "Approved Arrangements"[45] (formerly "Quarantine Approved Premises") biocontainment. There are no specific regulations for more general animal research equipment or facilities except for requirements relating to building maintenance, hygiene, and OHS (discussed earlier in this chapter), compliance with license requirements for scientific use of animals,[4–11] or to operate diagnostic radiography units, annual certification of autoclave chambers, and so on.[12]

EQUIPMENT AND FACILITIES—NEW ZEALAND

The *Good Practice Guide for the Use of Animals in Research, Testing and Teaching*[37] has a section on facilities (Section 4), which requires that

> *"Investigators, Animal Ethics Committees and the institutions must ensure that facilities [indoor and outdoor] are appropriately designed, constructed, equipped, staffed and maintained to ensure the health and welfare of the animals and to fulfil scientific requirements. The design and management of facilities will depend on the type of animals to be kept and the experiments to be undertaken. The overall condition and management of facilities should permit effective maintenance and servicing and be compatible with maintaining good health and welfare."*

Special considerations for farm animals are included.[37]

REFERENCES

1. Russell WMS, Burch RL. *The principles of humane experimental technique*. London UK: Special edition published by Universities Federation for Animal Welfare—UFAW; 1959/1992.
2. Australian Government Department of Agriculture, Water Resources. *Animal welfare in Australia*. 2016. Available from: http://www.agriculture.gov.au/animal/welfare/animal-welfare-in-australia.
3. Australasian Legal Information Institute. 2016. Available from: http://www.austlii.edu.au/au/.
4. New South Wales Government. *Animal research act 1985*. 2016. Act no. 123 of 1985. Available from: http://www.austlii.edu.au/au/legis/nsw/consol_act/ara1985134/.
5. Victorian Government. *Prevention of cruelty to animals act 1986*. 2016. Act no. 46 of 1986. Available from: http://www.austlii.edu.au/au/legis/vic/consol_act/poctaa1986360/.
6. *Australian Capital Territory, Animal welfare act 1992*. 2016. Act no. 45 of 1992. Available from: http://www.austlii.edu.au/au/legis/act/consol_act/awa1992128/.
7. Northern Territory Government. *Animal welfare act 1999*. 2016. Act no. 44 of 1999. Available from: http://www.austlii.edu.au/au/legis/nt/consol_act/awa128/.
8. South Australian Government. *Animal welfare act 1985*. 2015. Act no. 106 of 1985. Available from: http://www.austlii.edu.au/au/legis/sa/consol_act/awa1985128/.
9. Tasmanian Government. *Animal welfare act 1993*. 2016. Act no. 63 of 1993. Available from: http://www.austlii.edu.au/au/legis/tas/consol_act/awa1993128/.
10. Western Australian Government. *Animal welfare act 2002*. 2016. Act no. 033 of 2002. Available from: https://www.slp.wa.gov.au/legislation/statutes.nsf/main_mrtitle_50_currencies.html.

11. Queensland Government. *Animal care and protection act 2001*. 2016. Act no. 64 of 2001. Available from: https://www.legislation.qld.gov.au/Acts_SLs/Acts_SL_A.htm.
12. National Health and Medical Research Council 2013. *Australian code for the care and use of animals for scientific purposes*. 8th ed. Canberra: Australian Government; 2013. NHMRC reference no. EA28. Available from: http://www.nhmrc.gov.au/guidelines/publications/ea28.
13. Australian Government Department of Agriculture. *Fisheries and Forestry, NCCAW position statements*. 2004. Available from: https://web.archive.org/web/20060208131840/http://www.affa.gov.au/content/output.cfm?ObjectID=D2C48F86-BA1A-11A1-A2200060B0A00816.
14. Australian Government Department of Agriculture and Water Resources. *Australian animal welfare standards and guidelines (model codes of practice)*. 2016. Available from: http://www.agriculture.gov.au/animal/welfare/standards-guidelines.
15. Australian animal welfare strategy. 2016. Available from: http://www.australiananimalwelfare.com.au/.
16. Animal Health Australia. *Animal welfare standards and guidelines*. 2016. Available from: http://www.animalwelfarestandards.net.au/.
17. Australian Government Department of Agriculture and Water Resources. *Animal welfare issues*. 2016. Available from: http://www.agriculture.gov.au/animal/welfare/animal-welfare-issues.
18. National Health and Medical Research Council. *Animal research ethics*. 2016. Available from: http://www.nhmrc.gov.au/health-ethics/animal-research-ethics.
19. National Health and Medical Research Council Animal Welfare Committee. *Guidelines for the generation, breeding, care and use of genetically modified and cloned animals for scientific purposes*. Canberra: Australian Government; 2007–14. NHMRC reference no. EA17. Available from: https://www.nhmrc.gov.au/guidelines-publications/ea17.
20. Primary Industries Standing Committee (PISC/SCARM). *Primary industries report series*. Collingwood Victoria: CSIRO Publishing; 2016. Available from: http://www.publish.csiro.au/books/series/11.
21. Animalwelfarestandards.net.au. *Canberra: Animal Health Australia. Australian animal welfare standards and guidelines*. 2016. Available from: http://www.animalwelfarestandards.net.au/.
22. New South Wales Animal Research Review Panel. *Policies and guidelines*. 2016. Available from: http://www.animalethics.org.au/policies-and-guidelines/.
23. New South Wales Department of Primary Industries. *Animal Welfare Branch fact sheet 21: supply of dogs and cats for use in research*. 2016. Available from: http://www.dpi.nsw.gov.au/agriculture/livestock/animal-welfare/research-teaching/factsheets/aw-fact21.
24. Victoria Agriculture. *Code of practice for the housing and care of laboratory mice, rats, guinea pigs and rabbits*. 2016. Available from: http://agriculture.vic.gov.au/agriculture/animal-health-and-welfare/animal-welfare/animal-welfare-legislation/victorian-codes-of-practice-for-animal-welfare/code-of-practice-for-the-housing-and-care-of-laboratory-mice,-rats,-guinea-pigs-and-rabbits.
25. South Australia Department of Environment, Water and Natural Resources. *Policies and guidelines for wildlife research*. 2014. Available from: http://www.environment.sa.gov.au/Plants_Animals/Animal_welfare/Policies_and_guidelines_for_wildlife_research.
26. ANZCCART. *Australian and New Zealand council for the care of animals in research and teaching Australia*. 2016. Available from: http://www.adelaide.edu.au/ANZCCART/resources.
27. New Zealand Government. Animal Welfare Act 1999. Available from: http://www.legislation.govt.nz/act/public/1999/0142/latest/versions.aspx.
28. New Zealand Government. *Animal welfare amendment act (No 2)*. 2015. Available at: http://www.legislation.govt.nz/act/public/2015/0049/30.0/DLM5174807.html.
29. National Animal Ethics Advisory Committee. *Guide to the preparation of codes of ethical conduct*. 2012. Available at: https://www.mpi.govt.nz/document-vault/1471.

30. National Health and Medical Research Council. *Principles and guidelines for the care and use of non-human primates for scientific purposes*. Canberra: Australian Government; 2016. NHMRC reference no. EA15. Available from: https://www.nhmrc.gov.au/guidelines-publications/ea15.
31. Australian Government. *Gene technology act 2000 – C2016C00792*. 2016. Act no. 169 of 2000. Available from: https://www.legislation.gov.au/Details/C2016C00792.
32. Australian Government. *Gene technology regulations 2001 – F2016C00615*. 2016. Statutory Rules 2001 no. 106. Available from: https://www.legislation.gov.au/Details/F2016C00615.
33. Department of Health Office of the Gene Technology Regulator. *Guidelines for the transport, storage and disposal of GMOs 2011*. 2012. Available from: http://www.ogtr.gov.au/internet/ogtr/publishing.nsf/Content/tsd-guidelines-toc.
34. New Zealand Hazardous Substances and New Organisms Act 1996. Available from: http://www.legislation.govt.nz/act/public/1996/0030/latest/DLM381222.html.
35. New Zealand Biosecurity Act 1993. Available from: http://www.legislation.govt.nz/act/public/1993/0095/latest/DLM314623.html.
36. *New Zealand Biosecurity Authority, Standard 154.03.03 containment facilities for vertebrate laboratory animals*. 2003. Available from: http://www.epa.govt.nz/Publications/154-03-03-MAF-ERMA-Std-2002.pdf.
37. New Zealand Government. *Good practice guide for the use of animals in research, testing and teaching*. 2010. Available from: https://www.mpi.govt.nz/document-vault/3660.
38. Health and Safety at Work Act 2015. Available from: http://www.legislation.govt.nz/act/public/2015/0070/latest/DLM5976660.html.
39. New South Wales Animal Research Review Panel, New South Wales Department of Primary Industries Animal Welfare Branch. *Animal ethics Infolink*. 2016. Available from: http://www.animalethics.org.au/.
40. Victoria Agriculture. *Training opportunities for Animal Ethics Committee members and governance staff*. 2016. Available from: http://agriculture.vic.gov.au/agriculture/animal-health-and-welfare/animals-used-in-research-and-teaching/training-opportunities.
41. Queensland Department of Agriculture, Fisheries. *Animal ethics training*. 2015. Available from: https://www.daf.qld.gov.au/animal-industries/welfare-and-ethics/using-animals-for-scientific-purposes/ethics-training.
42. University of Melbourne Office for research ethics, integrity. *Online resources*. 2016. Available from: http://research.unimelb.edu.au/office-for-research-ethics-and-integrity/animal-ethics.
43. Animal Welfare Science Centre. *Scientific seminars*. 2016. Available from: http://www.animalwelfare.net.au/article/scientific-seminars.
44. Australian Government. *Australian animal welfare strategy*. 2016. Available from: http://www.australiananimalwelfare.com.au/.
45. Australian Government Department of Agriculture, Water Resources. *Biosecurity in Australia*. 2016. Available from: http://www.agriculture.gov.au/biosecurity/australia.
46. CITES.org. Geneva; Convention on International Trade in Endangered Species of Wild Fauna and Flora (CITES). Available from: https://www.cites.org/.
47. IATA.org International Air Transportation Association. *Live animal regulations 2016*. 2016. Available from: http://www.iata.org/publications/store/Pages/live-animals-regulation.aspx.
48. National Animal Welfare Advisory Committee. *Animal welfare (transport within New Zealand) code of welfare*. 2011. Available at: http://www.mpi.govt.nz/protection-and-response/animal-welfare/codes-of-welfare/.
49. National Health and Medical Research Council. *Guidelines to promote the wellbeing of animals used for scientific purposes: the assessment and alleviation of pain and distress in research animals*. Canberra: Australian Government; 2008; 2014. NHMRC reference no. EA17. Available from: https://www.nhmrc.gov.au/guidelines-publications/ea18.

50. Swiss Academy of Medical Sciences. Ethical principles and guidelines for experiments on animals. *Altern Lab Anim* May-June 1997;**25**(3):379–84.
51. New Zealand Government. *Animal use statistics*. 2013. Available at: https://www.mpi.govt.nz/mpisearch/?site-search=animal+use+statistics&action_doSimpleSearch=.
52. Australian and New Zealand Council for the Care of Animals in Research and Teaching. In: Reilly JS, editor. *Euthanasia of animal for scientific purposes*. 2nd ed. Adelaide: ANZCCART; 2001 (Currently under revision) Available at: https://www.adelaide.edu.au/ANZCCART/docs/euthanasia.pdf.
53. American and Veterinary Medical Association. *AVMA guidelines for the euthanasia of animals*. 2013 ed. Schaumberg, IL: AVMA; 2013. Available at: https://www.avma.org/KB/Policies/Documents/euthanasia.pdf.

Index

'*Note*: Page numbers followed by "f" indicate figures and "t" indicate tables.'

A

C

F

G

H

I

J

K

L

M

O

U

V

W

Z

Printed in the United States
By Bookmasters